Agricultural
Bioethics

Agricultural Bioethics

Implications of Agricultural Biotechnology

EDITED BY

Steven M. Gendel
A. David Kline
D. Michael Warren
Faye Yates

Iowa State University Press / Ames

First edition, 1990

Library of Congress Cataloging-in-Publication Data

Agricultural bioethics: implications of agricultural biotechnology / edited by Steven Gendel . . . [et al.] — 1st ed.

    p.   cm.
    ISBN 0–8138–0129–X
    1. Agricultural biotechnology — Moral and ethical aspects. I. Gendel, Steven.
    S494.5.B563A37   1990
    174'.963 — dc20   89–15598
                       66940   CIP

Contents

I | Safety and Regulatory Issues

II | Impact on Scientific and Industrial Communities

v

III | Public Perceptions

IV | Economic Prospects

V | Social Considerations

VI | Ethical Dilemmas

Foreword

DANIEL J. ZAFFARANO

In fall 1985, in response to scandals described in newspaper articles, I convened a small group of faculty members to consider the question of how Iowa State University might treat a case of falsified research data if the need should ever arise. Our discussions developed into more general arguments about what constitutes ethical behavior in research.

In winter 1986, the Iowa State Legislature developed a bill to provide new funding for biotechnology research at ISU. During the session, Representative Paul W. Johnson, a farmer from Decorah who has done graduate work in biological sciences, asked us if we would be interested in studying ethical questions, such as the potential impacts of biological innovations on the quality of life for producers and consumers of agricultural products, which might emerge from agricultural biotechnology research. We responded enthusiastically, and the final appropriations bill for $17 million contained his amendment directing support for bioethics research at ISU for fiscal year 1986–1987. Subsequently, our applications to the Northwest Area Foundation (Minneapolis) and the Joyce Foundation (Chicago) for matching funds were approved.

The Bioethics Committee, now with financial support and endorsement of administrators of the biotechnology research funds at ISU, began to organize a long-range study program to parallel and complement the technical research plans. Four groups were formed to consider questions such as the following:

1. What are the implications for the university? Will the reward structure for faculty be affected? Will the special emphasis on biological research distort the mission of the university?

2. What will be the effect of biological advances on the farmer and the consumers of agriculture products? Will the quality of life be affected? What are the ecological implications?

3. Will economic values overshadow ethical considerations? Who will benefit? Who will lose?

4. Can we organize a national or international conference on agricultural bioethics to bring together those who have publicly written or spoken about the pros and cons of coming biological advances?

The answer to the last question is obviously *yes,* a fact to which we all can attest today. One graduate student research assistant was assigned the task of providing an annotated bibliography of everything published on bioethics in English language journals within the last five years. Another was asked to write a history of the use of stilbestrol as a feed additive during the last two decades. An external advisory committee was formed with representatives from agro-industry, state government, and rural and consumer advocacy groups. A monthly meeting in the ISU Memorial Union was established at which bioethics questions were opened for general discussion with students, faculty, and townspeople.

The impact of developments in agricultural biotechnology may be as far reaching as the development of nuclear energy sources during and after World War II. Whereas the technical aspects of production of nuclear energy were solved first and the environmental and social impacts were debated later, we have the advantage in biotechnology that the dialogue on ethical impacts has begun in parallel with the development of the technology. We hope that this will lead to a better understanding by the public of what can be expected and what we may gain or lose as a result of new technologies. As additional campus projects in research are proposed, the Bioethics Committee will attempt to elicit some advice and guidance in these matters from the researchers themselves and will try to explain and evaluate these considerations for people who will be affected by these developments.

In the best expression of the land-grant tradition, our faculty is dedicated to understanding new knowledge produced both here and elsewhere through basic research, evaluating its usefulness to Iowa agriculture, and transmitting our findings as rapidly as possible to the people who produce and use agricultural products. Thanks to the support we have received for studies in bioethics, we should be able to present with the new technical procedures and data a thoughtful analysis of the short-term and long-term implications of these developments in biotechnology on their quality of life.

Funding for the Iowa State University Agricultural Bioethics Symposium and several of the studies presented at the symposium was provided by the State of Iowa, the Northwest Area Foundation and the Joyce Foundation, for which we are most grateful.

Introduction: *Agricultural Bioethics and the Control of Science*

A. DAVID KLINE

It is commonly acknowledged that biotechnology is intellectually exciting and economically promising.[1] But what is special about it from the moral point of view? Biotechnology has been the object of intense social and ethical reflection both from within and without the molecular biology community. Why? The microelectronics industry was born and thrives with little moral examination. The superconductor revolution, which appears to be in the final stages of labor, has received nothing worse than impatient anticipation. Is there something unique about biotechnology that gives rise to moral questions which are different in kind from those raised by other technologies?

Focusing on the medical applications with the dust raised by Mary Shelley's images actually distorts the deeper considerations. It will be argued that the core issue, which gave rise to the debate and which sustains it, is a very old one, viz., who should control science? What is special about biotechnology is not that it raises new issues but that it raises an old issue in a particularly dramatic form—a form that jars scientists from naive political isolationism and shakes farmers out of their complacent trust in agriculture science. The r-DNA debate raises the issue of who should control science, and it forces us to see that we all have a stake in the outcome.[2]

I shall begin with two topics that are readily recognizable as central to the debate. The first I call the safety issue; the second, the issue of the common good. I will show that these topics lead to the problem of control. I will try to provide some understanding of the issues by making clear that at a certain level of abstraction there is a unity to them. Most questions at a lower level of abstraction are accommodated by the taxonomy but will not be discussed individually. For example, the area of appropriate federal regulations pertains at least to the safety issue. What sort of regulations will protect our health and the environment *and* will not unduly hinder the growth of science? The patenting issue is an instance of worries about the common good. Whose interests are served by patents in the area of biotechnology? How can we write patent laws that will encourage biotechnological research and guarantee long-term benefits to society?[3]

The Safety Issue

Concerns for safety initiated the debate over biotechnology. While the concerns have evolved into new forms, safety is still a live issue. The original worries pertain to what we can label *laboratory safety.*[4] Most commentators designate 1974 as the beginning of the controversy. In 1973 a special session on gene-splicing techniques was held at the Gordon Conference on Nucleic Acids. As a result a communication was sent to the National Academy of Sciences requesting that a panel be appointed to study biohazards. Paul Berg, from M.I.T., headed the appointed panel. In 1974 a report, "The Berg Letter," was issued. It called for a voluntary moratorium on certain types of experiments.

While realizing it is not constructive to quibble over the precise starting point of the debate, I believe that it is conceptually and philosophically illuminating to consider the First Asilomar Conference (Pacific Grove, California, January 1973) as initiating the debate. At this early stage the central issue was presented with a crispness and elegance that was not to be regained throughout the debate.

In 1971 Paul Berg considered constructing a hybrid molecule. The components were segments of a bacterial virus and certain bacterial genes plus an animal virus (SV40). SV40 is a monkey virus known to cause some animal cells to become malignant. The details of the rationale for constructing the hybrid molecule are not crucial here. Berg believed that it could be used to obtain important knowledge related to a general understanding of cancer in humans. Despite the promise of the system, he thought that there was a very small but definite probability that the bacterial segment would take up the SV40 virus, spread to animals including humans, and then cause the development of cancers.

The First Asilomar Conference was organized by Berg and others in the realization that not enough was known about biohazards in general and SV40 in particular for rational decisions to be made as to the health risk of certain experiments. The conference brought together leading biologists from universities, institutes, and government. While the work considered was not technically r-DNA work, it was a very near relative.

Michael Oxman, who held a position at the Harvard Medical School, was one of Asilomar's co-organizers. During the meeting he put an argument to the group that, when taken seriously, foretells the subsequent debate. In the following passage he is responding to worries about the protection of laboratory personnel.

> Whereas an investigator may himself decide to assume certain risks, he does not have the right to make that decision for anyone else. In fact, it seems to me that the decision to assume a risk can only legitimately be made by the individual who will be in jeopardy. This principle of "informed consent" can readily be applied to other investigators, graduate students and technicians, as long as they are informed of the nature and magnitude of the risk and are free to decide whether or not they are willing to take it.[5]

Put precisely, Oxman's principle of informed consent (PIC) is as follows: It is morally permissible to subject some person, P, to risk, R, only if P knows and understands R and freely chooses to assume R.

PIC is a very plausible moral principle. It is not surprising that someone from a medical environment would urge it and that scientists from the biological disciplines would have sympathy for it.

What is surprising is that neither the conference participants nor even Oxman appreciated the significance of PIC's consequences. It is a very radical principle once imported into the r-DNA debate.

First, the principle demands an enormous amount of knowledge that was not at hand. Asilomar itself was a testimony to the scientific community's ignorance with respect to biological hazards. Obviously, much more had to be known if the relevant risks were to be rationally determined.

Second, how should the issue be handled with respect to janitors or secretaries or the public? Is not their consent needed? Oxman had thought about this issue: "[T]he principle of 'informed consent' cannot realistically be applied to glassware washers, secretaries, housekeeping personnel, workers in adjacent laboratories or the public at large. Consequently we must insure that under no circumstances will these people be exposed to risks as a result of our research."[6]

Oxman's solution is not to abandon PIC but to insist that the public be subjected to no risk. This seems quite naive. Surely the probability of risk cannot be reduced to 0. Oxman's strategy reveals a tendency on the part of scientists—a tendency that has existed throughout the r-DNA debate—viz., to see the issues as narrow scientific issues best discussed only within the scientific community. Oxman's disposition is to remove potential risks caused by science by doing more science.

It should be quite clear that this insular attitude is not tenable. PIC and the recognition that the risks are not 0 lead directly to the necessity of the consent of the public. The later struggle to form regulatory policies is the struggle to not hinder the growth of science while gaining the consent of the public.

Perhaps one should be more sympathetic with the desire to keep the debate within the scientific community. After all, the public cannot be represented through the process of voting. Most people, supposing that they are interested, know little about the scientific aspects of the problem.

A way to conceptualize the problem is to ask what the ideally rational person would do. The practice of performing a risk/benefit analysis is a common way of making the idealization substantial. The structure and methodology of risk/benefit analyses are complex.[7] But it is not difficult to see that parts of the analysis are purely scientific projects while others are not. For example, determining the probability of a certain outcome is a scientific endeavor. Enumerating the relevant risks and benefits is not. To the extent that one thinks of the risks in narrow medical terms, this will be obscured. Even more obviously, the project of assigning weighted values to the risks and benefits is not a project for which training in cell biology is particularly helpful. Furthermore, one would think that those who are likely to bear the risks and benefits should have some say in the weighing. We return to the need for the public participation in the control of science. Direct participation is admittedly unmanageable, but it is crucial to see that risk/benefit analyses are not purely scientific processes. The necessity and manner of incorporating the public's interest remain an issue.

Subsequent to the Asilomar I conference, meetings and discussions of the laboratory safety issue continued. In 1976 the National Institutes of Health (NIH) issued guidelines for r-DNA research. Some experiments were thought to pose hazards potentially significant enough that they should not be performed. The recommendations included prescriptions for less hazardous experiments.

Immediately following the issue of the guidelines, critics argued that they were too inhibiting. By 1981 the guidelines were weakened sufficiently to be of little significance.

This five-year period in which the scientific community moved from raising serious concerns to dismissing the worries is a curious one. Putting aside whether the change was a rational one, many of the original scientific worries do not appear to have been laid to rest by the weight of *scientific argument*.

That many people stopped worrying is clear, but why they stopped worrying is not. One example is Robert Sinsheimer's warning that using r-DNA techniques to cross species barriers could have dramatic and potentially dangerous effects on evolutionary patterns.[8] It is not that people did not respond to Sinsheimer. The problem is that the responses seem far less than conclusive. Nevertheless, by the early 1980s the laboratory safety issue was essentially dead.

Like Lazarus, the safety issue was not put to rest for long. But this time misgivings were not, at least originally, forced by the scientific community itself. Furthermore, the new safety issue was importantly different from that of laboratory safety and much more directly related to agriculture.

Research had been taking place on various organisms whose *deliberate release* would be useful. Critics maintained that we do not know enough to anticipate the possible negative consequences of such events. A particularly vivid and now celebrated confrontation took place in 1983–1984.[9]

Steven Lindow and Nicholas Panopoulos through r-DNA techniques developed a strain of *Pseudomonas syringae* that would help prevent frost damage in plants. They eliminated the gene which codes for the ice-nucleating protein. If the new strain of bacterium populated a plant, ice would be formed at a lower temperature than it would be in the case of its more mundane relatives.

Though the experimental testing of the new strain had been blessed by NIH, the test was blocked by Judge John Sirica. Sirica was responding to a request for a temporary injunction prepared by Jeremy Rifkin. (Rifkin is an attorney and a long-standing critic of r-DNA research.) The legal details are not important here. The contention behind the suit was that deliberate release experiments should be halted until appropriate procedures were developed to show the safety of the tests. The basic worry on the part of critics was that the deliberate release of organisms might cause seriously detrimental ecological damage.

Since 1984 there has been a healthy, and at times testy, debate within the scientific community on the new safety issue. At present the issue appears to be far from settled.[10] An interesting dimension of the scientific controversy is that researchers are roughly divided along discipline lines. Molecular and cell biologists regard worries over safety as unfounded speculation. Bacterial ecologists urge caution in the light of what they see as a lack of needed knowledge.

It is curious that, even though the issue arose from without the scientific community, scientists, intentionally or not, have encouraged the view that the debate is really a wholly scientific one. Those who want to push immediately ahead take themselves to have shown not that there is little risk in the technology but rather that there is no risk. This is the same tendency that narrowed the debate in the case of laboratory safety.

It strikes a disinterested observer as a case of hubris. Surely the presently calculable probability of harmful effects is not 0. The ecologists may turn out to be quite wrong. But their arguments at this historical moment are not wholly without merit. A rational approach again seems to be to conduct a risk/benefit analysis. This, as was previously argued, requires representation from outside the scientific community.

There still remains the problem of determining the probability of certain outcomes or scenarios in the analysis. How can this be done when experts disagree? Isn't the rational attitude to be cautious?

Regardless of how one answers these questions, we should agree that the questions are not scientific ones. They are science policy questions and clearly have a normative or moral dimension. So again, we are led to the realization that the issue of the deliberate release of organisms should not, on pains of self-deception, be construed as a narrow scientific issue. It raises fundamental questions about how to properly direct science.

The Common-Good Issue

Two examples, herbicide-resistant crops and bovine somatotropin, have become symbolic of the moral difficulties with agricultural biotechnology. The examples are just that — examples — but they illustrate dilemmas that will likely be played out time and again as biotechnology develops. It is important to appreciate the concrete examples, but even more crucial are the general issues that lie behind them. These cases, unlike the safety issue, do not rest on potentially unforeseen health and environmental consequences. They rest on consequences of biotechnology that give rise to a tension between narrow economic interests and the common good.

Herbicides, as we all are aware, are big business. Monsanto alone sold $1 billion worth in 1982. Not every crop tolerates a given herbicide equally well. Triazine compounds, a family of particularly powerful chemicals, work well on corn. Unfortunately, they are lethal to soybeans, a crop that is often rotated with corn. A field of soybeans which was immediately preceded by corn treated with a triazine herbicide is likely to suffer a reduced yield.

In the early 1980s, biotechnology came to the rescue. The goal was to develop crops that were more resistant to herbicides. Not only was the effort to provide a variety of soybeans that would be resistant but also to develop a variety of corn that could withstand larger doses of herbicide.[11] *Chemical Week* in July 1982 announced the new possibility. "The theory is that farmers would then be willing to use even more of the weed killers, safe in the knowledge that their crop won't be damaged."

Such developments from a certain corporate point of view are not a bad strategy. Most seed corn companies are owned by large chemical companies. (Only Pioneer Hi-Bred remains a large independent seed company.) Varieties of seed that can withstand larger doses of herbicides will increase herbicide sales.

Herbicides are safe relative to insecticides but their long-term effects are not well known. Furthermore, some are known to cause chromosome breakage and, in fact, are mutagenic. There is an increasing problem with ground water pollution, especially in several midwestern states. Herbicides are part of that problem. From a moral point of view, we have a situation where the union of science and business is urging the increased use of chemicals that are at least potentially harmful. This prompts the question, Why isn't the scientific genius of biotechnology turned to developing varieties of crops that are more tolerant of weeds rather than herbicides?

The agribusiness community has an answer to this criticism. It asserts that the development of herbicide-resistant crops will allow the farmer to use a new family of herbicides that is more lethal to weeds and less dangerous to humans than is the present family. In fact, a smaller volume of herbicide will be needed.[12] This may be progress, but the question of weed resistance still remains. Furthermore, some work in the area is clearly designed to allow more use of familiar herbicides such as atrazine. Even if the industry as a whole were to develop a satisfactory response, the example is still of moral significance. Nothing in "the system" guarantees the wise use of science.

So we have a case where technological genius does not appear to be aimed at the general or common good. As this case illustrates, there does not appear to be anything intrinsic to technology or the market that will, perhaps invisibly, guard our welfare. We are returned to the issue of how to control technology.

The early 1980s also spawned our second example. Bovine growth hormone (bGH) occurs naturally in dairy cattle. But by using genetically modified bacteria, large quantities of the hormone can be produced. Dairypersons can then use the manufactured bGH to stimulate their herds. Though actual products have not yet been licensed for use, government and private studies indicate an increased milk yield of 10 percent to 40 percent. Recall that this is at a time when the federal government is slaughtering one million cows in an effort to reduce the $1 billion cost of milk surpluses. In a time of gross oversupply, we will have more.[13]

The effects on the milk industry are expected to be dramatic. The number of dairy farms will be reduced by 25 percent to 30 percent. The dairy industry will shift geographically to the Southwest. The loss of farms and family farmers is morally significant. Without falling into romanticism, we can assert that the farm life is of value. This is especially obvious when one considers the kinds of jobs likely to be created in the "new milk industry."

So, given that we are awash in dairy products, we have the loss of valued activities for little reason. The union of technology and business will likely, at least in this case, prove blind to the general good.

The promoter or defender of bGH is not without a response: (1) bGH will reduce the cost of milk to the consumer. Isn't that aiding the common good? (2) What is so privileged about family farms? What we really have is a simple case of technological progress! Part of technological progress often has the downside of the temporary displacement of workers. Family farms are no different from family hardware stores or mom and pop grocery stores. These are losses but most of us benefit from K-Mart and A&P. (3) bGH is a way to make farms more efficient. If we want to preserve any farms, we must do so by being efficient. Only then will our farms be able to compete in the international markets.

This is a tough response. The debate must be engaged in all its economic, political, social, and philosophical detail. I, of course, cannot do that here. I do, however, have two general points. First, writers often encourage the view that their claims are straightforwardly true as a simple result of the working of the laws of economics. As such, the rational attitude is to tuck one's chin and get behind the program. But one must remember that what happens to dairy farms is as much a product of the boundary conditions as it is the laws of economics. Tax laws and other artificial incentives have an enormous effect on the economic "facts." We shape the tax laws supposedly on the basis of our ideals of a fair, just, and compassionate society. It may be an economic fact of life that the family farm is dead. But that fact, if it is a fact, is in part crafted by deliberate choices.

Secondly, I do not know how the bGH debate will come down. But it is at least possible that we will decide that the bGH technology and its implementation were a mistake. Given the present management of science, all we could then say, perhaps with sufficient Greek courage, is that the situation was tragic. It seems clear that a rational society needs a policy for managing its scientific might which will not leave one in Hektor's condition. Again we return to the question of how to manage science.

Both the herbicide and bGH cases illustrate a tension between what I have called narrow economic interests and the common good. Four shared features of the cases deserve comment.

First, the discussion has implicitly limited the common good to the good of the citizens of the United States. In many agricultural policy discussions, the common good is treated even more parochially. This is a mistake. From a moral point of view, it is very difficult to justify many regional, even national, distinctions. How far one lives from you does not seem to be terribly relevant, morally speaking. A fully developed agricultural biotechnology policy will need to consider the interests of Third World peoples. That the promised increases in production will be a blessing for the hungry and poor of the Third World is not at all obvious.[14]

Second, what has been called the narrow economic side is an amalgam of science and business. One might argue that the moral problems are really a consequence of the actions of business. The scientific and business components should not, if one wants to avoid encouraging confusion, be run together. The problem is with the *use* of science. It is not science that needs managing; it is the exploitation of scientific knowledge by narrow business interests.

This often-repeated story is used to support a laissez-faire attitude toward science. Despite its frequency of appearance, the argument has a very serious problem.

Perhaps one could maintain the science/business distinction with some clarity for pure science, supposing that we know what *pure* science is. But the present context is technology. Technology, including biotechnology, is driven by an end or goal. It is good technology to the extent that it efficiently, simply, economically, morally achieves the goal. The point is that we cannot sever technology from its use. The use is built into the technological problem. So it is incoherent for one to argue that it is not technology which needs managing but the use of technology.

Third, it might be attractive to suppose that the proper way to deal with our examples is to put more responsibility on the farmer. The farmer can refuse to use high levels of herbicide or bGH. A little restraint by the producer and the workings of the market will solve the problem.

To see what is wrong with this tidy solution, one must appreciate that it is rational for the farmer to use more herbicide and bGH. This is the case despite the fact that it is not in the farmer's long-term interests. This paradoxical situation is not an abstract puzzle but is endemic to the agricultural sector. New technological innovations, whether they support the communal good or not, must be adopted if they give the adopter a short-run economic advantage. Otherwise the producer will be defeated by the market. The farmer may not want to adopt the innovation. But since it would be an immediate economic advantage to do so, he had better! There is a justifiable fear that others will adopt the technology. We should not rely on the altruism of farmers to handle these kinds of cases.[15]

Fourth, there is a moral dimension of these cases that has not been considered. Many of the advances in biotechnology and the fundamental science that underlies biotechnology have been financed by individual states and the federal government. The above examples have been heavily supported by land-grant institutions. Charles Arntzen at Michigan State University was influential in developing the capabilities for producing herbicide resistance. Cornell University, another land-grant institution, has played an important role in the development of bGH.

Michigan State and Cornell are partially supported by the citizens and farmers of their respective states. Given this financial relationship, farmers expect, and have a *right* to expect, that their collective interests will be guarded. In a very real sense, land-grant institutions should serve at the pleasure of the rural populations that support them.

Given that land-grant institutions play an important role in biotechnology and given that it is part of the character of those institutions to be concerned with the public's welfare, these institutions are an appropriate locus for insisting on the rational management of science. This introduction is not the place to discuss the form of that management. My point is a preliminary one: institutions already exist which have the might to produce biotechnological knowledge and the duty to do so in a manner which is safe and aimed at the common good.[16,17]

Readings

Chapters in this volume were originally delivered as papers at an Iowa State University symposium on agricultural bioethics in early November 1987. The papers by Buttel, Doyle, Harl, Hollander, Kingsbury, Rollin, and Wright were invited. The remaining essays were selected from the contributed papers. Given our goals of helping a very new field get started and encouraging interest in agricultural bioethics, the editors included as many contributed essays as possible.

Notes

1. We need not worry about a precise definition of *biotechnology*. I understand the notion to include as central cases the use of recombinant DNA (r-DNA) techniques.

2. Part of the explanation of the volume of interest in agricultural biotechnology is the activism of a group of economists and sociologists located at Cornell University and the University of Wisconsin. Their work is not typical of agricultural economists and sociologists.

3. For an introduction to the patenting question, see "Patenting Life," *Report from the Center for Philosophy and Public Policy* 5 (1985):13–14; and Frederick H. Buttel and Jill Belsky, "Biotechnology, Plant Breeding, and Intellectual Property: Social and Ethical Dimensions," *Science, Technology, and Human Values* 12 (1987):31–49.

One moral concern that I do not discuss and that is different in kind from the safety and common-good issues concerns the treatment of farm animals. Are there any reasonable prohibitions on the genetic modification of animals in the service of agricultural goals? See B. E. Rollin, "The 'Frankenstein Thing': The Moral Impact of Genetic Engineering of Agricultural Animals on Society and Future Science," Chapter 21 of this volume.

4. For an excellent set of essays on many aspects of the safety issue, see David Jackson and Steven Stitch, eds., *The Recombinant DNA Debate*, (Englewood Cliffs, N.J.: Prentice-Hall, 1979). An excellent social history of the debate is Sheldon Krimsky, *Genetic Alchemy*, (Cambridge, Mass.: M.I.T. Press, 1982). The present historical presentation draws heavily from Krimsky.

5. Sheldon Krimsky, *Genetic Alchemy*, p. 64.

6. Ibid., p. 65.

7. The epistemological and moral difficulties with risk/benefit analyses are discussed in the context of biotechnology in the following paper: Stephen Stitch, "The Recombinant DNA Debate: Some Philosophical Considerations," in *The Recombinant DNA Debate*, eds. Jackson and Stitch, pp. 183–202.

8. Robert L. Sinsheimer, "An Evolutionary Perspective for Genetic Engineering," *New Scientist* 73 (1977):150–52.

9. For a brief history see Rudy M. Brown, "Genetic Engineering Engulfed in New Environmental Debate," *Chemical & Engineering News* 62 (August 13, 1984):15–22.

10. For a window on the scientific debate, see Gina Kolata, "How Safe Are Engineered Organisms?" *Science* 229 (1985):34–35, and Robert K. Colwell et al., "Genetic Engineering in Agriculture," *Science* 229 (1985):111–12.

11. See Jack Doyle, *Altered Harvest* (New York: Viking, 1985), 214–20; and Constance Matthiessen and Howard Kohn, "Ice Minus and Beyond," *Science for the People* 17 (1985):21–26. As the title indicates, the latter also pertains to the deliberate release issue.

12. Another argument given is that introducing herbicide resistance in corn is a "model" problem. It is technically simple, hence an appropriate place to practice the technology. Supposing that that is the case, what justifies introducing the technology into the market place?

13. See R. J. Kalter, "The New Biotech Agriculture: Unforseen Economic Consequences," *Issues in Science and Technology* (1985):125–33; and R. J. Kalter et al., "Biotechnology and the Dairy Industry: Production Costs and Commercial Potential of the Bovine Growth Hormone," *A.E. Research 84–22* (Ithaca, N.Y.: Department of Agricultural Economics, Cornell University, 1984).

14. Martin Kenney, "Is Biotechnology a Blessing for the Less Developed Nations?" *Monthly Review* (1983):10–19; Martin Kenney and Frederick Buttel, "Biotechnology: Prospects and Dilemmas for Third World Developments," *Development and Change* 16 (1985):61–91.

15. See Loren W. Tauer, "Economic Changes From the Use of Biotechnology in Production Agriculture," *Proceedings of the Iowa Academy of Sciences* 95:27–31. The paradox is very close to the much-discussed prisoner's dilemma. See C. Dyke, *Philosophy of Economics* (Englewood Cliffs, N.J.: Prentice-Hall, 1981), 106–16.

16. For the development of this idea and some of its problems, see my "Biotechnology: A Dilemma for Land-Grant Institutions," *Proceedings of the Iowa Academy of Sciences* 95:32–34. Also see G. Edward Schuh, "Revitalizing Land Grant Universities," *Choices* (1986):6–10.

17. I want to thank Gary Comstock, Steven Gendel, Joseph Kupfer, Tony Smith, Faye Yates, and Laura Kline for their help with this project.

Acronyms

AAAS	American Association for the Advancement of Science
APA	Administrative Procedures Act
APHIS	Animal and Plant Health Inspection Service
ARS	Agricultural Research Service
ASA	American Society of Agronomy
bGH	bovine growth hormone
BSCC	Biotechnology Science Coordinating Committee
BST	bovine somatotropin
BW	biological warfare
BWC	Biological Weapons Convention
CB	chemical-biological
CBW	chemical and biological warfare
CRLA	California Rural Legal Assistance
CSRS	Cooperative State Research Service
DHHS	Department of Health and Human Services
DOD	Department of Defense
DOE	Department of Energy
EEC	European Economic Community
EIA	environmental impact assessment
ERS	Economic Research Service
EPA	Environmental Protection Agency
ESCOP	Experiment Station Committee on Organization and Policy
FAO	Food and Agriculture Organization
FDA	Food and Drug Administration
GAO	Government Accounting Office
HFCS	high fructose corn syrup
ICGEB	International Center for Genetic Engineering and Biotechnology
INAD	investigational new animal drug
IRB	Institutional Review Board
LDC	lesser-developed country

NADA new animal drug application
NAS National Academy of Sciences
NEPA National Environmental Protection Act
NIH National Institutes of Health
NSF National Science Foundation
OECD Organization of Economic Cooperation and Development
OGPS Office of Grants and Program Systems
OTA Office of Technology Assessment
PGH porcine growth hormone
PIC principle of informed consent
RAC Recombinant DNA Advisory Committee
RAF Rural Advancement Fund
RAFI Rural Advancement Fund International
SAES state agricultural experiment station
SERA Socioeconomic Effects Regulatory Agency
TSCA Toxic Substance Control Act
UIR university-industry relationship
UNCTAD UN Conference on Trade and Development
UNIDO UN Industrial Development Organization
USDA U.S. Department of Agriculture
WHO World Health Organization

PART

1

Safety and Regulatory Issues

1

The Release of Bioproducts for Agriculture: Environmental and Health Risks

ROBERT GROSSMAN and BRUCE KOPPEL

Introduction

Science and scientists are periodically critiqued for social and ethical waywardness. The critiques are often valid admonitions of actual and potential consequences of technological change, but should the critiques be directed exclusively at science and scientists? Does such emphasis constitute an oversimplification of problems and, worse, a distraction from more important problems in the social management of science and technology? For example, is contemporary agricultural science so autonomous as a cultural or political system that canons of social, political, or environmental accountability that are at least acknowledged (if not actually practiced) elsewhere in society are simply not perceived? This may be a valid characterization—it certainly is a premise of some of the more well-known critiques—but in terms of what the characterization reveals about the deeper complexity of what agricultural science is and does, it may actually only facilitate marginal insights.

Increasingly we are coming to recognize that relations among science, business, government, universities, and "the public" are pervasive, diffuse, and complex. In this context—where responsibility and accountability can virtually be lost—issues of interests, values, and ideology need to be continually and critically clarified. Biotechnology is a preeminent example. The rising attention being given to biotechnology and the somewhat unique course that debates about biotechnology have followed all testify to the urgency of better understanding (1) the complex relationships that constitute the biotechnology picture; (2) the interests these relationships are ac-

3

tually expressing and serving; and (3) when, why, and how these relationships avoid the detection that public accountability would appear to require. This chapter addresses these issues through an examination of an important component in the biotechnology debate: assessing the environmental and health risks that may be associated with the intentional release of biotechnology-generated products for agriculture. This is an excellent place to pursue these issues because it is an explicit intersection of stakes in scientific advance and the commercialization of biotechnology products, diverse "private" interests in the possible consequences of commercialization, and a policy system that purportedly is steward for a broader and longer-range public interest.

The Intentional Release of Genetically Modified Organisms

The risks that need to be considered to evaluate the intentional release of bioproducts are frequently discussed relative to the previous debate over the NIH guidelines for basic research involving recombinant DNA. This transference of the risk debate from the 1970s to the commercialization of the 1980s has led to a degree of complacency. For example, the biosafety committee at the University of Hawaii disbanded in the early 1980s. Today there is no local mechanism for peer review of experiments or proposed introduction of bioproducts.[1] Specific questions of risk assessment for the intentional release of bioproducts are different from earlier considerations and will be discussed later.

The intentional release of genetically engineered plants, animals, and microorganisms raises scientific and social concerns different from those discussed for basic research. The guidelines established by the NIH were aimed at establishing socially acceptable levels of containment based on assumed hierarchies or risk, whereas the intentional release of recombinant DNA bypasses any concept of containment. To understand this, it is necessary to reexamine the concept of environmental risk that was incorporated into the guidelines.

The NIH identified three assumptions in a decision to call recombined DNA a unique risk: (1) that a unique organism, never found in nature, might be constructed by recombinant DNA techniques; (2) that such a unique organism might be able to establish itself in the environment outside the laboratory; and (3) that such an organism established in the environment and possessing unique properties bestowed upon it by the recombinant DNA techniques, might be harmful to man, animals, or plants.[2] If one of the above assumptions could be proved false, then no special precautions would be needed because there would be no harm.

Two fundamental questions linger with the deliberate release of genetically engineered organisms. First, are the recombined organisms unique? And second, what are the possible avenues for harm? Although there is evolutionary evidence to show that genetically engineered organisms are likely to be at a comparative disadvantage once introduced into the environment, there is also anecdotal evidence from previous epidemics (e.g., the Irish potato famine of 1845) in which a substantial selective advantage among natural populations of pathogens occurs. However, there is little empirical evidence based on population dynamics or coevolutionary theory.

Proponents of reduced guidelines for the release of genetically engineered organisms feel that the question of uniqueness (novelty) is not a serious barrier. It is reasoned, in part, that the level of risk has been reduced through techniques which transfer more defined segments of genetic material (perhaps even synthesized) or gene deletions, so there is greater reassurance that "unknown baggage" is not being transferred. This is in contrast to the initial "shotgun" experiments, where gene uptake was random. However, it is the lack of control of these genes, because of the potential movement through the environment by horizontal transfer, that makes others feel uncomfortable. For example, bacterial plasmids can rapidly transport genes across species barriers, and some genetic material can be picked up as naked foreign DNA even after an organism is no longer living. There are still lingering questions about the enhanced potential for gene transfer from lower to higher life forms.

One of the original assumptions—that the biological barrier between lower and higher organisms would restrict the flow of genes inadvertently released into the environment—may need to be rethought. The regulation and expression of DNA (translation and transcription) may not differ to the degree originally thought. Laboratory experience over the last ten years has shown that DNA from any source can be introduced into any other host. The limitations appear to be the amount of genetic information that can be stably inherited and expressed. Genetic engineering increases the mechanisms for maintaining new sequences even in the absence of homology.[3] There are cases in nature, such as the Ti plasmid of the *Agrobacterium tumefaciens,* where there is genetic exchange between eukaryotes and prokaryotes. Even the presence of self-replicating organelles (e.g., chloroplasts) is evidence of possible incorporation of genes. The rapid increase of antibiotic resistance is evidence of the spread of genetic information through transfer of bacterial plasmids and the resultant adverse impact on human health.

The applications of genetic engineering techniques are different from conventional breeding because they can bypass breeding barriers by, for example, inserting genes from microorganisms into plants and animals. Such innovations include glyphosate-tolerant tobacco, superfixing inoculants, plants engineered for increased tolerance to harsh environments, engineered microbial pesticides, and the insertion of growth hormones into domestic animal breeds. The range of possible risks are, perhaps, as broad as the diversity of the experiments.

Examples where genes have been introduced raise other more difficult questions about pathogenicity and biological or ecosystem stability. Other factors, such as increased genetic homogeneity, may be important because of how the genetic innovations will be used in production systems. Therefore, it is essential to give some scope to the innovations and breakthroughs, pending or likely, and how they fit. A review of biological innovations for U.S. agriculture reveals the links between structure and increased risk. For instance, superfixing strains of *Rhizobium* may be engineered to fix atmospheric nitrogen even in the presence of soil-borne nitrogen. This could potentially increase the amount of nitrogen runoff from fields into water sources depending on the current use of inorganic fertilizers in a given agricultural system. The use of nitrogen fixation genes in combination with other biological innovations could further increase the level of nitrogen in ecosystems.

Some scientists have argued that although there is an abundance of gaseous nitrogen in the atmosphere, there are insufficient carbohydrate sources; hence there is a natural limiting factor. Others have argued that limited carbon sources could be overcome for agricultural systems through higher photosynthetic efficiency, artificial substrates, or intercropping with "leaky" legumes. There is little doubt that biological innovations will be used in combination. Therefore, risk must be evaluated in different agricultural systems and with combinations of technologies.

Genetic homogeneity in crops is another area of serious concern.[4] The corn blight of the 1970s, which struck the male sterile lines used to reduce the cost of hybrid seed, is a good recent example. By the late 1960s, some 90 percent of U.S. corn had a common genetic link (a trait located in the mitochondrial genome) which in unfavorable weather conditions became the weak link. This practice in foundation seed production increased the overall vulnerability of the corn crop to pathogens. It is likely that elite varieties produced by biotechnology will rapidly increase the levels of nuclear and cytoplasmic homogeneity. In conclusion, the concept of risk should include considerations of vulnerability for production systems.

Assessing Risk

There is consensus among scientists that the possibility of risk does exist. The disagreement is over the probability of damage occurring. Since an element of risk exists, society must decide whether the burden of proof rests with those who are looking for a level of safety or with those demanding evidence of harm.[5] These two different theoretical approaches to risk are evident in the legal battles involving the adequacy of environmental impact assessments (EIAs) under the National Environmental Protection Act (NEPA). The battles—which have slowed down the commercialization—have led to a restructuring and redistribution of the federal mandates for regulation of genetic engineering. This change marks a phase of deregulation. Evidence suggests that the Executive Office of Science and Technology Policy and the Biotechnology Science Coordinating Committee (BSCC) played a role in promulgating these shifts. The debate, involving the level of risk society is willing to take, has been narrowed to economic considerations. However, the assumption that entire categories of bioproducts, such as gene deletions, can be released without regulatory scrutiny needs to be questioned because of what is known and not known about the horizontal transfer of genetic information introduced into the environment.

The regulatory restructuring has raised the question of conflict of interest in such agencies as the USDA and sparked litigation over whether the change in rules in the Coordinated Framework also violated the Administrative Procedures Act (APA). Although the possibility of risk may be small, the damage to the environment, should it occur, could easily outweigh the benefits.

Trying to Understand the Current Regulatory Framework

In May 1984, a working group on biotechnology was created by the White House Cabinet Council to review regulation by federal agencies. In addition to NIH, the major regulatory agencies include the FDA, EPA, and USDA. The National Science Foundation, under the agreements of the multiagency framework, will examine the environmental impacts of basic research. In addition, each agency committee reports to the Biotechnology Science Board, which was formed in August 1985. This orchestration has been charted by the Executive Office of Science and Technology Policy. It is interesting to note that neither the Biotechnology Science Board's meetings or minutes were open to the public.

Millions of dollars have been spent in basic research on biotechnology, but methodologies do not exist to fulfill NEPA requirements with respect to environmental assessment. The industry cannot move forward without experimental use permits to field-test their bioproducts. One of the main barriers is a lack of risk-assessment methodology. The NSF has taken a key role in addressing this problem.

In August 1985, the NSF published a report to the Office of Science and Technology Policy of the Executive Office of the President titled "The Suitability and Applicability of Risk Assessment Methods for Environmental Applications of Biotechnology." Risk assessment is defined in the report as "the process of obtaining quantitative or qualitative measures of risk levels, including estimates of possible health and other consequences, as well as the uncertainty in those consequences." The framework is composed of a five-stage process involving (1) risk identification, (2) risk-source characterization, (3) exposure assessment, (4) dose-response assessment, and (5) risk estimation. The fifth stage integrates exposure and dose-response assessments to give an overall summary of environmental, health, or safety risks. Methodologies for risk estimation include statistical analysis, worst-case analysis, sensitivity analysis, confidence bounds, and probability distributions (e.g., Monte Carlo, event tree, or probability tree).

Ecological assessment is less well developed than epidemiological studies for human health, although there has been application of similar studies for specific agricultural systems. Proliferation of genetic information in ecosystems is difficult to measure because of numerous "feedback loops"— for example, changes in the environment that can accelerate establishment or transfer of the genes to other organisms with different habitat requirements. Generally, ecosystem interactions are difficult both to describe and to predict. Pathogenic properties may also greatly enhance persistence and increase risk. Even nonconventional methods for testing persistence, such as plasmid analysis, protein electrophoresis, and DNA/RNA probes, depend on quantitative nucleic acid recovery. This is not always feasible. The key to risk assessment is prediction, and, logically, the vast permutations of information that are constantly changing make comprehensive assessment questionable. It is, therefore, a question of trade-offs, for there will always be some risk. Still, there is the underlying question of whether economic pressures are principally controlling the regulation of this power technology.

For example, the Office of Technology Assessment (OTA) has used media and the apparent legitimacy of "official" government reports to foster the perception that biotechnology must be commercialized if the United States is going to maintain its "competitive edge" within the international biotechnology community.[6] This frames the risks that may be associated with an aggressive program of intentional release of biotechnology products within the context of an idealized national goal (rather than a narrower context of environmental and health relationships). An example is *Public Perceptions of Biotechnology,* an especially egregious manipulation of public opinion in which OTA sells rather than gauges perceptions of the need to commercialize biotechnology as a "social good."[7] Although we are told that only 35 percent of those surveyed have "heard or read a fair amount about the topic," the executive summary confidently ends as follows:

> As in other areas of science and technology, people favor the continued development and application of biotechnology and genetic engineering because they believe the benefits outweigh the risks. And, while the public expects strict regulation to avoid unnecessary risks, obstruction of technological development is not a popular cause in the United States in the mid-1980's. This survey indicates that a majority of the public believes the expected benefits of science, biotechnology, and genetic engineering are sufficient to outweigh the risks. (P. 5)

The OTA and NSF Workshop on the Deliberate Release of Genetically Engineered Organisms: Horizontal Transfer and Genetic Testing

On November 21 and 22, 1985, OTA and NSF jointly hosted a workshop to explore the horizontal transfer and genetic stability associated with the deliberate release of genetically engineered organisms. The workshop had several goals, but its central focus was the exploration of horizontal transfer of novel genetic material. The chair was highly motivated to develop consensus on generic rules for distinguishing between situations of high and low frequency of secondary transfer of genetic information. Included in this effort was a discussion of ways to reduce the frequency of horizontal transfer through modification of vectors, conditional lethal genes, and marker genes. Such approaches would require experiments that can detect such transfer in the field. To date, this methodology is limited. So what were the points upon which there was scientific agreement?

The scientists had consensus on six key "facts." They agreed that (1) most microorganisms have never been identified and only 10 percent can be cultured; (2) transposons and plasmids are genetically unique; (3) plasmid location can result in more active donor sites; (4) population density is critical; (5) there is no concept of "recall" once an organism is placed into the environment because even at only small numbers, it will persist; and (6) it is important to know whether the gene(s)/vectors will be replicated.

Perhaps more important was what the group of key scientists were unable to agree upon. There was no consensus on a way to approach the issue of deliberate release that would yield generic rules for distinguishing between high and low frequency or to measure range of horizontal transfer of genetic information. How can one predict what the secondary transfer of genes is likely to be when so few microorganisms are known? What do we know about cryptic genes, especially those maintained through infectious transfer? How much free DNA survives on clay particles and for how long? What are the mechanisms which maintain the integrity of species? These were some of the unanswered questions. What was evident is that environmental assessment will need to proceed on a case-by-case basis. Since there was little room for generic approaches to risk assessment, deregulation was achieved through the introduction of a Coordinated Framework of Regulation of Biotechnology. Such changes in the "rules" have been questioned by new legal action filed by the Foundation on Economic Trends, which filed suit on July 15, 1986, in the U.S. District Court of the District of Columbia.

The 1985 Washington Conference on Biotechnology held in Arlington, Virginia, (April 22–23) dealt almost entirely with regulation. One key seminar, which focused on "The Economic Impact of Regulatory Delay," asked for a clearer understanding of the economic ramifications of further governmental delay, and what the government can do to speed up clearance of testing and release of biotechnology products. With concern for maintaining international competitiveness running so high, it is not surprising that the U.S. Department of Commerce played a role in this conference and in limiting the export of biotechnology on grounds of national interest (often linked to national security).

The analysis of U.S. competitiveness was a focus of OTA's 1984 report to Congress entitled *Commercial Biotechnology: An International Analysis.* One key conclusion of the report was that

> The United States could have difficulty maintaining its competitive position in the future if several issues are not addressed. Additionally, clarification and modification of certain aspects of U.S. health, safety, and environment regulation and intellectual property law may be necessary for the maintenance of a strong U.S. competitive position in biotechnology. (P. 8)

Regulation is seen as influencing where companies will locate their production facilities, especially for products with recombinant DNA. The summary goes on to say

> At the current time, public perception is not an important factor in the commercialization of biotechnology. However, the volatility of a potential public

response must be noted. Were there to be an accident due to commercial biotechnology, the public's reaction could be extremely important to the future of biotechnology. (P. 20)

Is There a Conflict of Interest between Promoters and Protectors?

In 1985, the GAO issued a briefing report, "The U.S. Department of Agriculture's Biotechnology Research Efforts," to the chairman of the House Committee on Science and Technology.[8] The GAO found that the USDA was funding, in whole or in part, 778 biotechnology research projects. The amount of USDA funding spent on these projects in fiscal year 1984 by The Cooperative State Research Service (CSRS) and the Office of Grants and Program Systems (OGPS) or planned to be spent in fiscal year 1985 by ARS totaled $40.5 million. It is evident, although not discussed in the GAO report, that there is a fundamental conflict of interest in having the USDA regulate the biotechnology industry for products while the agency is a promoter of the technology.

The great majority (84 percent) of the eighty-seven USDA scientists involved in the r-DNA experiments who were surveyed by the GAO felt there was little risk involved in the deliberate release of their products and that any problem associated with the release would be either self-correcting or controlled with little effort. It is uncertain at this time whether the USDA will be able to self-regulate its own experiments and products. This was recently the case with the USDA's sale of Omnivac-PRC, a genetically engineered vaccine against a pseudorabies virus. The USDA had licensed Biologics Corporation to put the vaccine on the market but halted sales after a petition was filed with the USDA to suspend the license because the department had failed to undertake an environmental impact assessment. The USDA justified the lack of an EIA statement on grounds that the product was not significantly different from other marketed products. The issue of who regulates whom and the correlated issue of conflict of interest exist not only for the USDA but also for the NSF, NIH, DOD, and DOE.

Conclusion

Ideology is maintained and driven by powerful symbols. In *The Symbolic Uses of Politics,* Murray Edelman distinguished between referential symbols which are "economic ways of referring to the objective elements in objects or situations" and condensation symbols that "evoke the associated emotions."[9] Such use of symbols by those interested in rapid commercialization of biotechnology has masked important questions of risk. This results from a false bounding which narrowly defines public policy and confuses the meaning of "public interest." One way this is accomplished is through manipulation of public perceptions. How does this masking operate? According to Edelman, "To define beliefs as public opinion is itself a way of creating opinion, for such a reference both defines the norm that should be democratically supported and reassures anxious people that authorities respond to popular views."[10]

There is disturbing evidence that precisely this is happening in the evolving cycle of biotechnology development, testing, and commercialization. There are debates, but it is essential to recognize how these debates are managed and what issues are actually being evaluated. In effect, the debates are not really about environmental and health risks that can plausibly be associated with bioproduct release but rather are better described as canvassing about why, whatever those risks are, they are outweighed by certain higher goals — most notably, the "freedom" of the scientific enterprise from social management and the competitiveness of the American economy.

Several final points can be made.

1. Discussion of risk in recent years has been orchestrated through the President's Office of Science and Technology Policy, the BSCC, and the redefining of regulatory roles for EPA, USDA, FDA, and NIH. With what is known and not known about horizontal transfer of genes in the environment, there is no scientific consensus on a generic regulatory approach for field testing and wide-scale application of bioproducts for agriculture. Decisions will still need to be made on a case-by-case basis.

2. The concept of risk must be broadened to include the genetic vulnerability of agricultural systems. Levels of genetic homogeneity should be monitored not only for genes within the nucleus of plants, but also in organelles within the cytoplasm. In the future, it is likely that genetic innovation (gene segments) will be introduced simultaneously into more than one crop (e.g., the use of nitrogen-fixing genes).

3. It is important that questions of risk involving the intentional release of genetic engineering be delineated from the debate over the revising of the NIH guidelines. Economic concerns over regulatory delays and loss of international competitiveness should not overshadow the need for more stringent debate over the lack of risk-assessment methodologies and liability issues. The various ideological components that are currently masking

many of the risk questions should be exposed. General statements about "public perception" and broad social benefits should be questioned.

Notes

1. Hawaii State Legislature. House Resolution 193, 1987.
2. National Institutes of Health, "Recombinant DNA Research: Proposed Revised Guidelines," *Federal Register* (December 4, 1981):59368–59425.
3. Ibid.
4. National Academy of Sciences, *Genetic Vulnerability of Major Crops* (Washington, D.C.: NAS, 1972); General Accounting Office, *The Department of Agriculture Can Minimize the Risk of Potential Crop Failures* (Washington, D.C.: GAO, 1981).
5. U.S. Congress, Office of Technology Assessment, *Impacts of Applied Genetics* (Washington, D.C.: GPO, 1981):203–204.
6. U.S. Congress, Office of Technology Assessment, *Commercial Biotechnology: An International Analysis* (Washington, D.C.: GPO, 1984).
7. U.S. Congress, Office of Technology Assessment, *New Developments in Biotechnology: Public Perceptions of Biotechnology* (Washington, D.C.: GPO, 1987).
8. General Accounting Office, *Biotechnology: The U.S. Department of Agriculture's Biotechnology Research Efforts* (Washington, D.C.: GAO, 1985).
9. Murray Edelman, *The Symbolic Uses of Politics* (Urbana: University of Illinois Press, 1967).
10. Murray Edelman, *Political Language* (New York: Academic Press, 1977):49–50.

2

The Development of the "Coordinated Framework for the Regulation of Biotechnology Research and Products"

DAVID T. KINGSBURY

On June 26, 1986, the final part of the "Coordinated Framework for the Regulation of Biotechnology Research and Products" appeared in the *Federal Register*.[1] This achievement resulted from more than two years of work by a large number of people representing more than eighteen federal agencies and executive offices. The goal in developing the Coordinated Framework was to explain to the American public that, for questions involving the products of biotechnology, human health and the health of the environment were adequately protected. The policy guidelines are based on generally accepted scientific principles and, therefore, provide a rational yet stringent basis for regulation.

The process of development of the Reagan administration's policy began in April of 1984 within a cabinet council working group. In a December 1984 *Federal Register* notice, that working group published a proposed coordinated framework.[2] Each agency attempted to define what it saw as the regulatory challenges of this technology and how the agency might respond. Each research and regulatory agency was asked to identify the most critical elements of their role in biotechnology and to focus on how that factor interfaced with the other affected agencies. The NIH and NSF had been working for many years under the *NIH Guidelines for Research Involving Recombinant DNA Molecules* and felt confident that no significant policy changes were necessary, although NSF, after discussions with the director of NIH, proposed a modified review procedure for some NSF proposers.

EPA felt that their most critical element was a definition of a "new" organism to fit with regulation under the Toxic Substance Control Act

14

(TSCA). The proposed definition was process-based. A microorganism would be considered "new" if significant human intervention had been used to develop it. Microorganisms developed by recombinant DNA (rDNA) and cell fusion techniques were presumed to be new because they involved significant human intervention. The December 1984 notice also included a statement from the USDA regarding what constituted a "regulated article," which (similar to the way the EPA definition facilitated regulation) was necessary to define those items to be covered under USDA statutes. In addition to defining the role of each agency, the working group proposed a strong central "biotechnology science board" to oversee all of the government's regulatory activity.

Those proposals are now history. In response to the December 1984 notice, the working group received hundreds of comments about all aspects of the policy. While each agency was responsible for dealing with its own policy statement, the working group served as a central policy-coordinating forum. The working group eventually voted to establish the Biotechnology Science Coordinating Committee (BSCC). The BSCC was constituted to include the commissioner of the Food and Drug Administration; the director of the NIH; the assistant secretary of agriculture for marketing and inspection services, the assistant secretary of agriculture for science and education, the assistant administrator of EPA for pesticides and toxic substances, the assistant administrator of EPA for research and development, and the assistant director of the NSF for biological, behavioral, and social sciences.[3]

The committee was to operate as a *scientific* coordinating committee between the agencies, within the context of its charter and a "memorandum of understanding" signed by each of the members. It is important to recognize two factors about the BSCC. First, its responsibility was to focus on scientific and not policy questions; and, second, it was to act as the vehicle to coordinate agency interaction and help establish a rational basis for establishing lead agency responsibility in those areas where possible overlap of jurisdiction exists.

There were many public comments to the EPA definition as it appeared in the December 1984 notice. Upon consideration of the comments, EPA sought assistance from additional scientists both in and out of the government. The result was to refine the process-based definition from "human intervention" to include organisms that would not be "naturally occurring" or "pathogens." It was felt that microorganisms that would combine genetic material among source organisms from different genera should be considered "new." They were new because of the degree of human intervention required, the significant likelihood of creating new combinations of traits, and the greater uncertainty regarding potential risks of such organisms.

Thus, an *intergeneric* organism would be a "new" organism. Other organisms in the same taxonomic genus created by genetic engineering would not be considered new organisms. However, under the related definition of *pathogen* that EPA was developing, even these would be regulated if they were pathogenic.

While EPA was refining the definitions of intergeneric organism and pathogen, USDA's Animal and Plant Health Inspection Service (APHIS) was also reviewing the comments it received from its December 1984 *Federal Register* notice.[4] APHIS and USDA's researchers under the assistant secretary for science and education were seeking to develop those categories of products or research that should be subject to USDA review. USDA staff were benefitting from their own expert analysis and the refinements being developed by EPA.

The issue of definitions eventually came before the BSCC as the members represented the particular agencies developing them. The BSCC felt that uniformity across agencies, to the extent possible and appropriate, would strengthen and simplify the regulatory structure. Uniformity would also tend to discourage "agency shopping" by the regulated community seeking to find a sympathetic regulatory agency.

At the time of publication of the June 26, 1986, notice, the formulators of the Coordinated Framework felt that we had finally achieved the appropriate regulatory balance necessary to deal with a new and rapidly developing industry. Furthermore, we hope that the principles established in this policy will be suitable as a model in other countries.

The Coordinated Framework is a broad and complex policy which explains the application of existing statutes to the regulation of biotechnology and outlines the approach to interagency coordination, which is so vital to this field.

The June 26, 1986, policy consists of six elements: the preamble and statements of policy from the Food and Drug Administration, the Environmental Protection Agency, the U.S. Department of Agriculture, the Occupational Safety and Health Administration, and the National Institutes of Health. Also published at the same time were proposed USDA rules governing the "Introduction of Organisms and Products Produced through Genetic Engineering Which Are Plant Pests" and an "Advanced Notice of Proposed Guidelines for Biotechnology Research," also from USDA.

The *Federal Register* policy statement shows how the separate agencies' policies form a coordinated framework, yet each of the agency statements stands alone; each describes the agency's separate and respective policy. The preamble is essentially an executive summary, providing an overview of what is biotechnology, what are the agencies' separate policies, and how they work together. Biotechnology is not a unitary entity. Instead it is an enabling technology; it has broad applications in many diverse aspects of industry and commerce. The sharing of different agency viewpoints and approaches is essential in this new field where generic review standards are not well established and each product must be reviewed on a case-by-case basis.

The preamble describes the BSCC efforts to coordinate scientific aspects of agency activities in biotechnology. One approach is to coordinate, as appropriate, the definitions used by each of the agencies for the items to be regulated. The BSCC section also mentions concern regarding "release," and requests comments on ways to define it. The preamble also mentions international policy and expresses the OECD principles relating to rDNA safety and regulation.

Government regulation is not organized around technological processes, but rather the government tends to be structured around products developed by various technologies for specific intended purposes. Therefore, one critical element in a coordinated regulatory framework is the common definition of the nature of the products subject to particular types of regulatory oversight.

The BSCC in its role as a science coordinating body recognized that the EPA's definitions of *intergeneric organism* and *pathogen* would take effect immediately upon publication of the *Federal Register* notice as it would be EPA's final notice. This immediate effect was important to insure that "new" microorganisms would become subject to review on the date of the notice. The committee worked with EPA scientists and their advisors as well as with the other agencies to insure that the definition of a regulated article would be as consistent as allowed by statute. The BSCC believed that it would be useful to specifically solicit the views of the public and thus asked for comments. The comments received have been reviewed by the BSCC, and the BSCC shared those comments with the affected agencies, which are considering what further steps, if any, should be taken to refine the definitions. It will be up to each agency itself to make any changes to its own definitions. The BSCC continued to work to promote consistency and coordination.

The principal focus of the policy is environmental release of new organisms. There has been general acceptance of the regulation of nonliving products of biotechnology, and, in fact, the new policy reiterates that the past regulatory practices will be maintained. The new policy explains the application of certain statutes over genetically modified organisms and in some cases imposes an abbreviated review over unmodified organisms applied to environmental uses. There is a clear policy established requiring review of genetically engineered microorganisms prior to release into the environment, with some organisms subject to an abbreviated review. (For reference see the summary table in the EPA policy section.[5]) This policy has been in effect since June 1986; however, additional rule making is needed for full implementation. Until the rules are final, EPA expects persons introducing such organisms into the environment for nonagricultural uses to report to EPA voluntarily. In the unlikely event of a problem arising in this period of time, EPA could use its authority under Section 7 of TSCA to immediately limit or prohibit the manufacture, processing, distribution, or use of the product. In addition to the EPA activity, USDA will review all genetically engineered plant pests and animal pathogens.

Of equal significance to the current policy is the future implementation of the policy and the continued coordination of agency activity. While it is clearly the statutory responsibility of each agency to carry out its function, it has been the role of the BSCC to help in the coordination and development of full implementation. During its first two years of life, the BSCC has charted a full and active course of action. The committee has carried out the majority of its work through a subcommittee structure designed to provide a broad multiagency and, in some cases, nongovernmental representation. For example, the BSCC operated through subcommittees on risk assessment; greenhouse containment standards and definitions of containment and release; training; research and research needs; and public information and education.

A central coordinating body will continue to deal with the scientific questions that arise in the future, including questions associated with risk-assessment methodology and its further development and implementation. This central body anticipates dealing with questions of agency jurisdiction and the appropriate form of agency cooperation in the review process. Central coordination should also extend into questions of data submission requirements.

While we are proud of the Coordinated Framework as a document and a policy, we recognize that this is only the beginning. Much work remains ahead in implementation. A major emphasis is on informational and educational activities. The Biotechnology Science Coordinating Committee and the involved agencies have sought to provide information to the public. Agency officials have testified before congressional committees. The regulatory and research agencies are now implementing aspects of their policies. Some rule making will be required, and certain research guidelines will need to be developed, particularly for those actions involving the environmental introduction of certain biotechnology products.

The Coordinated Framework imposed new regulatory requirements, although a few unfortunately misinterpreted the approach as deregulatory. Part of this confusion may have resulted from an active and organized opposition by several small groups to the underlying enabling technology used in biotechnology. The challenge for researchers and regulators is to reaffirm their competence and judgment while allowing the emerging biotechnology industry to progress quickly in a safe and rational fashion. It is worth noting at this point that this industry has not had a safety problem in its more than ten-year history.

While it is not possible to predict precisely what impact biotechnology will have on the future of the world economy, most observers feel it will be substantial. Recent predictions state that the current situation in the United States of relatively low sales will change to several billion dollars of sales in the 1990s and as much as forty billion by the year 2000. The product mix by that time is expected to be pharmaceuticals, agricultural chemicals and pesticides, growth-promoting hormones for animals and plants, and a line of very-high-value-added industrial chemicals. Many observers predict that the use of specially engineered microorganisms to degrade pollutants and chemicals and biological waste-treatment processes are also highly probable products for the relatively near future.

It appears, to the concern of many observers, that two nonscientific issues may play a dominant role in the future of biotechnology development and implementation in the United States. First, the regulatory climate could, if not rational, present U.S. industry with an insurmountable problem for the eventual introduction of products being developed now, leading to a future withdrawal of activities in these areas. Second, concerns within the financial community about the long-term stability of companies doing business in environmentally regulated fields will lead to decreased company values and lack of capital for the continued development and testing necessary to overcome the regulatory requirements of safety and effectiveness.

Everyone associated with the regulatory questions related to the issues of biotechnology has a major responsibility to act quickly and definitively to balance the various opposing forces facing this technology. We cannot accept the notion that all biotechnology-derived products are too dangerous to introduce into the environment, yet we must acknowledge the legitimate concerns of the public and work to establish principles which govern the safe environmental use of these products and allow the underlying research to proceed. In such a fast-moving technological environment, it is necessary to regularly review the appropriateness of the scientific basis of existing regulation and to make any required adjustments in either the technology of regulation or the statutory basis for regulation. The interagency committees that worked on the current policy are very confident that the existing regulations cover the current and near-term biotechnology industry. As the technology changes, we must continue to monitor those changes and work together to keep our regulatory Coordinated Framework modern and effective for fulfilling the responsibility we have to the American public.

As the industry continues to grow and expand into nonmedical applications, it is clear that many of the new products of the industry are going to be the *living organisms themselves,* and that most will be applied in the environment. This recognition has fueled the latest wave of concern regarding the environmental consequences of the application of genetically engineered products. This concern emanates from many sources, including environmental groups, the Congress, and groups generally concerned with genetic manipulation of any type. Those who have expressed the most concern have focused on the possible negative environmental impact of the new products and have generally disregarded the potential positive environmental impact from, for example, the replacement of toxic chemical fertilizers and pesticides with new microbial products. While most of the current concern about the environmental application of genetically modified organisms ("deliberate release") comes from the environmental protection groups, these same groups have frequently failed to grasp the enormous potential of applied biology in solving, or at least minimizing, a number of environmental problems.

Resolving this paradox places a great challenge on our national leaders. A recent poll sponsored by the Office of Technology Assessment suggests that the majority of the American public does not trust government spokespersons or those from industry, but rather turns to the academic community for factual information.[6] This places on the universities the responsibility to try to bridge the gap between the commercial sector, often viewed with distrust by the environmentalists, and the environmental community, which needs to understand the potential applications of these techniques to the service of the environment. This is developing as one of the serious social concerns for the future development of nonmedical biotechnology. The government, too, has a heavy stake in this problem and responsible government leaders must take some initiative. Modern biotechnology is a direct product of many years of government investment in biological and biomedical research. The United States is now beginning to reap the benefits of that research. The promise of biotechnology is starting to be realized. However, we must continue to work to ensure that this past investment pays its expected dividends and that the benefits are broadly shared by society. One factor, which will be necessary for many years to come, is a continued commitment to research, both to basic research in biological processes (such as gene regulation, genetics, cell biology, and plant sciences) and to the environmental sciences (such as ecology, population biology, and systematics). We are entering an era when the pharmaceutical and chemical industries are calling for broader research in generic scale-up production technology which will enable them to more quickly and economically produce their products. These are important areas of research and are critical for positioning the United States as a competitor in the world marketplace. I would argue that it is also important that we commit an adequate level of research funding to the environmental sciences in order to continue to build the broad data base necessary to ensure the intelligent application of this powerful technology to the pressing problems of the environment and to *scientifically based* regulatory decisions needed to enhance progress in the future.

In retrospect, we can now ask questions about the wisdom of delaying the development of rDNA technology in the 1970s and examine the lessons learned for the future. In the early NIH rDNA guidelines one of the several classes of experiments which were highly restricted was the cloning of viral genes, especially those from the RNA retroviruses. This delay in the development of cloned viral genomes significantly slowed the accumulation of knowledge about this now "famous" group of viruses. Does this delay in knowledge generation resulting from a few people's uncertainty over the technology give us insight into today's debate about the environmental introductions, all of which have been proposed as *research* introductions and not product applications? This example of the delay in knowledge generation about the retroviruses and its potential impact on AIDS research may serve as a paradigm to introduce the concept of the risk of not implementing a new technology when it can potentially address serious public health or environmental problems. It is, of course, impossible to precisely estimate the cost of the delay in basic research on the retroviruses and the subsequent impact on today's knowledge base about HIV. Several workers in this field estimate that there was the loss of two to three years; others may believe this to be excessive. Regardless of the exact loss of time, there is agreement that the delays were costly. Today we are faced with a series of many other questions that can also be addressed by the new biology but are being slowed by public concern over this generally misunderstood technology.

Biological remediation of chemical spills and chemical dump sites has been proven to be highly effective in several instances. All of the successful examples have been limited to indigenous organisms taken from the site in question, propagated to large numbers, and returned to the site in question, usually with some additional site engineering to enhance their activity. These organisms have all had limited potential and often "microbial communities" were needed to achieve the required result. Academic and industrial investigators alike have declined to apply modern biological genetic alteration to these organisms for fear of provoking a public response that the "cure may be worse than the disease." This reluctance to innovate is having a direct effect on the rate at which we can consider biological remediation of such problems as the cleanup of the alarming number of toxic waste sites currently identified in the United States. As a medical microbiologist who has worked with some of the most dangerous microorganisms known to humankind, I cannot help but question the wisdom of our society's willingness to trade the fear of microbes for the known chronic debilitating effects of these toxic chemicals.

Likewise, in the western part of this country we are faced with the serious problem of erosion of our farmlands. This erosion has more effect than the loss of the topsoil on our farms since this topsoil, in most cases, is entering our rivers and silting in behind our many dams. These dams form a critical part of the system of energy generation and flood control. If their effectiveness for either or both of these functions is lost, the potential environmental impact is very great. Biotechnology alone cannot solve these problems; however, it can play a role as one part of the solution. For example, no-till farming is commonly cited as one partial solution to the erosion problem. This farming technique is aided by the ability to control weeds and residual growth from previous plantings through the use of herbicides. This requires herbicide-tolerant or -resistant crop plants, usually with a series of different herbicides for different crops. Many of the newer herbicides like glyphosate and sulfonylurea are rapidly degraded in the environment and some may be used at very low application levels. Despite this fact, there has been widespread opposition to the introduction of herbicide-tolerant plants by environmental groups. There are several good arguments for concern about herbicide-tolerant plants, and these have been widely discussed in recent scientific meetings. What seems deplorable is not that the environmental groups are concerned, but that such concern over the possible problems has paralyzed the potential for a rational and supportive interaction to assist in the development of approaches to the development of the appropriate products. Blind opposition to biotechnology development interferes with the ability to make positive contributions toward the most intelligent use of the new technology for constructive solutions to environmental problems. In short, the environmental groups are as much stakeholders in the new biology as are those needing the pharmaceutical products of biotechnology. As stakeholders, they must begin to share in a constructive dialog about the formulation of the ongoing policy regarding the implementation of the products of the new biology, rather than focusing on banning their use.

Notes

1. *Federal Register* 51 (June 26, 1986):23302-93.
2. *Federal Register* 49 (December 31, 1984):50856-907.
3. Ibid.
4. U.S. Congress, Office of Technology Assessment, *New Developments in Biotechnology: Public Perceptions of Biotechnology* (Washington, D.C.: GPO, 1987), 2.
5. *Federal Register* 51:23319.
6. *New Developments in Biotechnology.*

3

The Development of Regulations for Biotechnology: Comments

NEIL E. HARL

This session comes after almost a decade of deregulation in the United States. The winds of deregulation have been blowing since the late 1970s in finance, in transportation, in commerce, and in antitrust, to mention some of the major areas. The most ardent proponents of deregulation argue that the market is an arbiter far superior to regulations for resource allocation, income distribution, and economic activity in general. To a point, I am in agreement with that view. The market is a highly effective mechanism, although I do not worship at that shrine.

There are several areas that, in my opinion, should not be subject to deregulation. These include (1) passenger safety, (2) product safety, (3) safety and soundness in financial markets, and (4) environmental matters.

In the area of biotechnology, the principal justification for regulation is safety. To a substantial degree, that means environmental safety, but product safety is also involved.

Traditionally, the role of regulations in the U.S. legal system has been to fill in the voids and interstices left by the enactment of statutes, utilizing the expertise and experience of agencies to develop essentially the operating rules. Regulations, however, are limited by the governing statutes with the rule-making power indeed constrained by the guidance provided by the legislative body as approved by the executive branch.

In the area of biotechnology, regulations have come to assume a quite different role. Congress has not provided the kind of detailed guidance usually supplied to agencies. Rather, the problem of creating an appropriate institutional or legal framework out of the existing statutes for research, experimentation, and commercial introduction phases of biotechnology has been left largely to the scientific community. In all likelihood, a more detailed statutory framework will ultimately emerge, influenced by the form and substance of regulations. Thus, the usual roles of regulations and statutes have been modified, and, to a degree, reversed.

Perhaps the more exalted role for regulations in this area has been necessary if not inevitable. The Congress has not had the insight to sketch a regulatory regimen. To do so could have caused serious delays in all phases of biotechnology development.

The fundamental question with the regulatory approach taken in the case of biotechnology is whether the process is likely to produce a regulatory framework that protects adequately the public interest. In that regard, the key questions would seem to be, What is the public interest? Are those charged with proposing and approving the regulations likely to understand, appreciate, and protect the public interest? How can we evaluate whether the public interest is being protected?

Before we take up those questions, I would like to make an observation about the general advisability of relegating complex issues to scientific or discipline-based decision-making groups. For several decades, arguments have been made that antitrust issues are too complex for the court system to comprehend and so should be left to economists; family relations issues should be left to sociologists and psychologists; water issues should be left to hydrologists and others familiar with the dynamics of water occurrence and use; and technology issues should be left to scientists and engineers. In general, I have been highly skeptical of such proposals and have doubted, with all of the attendant shortcomings, that the existing decision-making groups in society should be bypassed on a routine basis.

The relevant question is, Who or which discipline is best equipped intellectually to balance the contributions of the various disciplines? For most major issues, including biotechnology, several disciplines are involved. In the case of biotechnology, problem resolution may involve (1) the core sciences, (2) economics, (3) sociology, and (4) philosophy. For a particular issue, even more disciplines could be involved in contributing to the resulting framework within which society expects the resulting technology to function. In general, the task of integrating research insights has been left to the decision-making bodies in society: (1) the legislative bodies (statutory law), (2) the courts (judge-made law), (3) the administrative agencies (administrative law), and (4) the people themselves (constitutional law).

If the task is to frame regulations for research and experimentation in biotechnology and commercial release of products, the result could be expressed as λ_a, which is a dependent variable with an array of independent variables.

$$\lambda_a = f(x_1, x_2, x_3, x_4 \ldots x_n)$$

where x_1 represents scientific aspects, x_2 economic considerations, x_3 sociologic aspects, x_4 philosophic considerations and x_n, prior law.

In transforming the relationship into a simple linear model, the relationship can be stated as

$$\lambda_a = B1_{x1} + B2_{x2} + B3_{x3} + B4_{x4} + \ldots + BN_{xn} + \epsilon$$

subject to

$$\sum_{i=1}^{n} B_i = 1$$

Functioning in isolation, a molecular biologist might develop an expectation that, as a matter of policy, B_1 should be set equal to one and all other B_i's to zero. An economist would probably be inclined to set B_2 equal to one with all other B_i's set equal to zero. In all likelihood, that outcome could be repeated for all contributing disciplines.

Society has created decision-making processes for establishing weights for the B_i's. From time to time, suggestions have been made for the creation of special decision-making groups—a special court for antitrust cases, a special domestic relations court for family problems, a water court for issues dealing with water as a resource, and a science court as a forum to deal with controversial scientific issues. The evidence is meager, however, that the overall quality of decision making would be improved.

If the present system is to be maintained, disciplinarians should be strongly encouraged to take steps to assure that the decision-making bodies are fully aware of (1) what each discipline has to offer and (2) the effects of various combinations of weights for the B_i's. For disciplinarians, the research task is not fully complete until usable decision-making information has been, or will be, disseminated in a clearly understandable form to decision makers or others as ultimate users of information. The present reward system does not encourage that type of information dissemination.

Biotechnology regulation and control are unique in the sense that legislative bodies are scarcely able to frame the issues and thus must depend in any event on scientists for insight. Although they essentially give scientists the rule-making power, it is incumbent upon legislators to assure that the public interests are protected. If the perception emerges that the public interests are not being protected, the Congress will almost certainly assert greater oversight even without much understanding of the problem. Indeed, that has been the case in other areas.

Public Interests

There would seem to be two major areas of public interest. One dimension of effects from the introduction of new biotechnological developments involves the various dimensions of safety. Under this rubric are effects on human health; effects on animal (nonhuman) health; effects on the natural environment (air, soil and water); and effects on property and other features of the environment other than for humans, animals, and the natural environment. For this category of effects, the market does not function very well. Economists refer to this area as one of "market failure."

Some individuals in society are quite concerned about this problem area. For example, plaintiffs in the lawsuit *Foundation on Economic Trends* v. *Heckler,* filed in 1983, challenged the adequacy of NIH guidelines for research involving recombinant DNA molecules and asked the U.S. District Court to grant an injunction against experiments carried out involving the deliberate release of any organism containing recombinant DNA into the environment pursuant to the NIH guidelines. The court on May 16, 1984, granted a preliminary injunction barring NIH from approving experimentation involving the deliberate release of recombinant DNA into the environment and barring the University of California from release of the bacteria in question into the environment. The suit involved release of *Pseudomonas syringae* that would displace natural ice-nucleating bacteria on the surface of plants and, consequently, protect the plants from frost injury. The suit alleged that NIH violated the National Environmental Policy Act by not preparing either an environmental impact statement or an environmental assessment when the NIH guidelines were revised in 1982 and 1983 to permit the deliberate release of recombinant DNA organisms into the environment. In issuing the preliminary injunction, the court indicated that evidence before the court to date was supportive of the plaintiff's position as to these matters. However, the court did not rule on the scientific aspects of the case. At stake were assurances that we will not, in our great cleverness, cause irreparable damage to the environment.

For this category of effects, it is anticipated that the major efforts will be to assure (1) an effective preassessment of human health and environmental effects before research is undertaken and before developments in biotechnology are released for field trials and for general use; (2) an assessment of the economic and social effects as soon as possible after a development has a reasonable probability of being introduced; (3) in some instances, compensation for those suffering unusually large compensable damages; and (4) the development of new technologies and other means to neutralize or to lessen the impact of negative externalities.

With respect to the preassessment function, a system is already in place to provide a review of proposed research in biotechnology. The centerpiece of that system, the Recombinant DNA Advisory Committee of the National Institutes of Health (RAC), focuses primarily on anticipated effects that would reasonably be expected to have an adverse effect on health. For agricultural research, of course, the problem is broader than human health. A review sufficient to insure consideration of all probable effects including those on human health, animal health, and the natural environment (air, water, and soil) is vital.

It is quite important that the review functions be conducted in such a manner as to have and maintain the highest level of confidence possible on the part of the public. As part of that objective, the review effort should involve the most capable and knowledgeable individuals from the disciplines and subdisciplines within the review jurisdiction.

The second dimension of the public interest involves the question of who benefits from the new technology and who bears the burdens.

New technologies are generally either cost decreasing or output increasing in nature, or both.

Once implemented, a new development in the area of production may lead to a change in the demand for inputs, a change in the supply of inputs, or a change in the productivity of inputs.

A major question is who ultimately benefits from new technology — farmers, consumers, input suppliers, or output processors — and in what proportions. In many instances, consumers tend to be the major beneficiaries in the long run with more output at a lower price. In a market characterized by highly elastic demand or by a rapid growth in demand, producers may retain a relatively large share of the gain from technical change. However, in a market characterized by relatively inelastic demand and by slow growth in demand, as is typically the case with food in high-income countries, most of the gains from technical change are passed along to consumers in the form of lower product prices. Only early adopters gain under competitive market conditions.

One likely result of research in the biotechnologies in agriculture is an increase in production from a given amount of inputs. Thus, the level of utilization of some inputs might decline, use might be made of lower quality inputs, or the level of inputs could remain constant with production levels increasing. If the societal interest is in maintaining a level of output, attention would be expected to focus on reductions in levels of inputs and in production or use of lower quality inputs in production. This could permit some easing of concern about soil conservation, for example, if an increase in production per acre of land under cultivation were to occur as a result of developments in the biotechnology area. If the national interest is in maintaining a particular level of production, the public interest would seem to be well served with investment in research in biotechnology with the potential for reducing the utilization level for inputs of greater present value than the present value of the research investment. Quite obviously, research would be needed on the trade-offs between expenditures of funds for research and development work in the new biotechnologies and expenditures of funds to maintain the productivity of inputs such as land.

In some instances, long-range effects may be felt on the structure of agriculture, depending upon the benefits from the new technology by size and type of firm. The new technologies may also affect risk and uncertainty with important implications for investment decision making and risk-sharing arrangements.

Another area of concern is the impact on Third World countries in terms of competitiveness and trade flows, ability to master the science needed to make effective use of biotechnology, and adequacy of compensation for use of their genetic pool.

The procedures being followed to develop and adopt regulations to govern biotechnology research, experimentation, and commercial introduction focus heavily on the safety aspects including health. The regulations make no effort to deal with the allocation of benefits and costs from the introduction of biotechnology. The question is whether the Congress will become involved in that set of issues. It is reasonably clear that the groups charged with adopting and implementing regulations are not constituted to provide such a review. Some, for ideological reasons, argue there should be no regulatory emphasis in that area. Others, with equal eloquence, argue that the area needs and deserves attention. The Congress will be the ultimate arbiter of the matter. My assessment is that the Congress will be sensitive to the losers but that sympathy is unlikely to leave that group economically whole.

Conclusion

A final word: It is a global economy and, to a considerable extent, a global environment. The challenge is to develop a rational, effective regulatory regimen on a global basis without the presence of a global government.

4 | Ownership of Living Tissues and Cells

GLADYS B. WHITE

Human Biological Materials—Questions for the Future

New developments in biotechnology hold great promise for advancing knowledge about various life forms and for improving human health. But with this promise come greater responsibilities for scientists and policymakers. Human biological materials—tissues and cells—can be used to develop commercial products (such as hybridomas and cultured cell lines) for diagnostic and therapeutic purposes. The use of human biological materials for therapy, research, and profit raises a number of important legal, ethical, and economic issues (see Case A).

Many of these issues are similar to those that have been raised concerning human organ donation, which is currently regulated as a result of the Uniform Anatomical Gift Act (National Conference of Commissioners on Uniform State Laws, 1968) and the 1985 National Organ Transplant Act (Public Law 98–507). But the use of human tissues and cells in biotechnology raises questions that have not been answered in previous public policy deliberations concerning the acquisition of human organs. Who owns a cell line—the human source of the original tissues and cells or the scientist who developed the cell line? Should biological materials be sold, and if so, what are the implications for equity of distribution? Should disclosure, informed consent, and regulatory requirements be modified to cope with the new questions raised by the increased importance and value of human biological materials? There are no easy answers. These issues are novel and complex, and no single body of law, policy, or ethics applies directly.

| CASE A | The Hagiwara Case: An Example of a Dispute over Human Biological Materials |

In early 1981, a researcher at the University of California, San Diego, was developing human hybridoma cell lines that would secrete antibodies to cancer cells. Learning of the project, Dr. Heideaki Hagiwara suggested the use of lymph cells from his mother, who was suffering from cervical cancer. The researcher agreed, and the Hagiwara cells were fused to an immortal cell line developed and patented by the investigator. A hybridoma that secreted an antitumor antibody resulted.

Without the investigator's permission, Hagiwara took a subculture of the hybridoma cell line with him to Japan and gave them to the Hagiwara Institute of Health, directed by his father. The university and the Hagiwaras subsequently executed an agreement that permitted the Hagiwaras to use the cell line for scientific research but forbade their transfer to any other party for commercial purposes.

Several months later, the Hagiwaras asserted rights to the cell line and antibody, claiming that they had tangible property rights in the original tissue and were therefore entitled to a pecuniary interest in the derivative cell line. In 1983, the parties reached an agreement under which the university retained all patent rights and the Hagiwaras received an exclusive license to exploit the patent in Asia.

Source: Office of Technology Assessment.

Sources of Human Biological Materials

Human bodies contain a number of elements that are useful in biomedical research. Healthy people continually produce a number of replenishable substances, including blood, skin, bone marrow, hair, urine, perspiration, saliva, milk, semen, and tears. Human bodies also contain non-replenishing parts, such as organs or oocytes. Organs may be either vital (e.g., heart) or to some extent expendable (e.g., lymph nodes or a second kidney). Finally, the body can also have diseased parts. While this report refers to all human parts—replenishing and nonreplenishing, living and nonliving, beneficial and detrimental—collectively as *human biological materials,* the report covers primarily those biological materials most frequently used in biotechnology tissues and cells. The terms *specimens, body parts, human tissue, fluids, bodily substances,* and *biologicals* are also used.

A Problem of Uncertainty

Uncertainty about how courts will resolve disputes between the human sources of specimens and specimen users could be detrimental to both academic researchers and the nascent biotechnology industry, particularly if the rights of a human source are asserted long after the specimen was obtained. The assertion of rights by human sources would affect not only the researcher who obtained the original specimen but other researchers as well, because biological materials are routinely distributed to other researchers for experimental purposes. Thus the original researcher and also scientists who obtain cell lines or other specimen-derivative products (such as gene clones) from the original researcher might be sued. Furthermore, because inventions containing biological materials can be patented and licensed for commercial uses, companies are unlikely to invest in developing, manufacturing, or marketing a product when uncertainty about clear title exists.

This uncertainty about rights could have far-reaching implications as research and development progresses. Research using human biological materials could be thwarted if universities and companies have difficulty obtaining title insurance covering ownership of cell lines or gene clones or liability insurance. Insurers would be concerned not only with suits by individuals who can be identified as the sources of specimens but also by the potential for class action lawsuits on behalf of all those who contributed specimens to a particular research project. Researchers generally claim that the pervasive use of human cells and tissues in biomedical research makes it impractical and inefficient to try to identify the sources of various specimens or to try to value their contributions. Regardless of the merit of these claims, however, resolving the current uncertainty may be more important to the future of biotechnology than resolving it in any particular way.

The Technologies

Three broad classes of basic biotechnological techniques are of particular relevance to this report. They are tissue and cell culture technology, hybridoma technology, and recombinant DNA technology. (A hybridoma is a hybrid cell resulting from the fusion of a particular type of immortal tumor cell line, a myeloma, with an antibody producing B lymphocyte. Cultures of such cells are capable of continuous growth and specific, monoclonal antibody production. A cell line is a sample of cells, having undergone the process of adaptation to artificial laboratory cultivation, that is now capable of sustaining continuous, long-term growth in culture.)

Tissue and Cell Culture

Cells are the basic structural unit of living organisms. A single cell is a complex collection of molecules with integrated functions forming a self-assembling, self-regulating entity. There are two broad classes of cells, *prokaryotic* and *eukaryotic*. Prokaryotes, generally considered the simpler of the two classes, include bacteria. Their genetic material is not housed in a separate structure (a nucleus) and the majority of prokaryotic organisms are unicellular. Eukaryotes are usually multicellular organisms; they contain a nucleus and other specialized structures to coordinate different cell functions. Human beings are eukaryotes.

Because eukaryotes are complex, scientists have studied these organisms by examining isolated cells independent of a whole organism. This reductionist approach, called *cell and tissue culture,* is an essential technique for the study of human biological materials and the development of related biotechnologies. Establishing human cell culture directly from human tissue is a relatively difficult enterprise and the probability of establishing a cell line from a given sample varies, ranging from 0.01 percent for some liver cells to nearly 100 percent for some human skin cells.

Cell cultures isolated from nontumor tissue have a finite life span in the laboratory and most will die after a limited number of population doublings. These cultures will age (the aging is called *senescence*) unless pushed into immortality by outside interventions involving viruses or chemicals. The type of donor tissue involved and culture conditions are important variables of cell life span. Long-term growth of human cells and tissues is difficult, often an art. Most established cell cultures have been derived from malignant tissue samples. Tissue and cell culture techniques have greatly increased knowledge about cell biology and set the stage for the development of hybridoma technology.

Hybridoma Technology

In response to foreign substances, the body produces a constellation of different substances. Antibodies are one component of the immune response and have a unique ability to identify specific molecules. Lymphokines, sometimes called *bioregulators,* are also produced during an immune response.

Cell culture technology provides the tools scientists need to produce pure, highly specific antibodies. By fusing two types of cells—an antibody-producing B lymphocyte with a certain tumor cell line (a myeloma)—scientists found that the resulting immortal hybrid cells, called *hybridomas,* secrete large amounts of homogeneous (or monoclonal) antibodies. Monoclonal antibodies have led to a greater understanding of the intricacies of the immune response, have become powerful and widely used laboratory tools, and have been approved for use as therapeutic agents. Although the production of human monoclonal antibodies has proven much more difficult than the production of rodent monoclonal antibodies, the increasing availability of large supplies of monoclonal antibodies is revolutionizing research, commerce, and medicine.

Lymphokines (such as interferon), for example, were previously available in minute and usually impure amounts—if at all. Hybridoma, cell culture, and recombinant DNA technologies now permit lymphokines to be isolated in pure form and in quantities facilitating further analysis or use. The increased production and availability of these molecules has significant therapeutic promise in the treatment of a spectrum of diseases because of their exquisite specificity and reduced toxicity.

Recombinant DNA Technology

Recombinant DNA technology, also referred to as *genetic engineering,* involves the direct manipulation of the genetic material (the DNA) of a cell. Using these techniques, it is now possible to speed the isolation, examination, and development of a wide range of biological compounds. As with the use of cell culture, the use of recombinant DNA techniques has shed further light on the details of many important biological processes. Gene cloning is a process that uses a variety of recombinant DNA procedures to produce multiple copies of a particular piece of genetic information. It is an important tool that accelerates the study or production of genes. All recombinant DNA methods require a suitable vector to move DNA into the host cell, an appropriate host, a system to select and cull host cells that have received recombinant DNA, and a probe to detect the particular recombinant organisms of interest.

Recombinant DNA techniques have done much to illuminate the regulation and control of important human processes. In addition, advances in this technology underlie many commercial ventures to isolate or manufacture large quantities of scarce biological commodities.

The Interested Parties

Although tissues and cells can be used for exclusively diagnostic, therapeutic, research, or commercial purposes, in fact the various uses of biological materials are usually intertwined, sometimes inextricably. This means that a variety of people, including scientists in the research community, universities, and industry, plus physicians and patient and nonpatient donors, share an interest in the acquisition and use of human tissues and cells. All would likely benefit from a resolution of the uncertainty surrounding the uses of biotechnology.

Commercial Interest in Human Biological Research and Inventions

The government has always maintained an interest in the legal, ethical, and economic implications of the research it is funding, and this interest is magnified when such research might result in inventions that are patentable under federal law. In addition to advances in technology, two events occurred in 1980 to precipitate the increasing research and commercial interest in human biological materials. First, the U.S. Supreme Court held for the first time that federal patent law applies to new life forms created by DNA recombinations—opening up the possibility that products containing human cells and genes might also be patentable. Second, Congress amended the patent statute to encourage patenting and licensing of inventions resulting from government-sponsored research.

Even though the government is the primary source of funding for basic biomedical research, no single patent policy existed for government-supported research until 1980. Instead, each agency developed its own rules, resulting in twenty-six different patent policies. Under this system, only about 4 percent of some thirty thousand government-owned patents were licensed. Furthermore, the government policy of granting nonexclusive licenses discouraged private investment, since a company lacking an exclusive license is reluctant to pay the cost of developing, producing, and marketing a product. Potentially valuable research thus remained unexploited. To resolve this problem, Congress passed the Patent and Trademark Amendment Act (Public Law 96–517) in 1980 to prompt efforts to develop a uniform patent policy that would encourage cooperative relationships between universities and industry, and ultimately take government-sponsored inventions off the shelf and into the marketplace.

The changing legal climate has provided a fertile medium for the growth of university biomedical research and development using novel biotechnologies. From 1980 through 1984, patent applications by universities and hospitals for inventions containing human biologicals increased by more than 300 percent compared to the preceding five-year period. The extent to which these and forthcoming patents will be of commercial value is difficult to assess.

Sources of Human Tissue

There are three major sources of human tissue specimens: patients, healthy research subjects, and cadavers. Patients are a source of both normal and atypical specimens and these individuals may or may not be research subjects. Patient-derived specimens may be "leftovers" obtained from diagnostic or therapeutic procedures and many human tissues or cells that find their way into research protocols are of this type. Patient-derived samples can also be obtained as part of a research protocol. Healthy volunteer research subjects may donate replenishing biologicals if specimen removal involves little or no risk of harm, according to generally accepted principles of human subject research. Cadavers are the only permissible source of normal and atypical vital organs (including the brain, heart, and liver, but excluding kidneys and corneas). They are also the only permissible source of healthy benign organs (such as corneas) destined for research rather than transplantation.

While these donor classifications may seem fairly straightforward, the human relationships involved are more dynamic than these categories suggest. In particular, the physician-patient relationship may change over the course of time into a researcher-subject relationship.

The Research Community

Research uses of human tissue are diverse and difficult to categorize. Generally, researchers study the characteristics and functions of healthy and diseased organs, tissue, and cells. Commercial products developed from human specimens are usually related to medical or research uses. The use of human biologicals is widespread; a recent survey conducted by the House Committee on Science and Technology found that 49 percent of the researchers polled used patients' tissues or fluid in their research.

The revolutionizing effect of biotechnology on the use of human specimens is principally due to three factors: (1) isolation of increasingly smaller amounts of important naturally occurring human biological factors (also known as *biopharmaceuticals, bioresponse modulators,* or *biological mediators*); (2) production of virtually unlimited quantities of these factors (usually found in the body in only small amounts) using recombinant DNA methods; and (3) discovery of techniques to create hybridomas, making it possible to generate large, pure supplies of specific antibodies.

At the most fundamental scientific level, human material is a source for studies designed to understand basic biological processes. From this basic research, commercial development may follow. However, the probability that any one person's biological materials will be developed into a valuable product is exceedingly small. Thus the issue of potential commercial gain from donated materials is relevant to a minority of patients and donors. However, in the future as biotechnology progresses, the importance of the issue and the number of people involved could increase. The potential for commercial gain, which to date is mostly a speculative consideration, may quickly become a reality. It is appropriate to consider these issues and the possible roles of the interested parties now, in advance of their becoming highly visible, so that public policy perspectives can be developed with wisdom and foresight.

Industry

The biotechnology industry is a major interested party in the controversy surrounding the use of human tissues and cells for financial gain. It is comprised of a variety of different types of organizations including the established pharmaceutical companies, oil and chemical companies, agricultural product manufacturers, and the new biotechnology companies. Of the more than two hundred commercial biotechnology firms in the United States actively engaged in biotechnology research and commercial product development, approximately 50 percent appear to be engaged in research to develop a human therapeutic or diagnostic reagent. There is a strong international component to the biotechnology industry, with numerous research and development arrangements and partnerships between American firms and firms in Japan and Europe.

Legal Considerations

United States law has long protected people from injury and damages. Much of this protection is afforded by the common law, the body of judge-made law built on judicial precedents. This body of legal principles has evolved over centuries as judges are called on to resolve disputes that have not been addressed by statute. Congress and state legislatures, however, have enacted numerous statutes to codify, modify, or overrule the common law, or to address larger societal issues that are inaccessible through the use of common law.

The common law does not provide any definitive answer to the questions of rights that arise when a patient or nonpatient source supplies biological materials to an academic or commercial researcher. Because neither judicial precedents nor statutes directly address this question, the court must do what common law judges have done for centuries: reason by analogy, using legal principles and precedent developed for other circumstances.

Three large collections of legal principles could prove relevant to the use of human tissues and cells: property law, tort law, and contract law. These three areas include a broad variety of statutes and precedents that might be relevant and thus this issue could arguably touch almost all facets of American law (see Case B). Overall, however, there is no discrete body of law that deals specifically with these human biological materials. Because common law reacts to damages only after they have occurred, it does not anticipate possible interests that have not existed previously. In the area of the use of human tissues and cells, technology in fact has advanced beyond existing law. It is not possible to predict what principles and arguments of law might actually be used as cases of this sort come before the courts.

Can Human Biological Materials Be Sold like Property?

No area of law clearly provides ownership rights with respect to human tissues and cells. Nor does any law prohibit the use or sale of human bodily substances by the living person who generates them or one who acquires them from such a person, except under certain circumstances unrelated to biotechnology research. In the absence of clear legal restrictions, the sale of tissues and cells is generally permissible unless the circumstances surrounding the sale suggest a significant threat to individual or public health, or strong offense to public sensibility. To date, neither deleterious health effects nor public moral outrage have occurred even though occasional reports of sales of replenishing cells have been publicized. But while the law permits the sale of such replenishing cells as blood and semen, it does not endorse such transactions and does not characterize such transactions as involving property. In this sense, either permitting or forbidding the sale of human specimens by patients and research subjects can be claimed to be consistent with existing law.

CASE B

A Hypothetical Case Study: How the Relationship between Doctor and Patient Might Change

The possible commercialization of human biological materials raises significant new questions about the proper transfer and use of human materials and how such transfers will affect the relationship between doctor and patient.

Ms. Doe, a 42-year-old female, visits her gynecologist once a year for a routine pelvic exam, Pap smear, and mammography. In recent years, her physician's premiums for malpractice insurance have soared. Ms. Doe suspects that this is one of the reasons why she must now sign a sheaf of consent forms and waivers concerning the possible use of her tissues and cells, the possible hazards of exposure to radiation, and a host of other topics. She has come to regard her relationship with her doctors as one in which she is both vulnerable and unprotected. She is determined not to be completely outsmarted and has purposely not waived her rights to any commercial interest that may result from the use of her cells or developed cell lines.

Dr. Ray is a 50-year-old obstetrician/gynecologist beset with the problems and conflicts associated with conducting what had once been a satisfying practice in the modern technological world. She has been sued 3 times in the course of her 25 years of practice, and in each case the charges have been dropped. Nonetheless, these experiences have caused her considerable stress and some humiliation. She no longer derives the satisfaction that she once did in her relationships with patients because she can no longer practice in a trusting fashion. When she first began her practice, she believed that her obligations were to benefit her patients and, above all, to do no harm. In recent years, she has had to change her approach from telling her patients what she thinks they may need to know to providing them with an abundance of information so they can make autonomous choices as recipients of health care.

Advances in biotechnology offer both Ms. Doe and Dr. Ray some potential recompense for what they both view as modern encroachments on what had heretofore been a relatively unencumbered and trusting relationship. Both have heard that it is possible and occasionally profitable to patent cell lines developed from unique cervical cells, and both now wonder if the yearly routine Pap smear could be an opportunity for financial gain.

It is not farfetched to consider the ways in which modern developments in biotechnology might transform the relationship between doctor and patient. It is now possible to obtain something of value in any medical procedure that involves collection of a patient's tissues or cells though the actual probability of profit is minimal. This possibility, however, seems to entail new obligations regarding informed consent. The nature of these obligations, however, is a subject of some debate.

Source: Office of Technology Assessment.

Informed Consent and Disclosure

Every human being of adult years and sound mind
has a right to determine what shall be done with his
own body.
                              —*Scholendorff* v.
                Society of New York Hospital, 1914

The fundamental principle underlying the need for consent for medical
or research purposes is respect for personal autonomy. Consent is a process
of communication, a two-way flow of information between caregiver/re-
searcher and patient/subject about the risks and benefits of the treatment
or research. Informed consent is necessary in both therapeutic and research
settings, but it is especially critical in human research because of ethical and
legal concerns about the rights and well-being of research subjects.

For consent to be valid, the patient or research subject must be given
an adequate amount of information with which to reach a reasoned choice.
Although there are differences from state to state, the information that
generally needs to be disclosed to obtain consent focuses on the nature and
purpose of the treatment or research; risk-benefit information; and the
availability of beneficial, alternate procedures or treatment. Consent in a
research setting, like consent in a traditional treatment context, must be
obtained in circumstances free from the prospect of coercion or undue
influence.

There are two main sources of federal regulations governing human
research. The Department of Health and Human Services (DHHS) and the
Food and Drug Administration (FDA) have promulgated regulations that
delineate the elements necessary for informed consent to research. The
DHHS regulations govern research conducted or funded by DHHS, including the National Institutes of Health. The FDA regulations govern clinical
investigations that support applications for research or marketing permits
for products such as drugs, food additives, medical devices, and biological
products. Where these federal regulations apply, disclosure requirements go
beyond the accepted norms and include disclosure regarding confidentiality, compensation for research-related injuries, and the right to withdraw
from research without incurring a penalty or loss of rights.

These federal regulations are a deliberate attempt to set ethical and
legal constraints on human research. A balance has been struck between the
needs of researchers and the rights and safety of human subjects. The
success of these regulations in achieving this balance is in no small measure
a function of the integrity of investigators and the diligence of institutional

review boards, which review proposed research projects for compliance with human subject research regulations.

Consent and the Prospect of Commercial Gain

The traditional view has been that in therapeutic settings, information disclosed to patients should be related to the risks and benefits of diagnostic tests or treatment, and it should include alternative procedures. Similarly, in the research setting the disclosure of information has focused on the nature of the study and its effects on subjects. Until recently, little thought had been given to disclosing information about the prospect for commercial gain, but with the advent of biotechnology and its potential use of human biological materials in valuable products, this issue merits consideration.

Arguments can be made both for and against the idea of including information about potential financial gain in the required disclosure of information to patients and research subjects.

ARGUMENTS FAVORING DISCLOSURE OF POTENTIAL COMMERCIAL GAIN

If the notion of personal autonomy and the right to decide what will be done with one's body is to be given full legal recognition, then the prospect of commercial gain should be disclosed because this information may help a person decide whether or not to take part in research. Indeed, the overall trend has been toward greater disclosure of information—details about the probable impact of a procedure on lifestyle, the financial costs of one procedure over another, even the length of disability. Requiring disclosure about commercial gain can be viewed as a logical extension of the consent process.

In fact, it can be argued that the federal regulations require disclosure of potential commercial gain because they require disclosure of "significant new findings developed during the course of the research that may relate to the subject's willingness to continue participation." Discovery of a commercially significant tissue or cell in a subject's body may constitute a "significant new finding."

ARGUMENTS AGAINST DISCLOSURE OF POTENTIAL COMMERCIAL GAIN

The primary argument against disclosing the prospect of commercial gain concerns the impact such information might have on the subject's ability to reach an informed choice free of undue influence. The prospect of financial gain stemming from marketable discoveries could hamper subjects from reaching informed decisions because attention to this highly speculative topic could distract attention from other important aspects of the consent process.

Disclosing information about commercial gain could sometimes jeopardize the health and safety of subjects as well as the validity of the research itself. The hope of gain, for example, might lead subjects to give less than candid answers to questions about medical or personal history that might otherwise disqualify them from the study. It might encourage them to expose themselves to risks they would otherwise consider unacceptable. In addition, because disclosure of potential gain is so speculative, it could set unreasonable expectations or be considered misinformation.

It can also be argued that federal human research regulations embody a philosophy that bans participation for inappropriate reasons. DHHS regulations, for example, make it clear that parole boards should not consider participation when making prisoners' parole decisions. DHHS might consider it improper for subjects to participate in research specifically because they might profit financially. Some people thus might argue that banning reference to the prospect of financial gain is necessary to safeguard subjects from undue influence on their decisions.

Are Changes Needed in the Consent Process?

The question of disclosing potential commercial gain related to diagnostic tests or treatment is one the courts or state legislatures will need to address. However, the federal government funds substantial amounts of human research and will also need to consider its regulations in light of this debate. Policymakers, institutional review boards, and researchers face these questions related to disclosure: Should potential commercial gain be disclosed? If so, what pertinent information is necessary? When is such disclosure best made? What safeguards need to be developed to minimize any detrimental impacts resulting from disclosure of probable commercial gain?

The prospect of financial gain is a troublesome issue in terms of voluntary consent and the use of human biological materials. It can be argued that to assure truly voluntary consent, research subjects should not be offered compensation for their time and inconvenience, let alone substantial financial gain. The counter argument is that the sources of human tissues and cells have rights or interests in marketable substances taken or developed from their bodies and so have a right to know about potential profits. Regardless of what decision is reached, care must be taken so research is not adversely affected because it becomes too complicated to get specimens.

Economic Considerations

The traditional relationships between donors and researchers, and among researchers themselves at various institutions, have been informal; both information and biological materials have been exchanged freely. Today, however, the techniques of biotechnology and the potential for profits and scientific recognition have introduced new concerns. At present, there is no widespread sentiment favoring a move toward a market system for human biological materials. However, a few types of human biological materials, such as plasma, and some patented cell lines are currently transferred within a market system. Future changes in the extent of profits generated from the biotechnology industry could force some changes in the current, primarily nonmarket system.

Two key factors probably will determine whether a change occurs in the current system of free donation of human biological materials for use in biotechnology research and commerce. First, a change could arise from judicial decisions in present or future cases under litigation. Second, a change could be initiated through greater public interest as the commercial applications of biotechnology increase and profits begin to be realized.

There are arguments both for and against payments for donations of human biological materials. Arguments over payments for human tissues and cells used in biotechnological research echo similar debates about markets in human organs. There are five principal issues in the debate: (1) the equity of production and distribution, (2) the added costs of payments to sources and costs associated with that process, (3) social goals (the merits of an altruistic system of donations versus a market system), (4) safety and quality (both of the donor and the donated materials), and (5) potential shortages or inefficiencies resulting from a nonmarket system or from changing from a nonmarket system to a market system.

The factors related to social goals, safety and quality, and shortages do not now offer compelling support either for or against paying the sources of human tissues and cells. But two of the issues are central to the debate, and they seem to argue in favor of opposing approaches. Issues of equity argue in favor of a payment system to human sources. On the other hand, the added costs of payments to sources argue against such a payment system.

Equity of Production and Distribution

The equity of a system can be considered from both the production and distribution sides. On the production side, one issue to consider is whether any of the participants are not receiving an equitable return for their services or products. On the distribution side, the main issue is whether there is adequate access to the goods by parties who seek them.

With respect to human biological materials obtained for research, it can be argued that sources are not entitled to the value of their donated materials because they do nothing to develop the materials into the valuable product. To a donor, replenishable tissue is often useless and diseased tissue is actually a threat. It is only the intervention of the researcher that gives value to these materials. Therefore, it is the researcher who should legitimately realize any economic gains from cell lines or other products developed from the original biological material.

With respect to distribution, researchers generally cooperate with each other in supplying biological materials. The main incentive to this cooperation is the scientific commitment to the free flow of ideas and materials, and to date the system has operated fairly efficiently. However, as biotechnological processes and products are commercialized, this free flow of information and materials is facing increasing constraints. Shortages of human tissues and cells for basic research could occur if the incentives to cooperate are insufficient to motivate researchers to go to the trouble of supplying fellow researchers.

Added Costs

Two types of additional costs would be incurred if human sources were compensated for their tissues and cells or if they shared in royalties accruing from licensing agreements concerning the transfer of developed cell lines: the actual compensation to the sources and the cost of administering the program (also called "transaction costs"). These costs could add significant burdens to the process of developing biotechnologies from human materials.

The actual compensation to the human source of original tissues and cells is unlikely to have a large economic impact on the use of human biological materials, but transaction costs are likely to dwarf the costs of payments to these individuals. Studies involving the development of cell lines can take years to complete and commercial application years longer, so the cost of keeping records of the origin of all the cell lines involved might be considerable. In addition, most of the cell lines studied are unlikely to have any commercial value so a large portion of the transaction costs would actually be unnecessary. Furthermore, under a payment system scientists would no longer exchange materials freely; they would have to negotiate over the transfer and value of property rights for cell lines and might hesitate to share materials at all. Such negotiations would further increase transaction costs.

Resolving the Payment Dilemma

From the point of view of equity, a market structure has a strong appeal because it eliminates the potential windfall realized by parties receiving the free donation. On the other hand, the magnitude of the transaction costs associated with payment to human sources may be sufficient to deter any forays into a market structure. Nonprofit organizations can play an important role in the procurement and distribution of human biological materials, just as they have played a key role in marketing blood and organs. At present, there does not appear to be movement toward a change in the existing system of free donations of human biological materials for use in research and commerce in biotechnology.

Ethical Considerations

Are the human body and its parts fit objects for commerce, things that may properly be bought and sold? There are three broad ethical grounds for objecting to or supporting commercial activities in human biological materials: respect for persons, concern for beneficence, and concern for justice.

First, the ethical principle of respect for persons is related to the idea that trade in human tissues and cells ought to be limited if the body is considered part of the basic dignity of human beings. To the extent that the body is indivisible from that which makes up personhood, the same respect is due the body as is due persons. If the body is incidental to the essence of personhood, trade in the body is not protected by the ethical principle of respect for persons.

The second ethical principle relevant to the acceptability of trade in human materials is beneficence—who would benefit? The basic question could be stated this way: Would commercialization of human materials be more beneficial than a ban on such commercialization? Marketing human tissues and cells might be justified if that would lead to only good results or to a preponderance of good results over bad. Those who hold differing ethical perspectives might consider different outcomes as beneficent.

A third relevant principle is justice. Would a market setting be equitable to all members of society, including those who are financially disadvantaged? Part of the public ambivalence about a market in human tissues stems from a sense that such a market would foster inequities.

The Moral Status of Bodies and Their Parts

Ethical and religious traditions do not provide clear guidelines about the ways in which human biological materials should be developed or exchanged. The absence of established customs regarding these materials is due to the relatively new potential for conducting and profiting from the development of human cells into cell lines. The debate about whether or not it is ethical for bodily materials to be bought and sold underlies all discussions about the commercialization of human biological materials. In addition, there are important questions about how justice should be preserved in the distribution of profits accruing from developed human biological materials.

Decisions about the commercialization of human biological materials depend in part on the ways in which human tissues are regarded or valued. Selected Western religious traditions offer some insights about the significance of the human body. Although there are significant variations among them, Jewish, Catholic, and Protestant traditions generally favor the transfer of human biological materials as gifts.

There are two main reasons why it is important to examine religious perspectives when the goal is to develop public policies in a pluralistic society. One reason is historical: many of the laws regarding bodies and their parts have been influenced by religious sources. A second reason is that religious traditions still shape the ethical values of many people and hence they influence whether some uses of bodily parts or materials will be viewed as ethically acceptable or unacceptable. It follows that religious organizations have to be considered when policymakers try to determine which policies are politically feasible.

Two major variables are present in these Western religious traditions that affect the use of human tissues and cells: the type of materials and the mode of transfer. The significance of different modes of transfer (or acquisition, if viewed from the viewpoint of the user) and different materials hinges on various ethical principles such as respect for persons; benefits to others; not harming others; and justice, or treating others fairly and distributing benefits and burdens equitably.

There is a distinction between ethically acceptable and ethically preferable policies and practices. Some modes of transfer and some uses may be ethically preferred — for example, tradition prefers explicit gifts and donations without necessarily excluding sales, abandonment, and appropriation in all cases. Western religious tradition prefers transfer methods that depend on voluntary, knowledgeable consent. Thus preferred methods recognize some kind of property right by the original possessor of the biological materials.

Tomorrow's Choices

Choices about how to handle transfers of tissues and cells from patients and research subjects to doctors, teachers, and researchers are important ethical decisions in two respects. First, these choices will characterize how individuals regard the human body. If certain human parts are "dignified," then social traditions suggest that they may be given, but not sold. Second, like the choice of how to obtain blood for transfusions, the system that is chosen for obtaining human biological materials will convey a sense of the symbolic weight modern society places on the human body and the use of human biological materials in order to relieve suffering and enhance human health.

The dispute between those who believe that commercialization of the human body is justified and those who think it is not is in part an argument between people who accept a philosophical view that separates the body (a material, physiological being) from personhood, identity, or mind (an immaterial, rational being) and those who do not.

Policy Issues and Options for Congressional Action

Four policy issues related to the use of human cells and tissues in biotechnology were identified during the course of this study. The first concerns actions that Congress might take to regulate the commercialization of human tissues and cells. The second involves the adequacy of existing regulations covering commercialization of cell lines, gene probes, and other products developed from human biological materials. The third concerns the adequacy of existing regulations covering research with human subjects. The fourth centers on whether present practice is adequate to ensure that health care providers disclose their potential research and commercial interests in the care of a specific patient or group of patients.

Associated with each policy issue are several options for congressional action, ranging in each case from taking no specific steps to making major changes. Some of the options involve direct legislative action. Others are oriented to the actions of the executive branch but involve congressional oversight or direction. The order in which the options are presented should not imply their priority. Furthermore, the options are not, for the most part, mutually exclusive: adopting one does not necessarily disqualify others in the same category or within another category. A careful combination of options might produce the most desirable effects. In some cases, an option may suggest alterations in more than one aspect of using human cells and tissues in biotechnology. It is important to keep in mind that changes in the one area have repercussions in others.

Issue 1: Should the Commercialization of Human Tissues and Cells Be Permitted by the Federal Government?

OPTION 1.1: TAKE NO ACTION

Congress may conclude that at the present the largely nonmarket basis for the transfer of human tissues and cells is appropriate. If a commercial market in human biological materials should arise, the lack of federal regulation might result in great variability in the amounts of money paid to the donors of the original tissues and cells. If no action is taken, it is unlikely that human patients or research subjects will be compensated for their tissues or cells in the near future.

OPTION 1.2: MANDATE THAT DONORS OF HUMAN CELLS AND TISSUES ARE COMPENSATED FOR THEIR DONATIONS

Some argue that in the interest of equity, the donors of human tissues and cells should be compensated. Congress could decide that human biological materials have a monetary value, even in their unimproved state, and that the donors of these materials have a right to this value. The amount and form of such compensation could vary. Donors could be paid for their time and trouble as opposed to payment for the actual specimen. Payment for service as opposed to substance is now standard practice in the case of sperm donation. Researchers argue that compensation for human tissues and cells in their unimproved form is impractical because the vast majority of these materials will have no ultimate value. Economists argue that the transactions costs of such compensation would outweigh any payment for the original biological material. In addition, many parties are concerned that any payment to the sources of human tissues and cells, no matter how small, would be so inefficient and inconvenient as to stifle research efforts in general. Lastly, some ethicists worry that any trade or market in human tissues and cells unacceptably alters the meaning and value of the human body.

OPTION 1.3: ENACT A STATUTE MODELED AFTER THE NATIONAL ORGAN TRANSPLANT ACT THAT PROHIBITS THE BUYING AND SELLING OF HUMAN TISSUES AND CELLS

Congress may conclude that at present, the existing situation in which human tissues and cells are largely either donated or abandoned for research purposes is satisfactory. If Congress concludes that any for-profit market in human tissues and cells should be stifled or avoided, it could prohibit the sale of these biological materials. Such a statute would prevent patients, research subjects, or other donors from making money from the donation of their tissues and cells. If the Congress enacted a statute modeled after the National Organ Transplant Act in particular, there would be a consistent line of federal reasoning concerning the transfer of human organs, tissues, and cells.

Issue 2: Should the Commercialization of Cell Lines, Gene Probes, and Other Products Developed from Human Tissues and Cells Be Permitted by the Federal Government?

OPTION 2.1: TAKE NO ACTION

At present, cell lines, gene probes, and other products developed from human tissues and cells are exchanged informally among researchers as well as by means of a market system. For the most part, profits are accrued in the form of royalties paid by those who want access to the developed products. If Congress takes no action, the use of patented inventions based on human biological materials will continue to be restricted to those who engage in licensing agreements for access to the patented products.

OPTION 2.2: AMEND CURRENT PATENT LAW SO PARTIES OTHER THAN INVENTORS (E.G., PATIENTS, RESEARCH SUBJECTS, OR THE FEDERAL GOVERNMENT) HAVE PROTECTED INTERESTS AND ACCESS TO ANY COMMERCIAL PRODUCTS DEVELOPED FROM THEIR DONATED TISSUES AND CELLS

Within the context of current patent law, the inventor has exclusive rights to patented material and this effectively bars access by the donors to their original biological material. Some argue that the patients or research subjects, particularly if they suffer from a disease, should have access to or some say in the use of patented products derived from their donated tissues and cells. At present, licensing agreements for the use of these patented materials do not commonly stipulate any protected interest for the original donor.

OPTION 2.3: ENACT A STATUTE PROTECTING THE RIGHTS OF PATIENTS OR RESEARCH SUBJECTS TO SHARE IN PROFITS ACCRUING FROM LICENSING AGREEMENTS FOR THE USE OF CELL LINES OR GENE PROBES DEVELOPED FROM THEIR ORIGINAL HUMAN BIOLOGICAL MATERIAL

The profitable features of patented cell lines and gene probes are the royalties that accrue from licensing agreements for access to these products. Congress may conclude that it is fair and equitable that the original donors of human biological materials should share in the derived profits. Such profit sharing could be in addition to or instead of a flat fee for the original unimproved tissues and cells. Some researchers argue that it is often impossible to identify the donor of the original material as cell lines and gene probes are developed. Many laboratory transformations over a long period of time separate the original donation from the patented invention. If Congress enacts a statute ensuring that the donors of human tissues and cells share in the profits accruing from licensing agreements, then an extensive and costly system of record keeping will be necessary to establish the identity and whereabouts of the original donors.

OPTION 2.4: MANDATE THAT ANY CELL LINE BE PRESUMED TO BE IN THE PUBLIC DOMAIN UNLESS IT HAS BEEN FORMALLY REGISTERED AT THE TIME THE TISSUE WAS EXTRACTED OR PLACED INTO CULTURE

The presumption that cell lines are in the public domain would bar anyone from claiming property rights to these products. While this would not directly compensate the donor or source of the unimproved tissues and cells or the researcher, it might relieve any sense of exploitation (that someone else has taken over that original property right). The patent and similar systems could still apply for further inventions made in developing applications of the cell line.

OPTION 2.5: MANDATE THAT THE FEDERAL GOVERNMENT SHARE IN PROFITS ACCRUING FROM INVENTIONS ORIGINALLY BASED ON GOVERNMENT-FUNDED RESEARCH

In many if not most cases, government-supported research serves as the basis for the existing commercial biotechnology industry as well as other industries and research programs. Some argue that it is fair and equitable for the government to share in the profits that accrue from products that are developed based on government-funded research. At present, most government-funded basic research takes place in universities. If the government were to share in profits accruing from inventions based upon government-funded research, it would transform present relationships among the government, industry, and the universities.

OPTION 2.6: ENACT A STATUTE PROHIBITING PARTIES OTHER THAN INVENTORS FROM SHARING IN ANY REIMBURSEMENT FOR, OR ANY PROFITS DERIVED FROM, THE USE OF PRODUCTS DEVELOPED FROM HUMAN TISSUES AND CELLS

Under the present market system, only those who have patent law protection or enter into a contractual relationship (e.g., a licensing agreement) realize commercial gain from developed tissues and cells. Congress may conclude that the sources should be barred from obtaining any reimbursement for products developed from their tissues and cells. Such action would affirm that commercialization of products developed through the use of human biological materials should be limited to the patent holder and licensees, and that patents and research subjects have no right to the value of their tissues and cells in their altered forms. While such an action might serve as an economic inducement for those who would obtain human tissues and cells for the purposes of developing new inventions, it is arguably contrary to current patent and contract law (which encourages commercial negotiation between willing parties) as well as to the concept of a person's authority over the use of bodily materials.

Issue 3: Are Guidelines on the Protection of Human Subjects (45 C.F.R. Part 46) Issued by the Department of Health and Human Services Adequate for the Use of Human Biological Materials in Biotechnology?

OPTION 3.1: TAKE NO ACTION

If no action is taken by the Department of Health and Human Services to alter the guidelines on the protection of human subjects, it will not be necessary for researchers to inform subjects about possible uses of pathological or diagnostic specimens. As a result, researchers can continue to use these materials as they choose without informing the patient (see option 3.2). In addition, if the guidelines are not altered, it will not be possible for subjects to specifically waive their interests in the uses of their tissues and cells when giving informed consent because of the existing ban on the use of exculpatory language (see option 3.4).

OPTION 3.2: DIRECT THE SECRETARY OF HEALTH AND HUMAN SERVICES TO REMOVE THE EXEMPTION REGARDING PATHOLOGICAL OR DIAGNOSTIC SPECIMENS (IF THESE ARE PUBLICLY AVAILABLE OR THE DONOR IS OTHERWISE UNIDENTIFIABLE) FROM THE REGULATORY REQUIREMENTS [S46.101(B)(5)]

According to existing DHHS guidelines for the protection of human research subjects, an exemption exists for research involving the collection or study of existing data, documents, or pathological or diagnostic specimens if these are publicly available or the donor is otherwise unidentifiable. Researchers are therefore not obliged to disclose their research interests to donors of specimens when this exemption applies. Congress could decide to remove this exemption so that it becomes necessary for research subjects to be informed about and have some say in the use of their cells and tissues. This option would assure that additional research subjects would be informed of the possible uses of biological specimens and related data and appears consistent with the general spirit of the guidelines to protect the interest of the research subjects. Implementation of this option could increase the administrative burdens of researchers who would need to be assured that consent had been obtained for data, documents, and pathological and diagnostic specimens that are currently exempted.

OPTION 3.3: DIRECT THE SECRETARY OF HEALTH AND HUMAN SERVICES TO AMEND THE GENERAL REQUIREMENTS FOR INFORMED CONSENT (S46.116) TO INCLUDE POTENTIAL COMMERCIAL GAIN AS A BASIC ELEMENT OF INFORMED CONSENT

Under the current DHHS regulations, certain information must be provided to each subject during the informed consent process. It could be decided to add a provision requiring that in seeking informed consent, a disclosure be made regarding the potential for commercial gain resulting from data, documents, records, pathological specimens, or diagnostic specimens obtained during the research. Such a requirement could be codified as a basic element of informed consent that shall be provided to each subject (S46.116a), or as an additional element of informed consent to be provided to each subject when appropriate (S46.116b). Such a requirement would make clear that potential commercial gain is an issue that would be reviewed by the Institutional Review Board (IRB).

OPTION 3.4: DIRECT THE SECRETARY OF HEALTH AND HUMAN SERVICES TO REMOVE THE BAN ON EXCULPATORY LANGUAGE AS IT PERTAINS TO COMMERCIAL GAIN (S46.116)

Under the current DHHS regulations, informed consent documents may not include exculpatory language which is used to make research subjects or their representatives waive or appear to waive any of the subjects' legal rights. The intent of this provision is to safeguard subjects and to make certain that they do not relinquish any legal rights. Some subjects may not want to reap financial benefits as the result of or as a byproduct of their participation in research, and some researchers and their sponsors may be deterred from conducting important research if they must share possible financial gain with research subjects. A change in the regulations could be made to modify the prohibition on the use of exculpatory language to permit research subjects to waive any rights to commercial gain. Such a provision would need to be clearly worded. Research subjects should understand exactly what rights are being waived and that they will not be denied treatment to which they are otherwise entitled even if they decide not to waive their rights. If the regulations are amended to permit the use of exculpatory language as it relates to potential commercial gain, the IRB will have a greater role.

OPTION 3.5: UNDER ITS POWER TO REGULATE INTERSTATE COMMERCE, CONGRESS COULD ENACT A STATUTE TO PERMIT AND REGULATE THE BUYING AND SELLING OF HUMAN TISSUES AND CELLS

The advantage of such a statute is that it would offer the possibility of financial compensation to the sources of human tissues and cells. In addition, such a statute would apply to the transfer of these materials from all sources and therefore go far beyond any alteration in guidelines for the protection of human subjects involved in federally funded research. The disadvantage of such a statute is that it permits commercialization of all human tissues and cells and extends federal regulation into a previously unregulated area.

Issue 4: Is Present Practice Adequate to Ensure That Health Care Providers Disclose Their Potential Research and Commercial Interests in the Care of a Specific Patient or Group of Patients?

OPTION 4.1: TAKE NO ACTION

Congress may decide that existing or altered DHHS guidelines concerning the protection of human subjects provide sufficient safeguards to ensure that individuals are aware of the purposes and methods of the research in which they are involved. At the present time, however, these guidelines only extend to research subjects participating in federally funded research. There are no protections for research subjects in privately funded research.

There are no guidelines to ensure that health care providers disclose their commercial interests in caring for a particular patient or group of patients. If Congress takes no action, physician/researchers will not be obliged to tell a patient about their intention to develop commercially valuable products from the patient's tissues and cells. Congress may decide that the commercial interests of health care providers do not necessitate new forms of disclosure in order for patients to be adequately informed.

OPTION 4.2: DIRECT THE SECRETARY OF HEALTH AND HUMAN SERVICES TO PROMULGATE GUIDELINES THAT REQUIRE HEALTH CARE PROVIDERS RECEIVING ANY FEDERAL REIMBURSEMENT TO DISCLOSE ANY RESEARCH OR COMMERCIAL INTERESTS THEY MAY HAVE IN THE CARE OF A SPECIFIC PATIENT OR GROUP OF PATIENTS

If the Congress acts to ensure that health care providers disclose their research and commercial interests in caring for particular patients, it will be necessary to discern what sort of commercial interests in particular merit disclosure. Physicians in private practice obviously have commercial interests in treating patients so their practice remains economically viable. It comes as a surprise to many people, however, to learn that their physician might also engage in research using a patient's tissues and cells and subsequently develop a profitable product based on these donated or abandoned materials. The relationship between physician and patient may be compromised if patients suspect that their caregivers may profit in unanticipated ways. The development of guidelines concerning this type of disclosure could promote greater trust between physicians and patients in the delivery of health care.

PART II

Impact on Scientific and Industrial Communities

5

The Scientific Implications of Biotechnology

J. M. ASPLUND

Nowadays there is scarcely a trade publication or the program of an industry or scientific society meeting that does not include an article or presentation extolling the limitless advantages of biotechnology. The only limit to the benefits to mankind extolled by the authors is the paucity of superlatives in the English language and the imagination of the extoller. Those superlatives are either Utopian or Orwellian, depending upon the persuasion of the author. The sad truth about actual biotechnological contributions, however, is that they have failed to fulfill the promises made for either human advancement or human catastrophe. In fact, despite several years of exciting and productive research and conceptual revolution, food production and veterinary and medical practice remain essentially untouched. Medical fields have changed the most, largely as a result of biotechnological advances in diagnostic tools and the production of certain biological therapeutic agents. Genetically engineered production of hormones and other naturally occurring molecules has made these more widely available and economically possible. In the production of animal foods, apart from the increased availability and decreased costs of some veterinary biologicals, the only concrete accomplishment is the use of bovine somatotropin to improve milk production. The basic biology of this process has been known for many years, but it has come to the fore recently because of the availability of the hormone through biotechnological manufacture. In agronomical food production, most of the traditional crops and cropping practices remain unaltered.

In short, the question needs to be raised concerning the actual impact of biotechnology as opposed to the theoretical accomplishments which have been so highly touted.

In the words of David Nash, director of biotechnology at the University of Alberta, "First, it is a good deal easier to imagine what one might create than it is to create what one might imagine." This author further gives a few tentative timetables for actual solid progress in a practical sense: "Characteristics like disease, drought and pesticide resistance are critically important to the farmer and biotechnology offers prospects of real advances on these fronts in the near future. . . . Radical genetic surgery on plants, animals, even on man, looks to be some years away."[1]

The question needs to be asked: Is the biotechnological revolution a reality; merely a figment of imagination; a product of rhetoric; or a concept that will, if properly managed, fulfill a real need for humankind?

I believe that, conceptually, many of the putative benefits of biotechnology are possible. I also believe that proponents are too sanguine, and the opponents too unreasonably suspicious of the practical realization of benefits. We are dealing with a concept before it is a reality. At present, biotechnology is a straw man for good or bad.

The failure of biotechnology to fulfill its promise (or its threats) to society is based on the imbalanced scientific approach which characterizes the current effort. The results of biotechnology will be felt only when a rational, permanent, sustainable, and significant change is made in some biological system. That change will have to be manifest in a whole organism existing in a particular ecosystem. In order to achieve this kind of change, therefore, it will be necessary to understand the whole organism and the whole ecosystem. Science has made startling and exciting progress in the understanding and manipulation of the genome and in cellular control mechanisms in general. However, the biotechnological potential of control alterations is achieved only when these mechanisms are applied to whole cells, organs, organ systems, organisms, and, eventually, to a specified ecosystem. As Nobel laureate Barbara McClintock is quoted as saying, "One must have a feeling for the organism. . . . One must understand how it grows, understand its parts, understand when something goes wrong with it."[2] Biotechnology, therefore, cannot be advanced simply by understanding the genome and cellular control mechanisms without concomitant progress in related total understanding of the whole organism and its environment. To attempt to do so would be analogous to expecting a revolutionary advance in home heating by building more sensitive and sophisticated thermostats without parallel progress in burner and insulation technology. Studies in control are not only useless but also dangerous unless they are accompanied by studies of the systems being controlled.

The promise of modern biotechnology is that we now have most of the tools and concepts to make rational alterations to biological systems. However, to fulfill this promise, the alteration must be rational and understood at all levels of the system—molecular, cellular, organic, organismic, and environmental. The major limitation to sound, rational change is not, at the present time, lack of understanding of the molecular mechanisms of genetic control, but it is in our understanding of the other aspects of biological change. We simply do not know enough about cellular and organic function, intraorgan transport and controls, intermediary metabolism, energy transduction, food intake control, animal behavior, or animal ecology at the present time to be able to make a fundamental, predictable and controllable change in the genome of a complex organism. David Nash, a noted geneticist, has said

> To this point, I have described, exclusively, prospects that involve rather simple modifications of microorganisms. Practically, any kind of genetic transfer we can envisage seems to work; but there are problems when it comes to operation of foreign genes in new host organisms. Basically, fundamental biotechnological research has so far concentrated on ironing out problems in micro-organisms; in more complex organisms, it's still hit or miss. Further, we do not yet fully understand the normal patterns of gene activity in most complex organisms. Hence, although we can transfer genes to man, for example, we cannot predict whether they would work properly, and indeed, would often not recognize proper operation if we saw it![3]

The insertion, deletion, or alteration of genetic material, however well understood at the molecular level, would, at the production level (the place where biotechnology becomes reality), have to be treated as a random mutation. Two alternative pathways of research would then be possible. The first would be to test the resulting animal for its survival fitness in its intended ecosystem and to evaluate its possible advantages and disadvantages in that system. This involves very empirical production research and will evaluate only phenotypic characteristics. Classical Mendelian genetics and metabolic and physiological studies would also be necessary.

The other pathway would be to examine the basic physiology and metabolism of the "mutant" to understand the phenotypic response. However, this would be wasted effort if the whole animal proved not to be fit or advantageous. Neither alternative constitutes rational, directed, controlled change. Rather it leads to reactive, post hoc research. In effect, it differs little from our past efforts at inducing or discovering mutations with subsequent screening of resultant phenotypes. How much better it would be to understand the whole organism well enough to be able to predict, yes, even engineer total change from the genetic code to the organism in the environment.

The explosive and heady revolution in molecular biology has, as has been stated, opened new vistas of opportunity, but current thinking and policy have created two serious limitations to the realization of ultimate biotechnological innovation. The first is the increasing political and conceptual isolation of the field of molecular biology from the rest of biological research. The other is the massive diversion of funds, facilities, and personnel to molecular biology and away from the other systemic components of total biology in both research and teaching. Molecular biology, therefore, cannibalizes the other research effort necessary for true progress. This decreases the chances for real advances and increases the danger of developmental accidents and environmental harm.

Throughout the country one sees departments, divisions, institutes, and centers of biotechnology being established. These facilities are extremely well funded and equipped. Recruitment of personnel is vigorous and competitive with salaries for new and untried molecular biologists approaching or exceeding those for senior scientists in other fields. Biotechnological effort appears to be centered entirely around molecular biology with little or no effort devoted to the upgrading of related disciplines. Indeed, no institutional or organizational liaisons are being considered with those disciplines. This thrust is perceived by the rest of biology as competition or even conquest, rather than cooperation for the advancement of the total program of biotechnology. Apart from the serious problems of morale and administrative difficulties that this causes, this relationship is causing major distortions in the overall scientific approach that is needed for rational biotechnological innovation. If molecular biology continues its isolation, self-imposed as well as conferred, it cannot make the contribution of which it is capable and will prove to be an obstacle to the progress of other disciplines and the integrated field of biology. The discipline is a major component of the team effort needed, but it is only one component and, just as the other disciplines are hampered and restricted without directed effort in molecular biology, molecular biology will be impotent without concomitant growth in all other aspects of biology.

The isolation of molecular biology from the rest of the biological sciences has manifold immediate impacts on all science. Educationally, it is causing major disruptions in curricula and syllabi. It is becoming difficult to get departments of biochemistry, microbiology, or biology to teach courses that are of value to students in anything but molecular biology. At the author's institution, the introductory course in zoology is at least two-thirds molecular biology. This replaces basic anatomy, physiology, taxonomy, ethology, or ecology. The course is offered without a biochemistry prerequisite so the student is without a basic background in the chemistry and biochemistry of the molecules involved in molecular biology. Thus, instruction even in molecular biology becomes impaired and vocational rather than truly scientific. Biochemistry courses, on the other hand, spend inordinate amounts of time on the mechanisms of gene action and gene expression at the expense of instruction in metabolic pathway, energetics, and enzyme kinetics. The consequences of ignoring proper pedagogical sequencing and disciplinary balance are twofold. First, there is the danger of producing molecular biologists with a distorted view of the place of their discipline in the general field of knowledge, molecular biologists who are even dangerously underprepared in disciplines directly and vitally related to their own field. One can easily imagine a scientist deeply involved in determining the genome of a given organism who is not aware of the structure of a nucleic acid or an amino acid. The second difficulty occasioned by this imbalance is the unavailability to scientists in other disciplines of the basic sciences which are prerequisite to their understanding of the phenomena they study. For example, nutritionists find it very difficult to find biochemistry courses which address intermediary metabolism, energy transductions, or amino acid metabolic pathways. As a result, many nutrition groups have begun to teach biochemistry courses for their own students. These courses are admittedly less rigorous and less on the cutting edge of biochemistry than if they had been taught by biochemists, but they are more pedagogically consistent than is the training available through formal biochemistry courses. One also wonders at the competence of future biochemistry teachers in anything but molecular biology. Thus, the basic disciplines, rather than providing a unifying, systematic basis for all biology, tend to fragment and stratify the field.

If a department wishes to hire a scientist trained in metabolic biochemistry, it has to choose someone underprepared as a biochemist or someone who is trained in molecular biology and whose background and interests will most probably lead him away from the intended program. In this way, traditional programs are becoming colonized and diverted to molecular biology. The isolation is self-perpetuating.

The isolation of molecular biology also causes a disruption of scientific communication within institutions. The eclectic nature of seminars and lectures is being lost and replaced with extremely restrictive and specialized presentations with resulting segregation of special interest groups. As traditional disciplines shrink and molecular biology grows, the scope of and interest in their seminars drop, further exacerbating the parochial nature of broad biology. Molecular biology thus loses the broad base upon which it must rely to be able to achieve measurable biotechnological advances.

Another danger that is evolving rapidly is the stratification of disciplines. There is already a strongly felt caste system among biologists with the Brahmins, the molecular biologists. The current feeling is that the quality of science is determined by the degree of sophistication of the field and not by the validity of the work. The author has heard molecular biology referred to as "good science" while studies of metabolic pathways or enzyme kinetics have been called "inferior science." When the quality and validity of scientific investigation are determined simply by the field one chooses, the whole fabric of science becomes imperiled.

The diversion of funds to molecular biology is a matter of record. However, the imbalance is even greater than it appears on first glance. For example, of the $40.5 million appropriated for competitive research grants under ESCOP of USDA, $19 million or 47 percent of all funds was designated for biotechnology. However, in all but two minor categories of the remainder, the *Federal Register*[4] specifies "work at the cellular or molecular level." Thus, the overwhelming majority of this funding is specified to go directly to molecular biology. An examination of almost every other major grant program in biology will reveal the same trend.

More parochially, many states are developing and funding programs in biotechnology which, as it turns out, are simply programs in basic molecular biology. Indeed, molecular biology may be too broad a definition for the kind of work envisioned at these centers. The bottom line seems to be that any project in molecular biology which can be conceivably connected to a biotechnological fantasy, however diaphanous, will be funded at a level adequate to create a major program, regardless of the quality or relevance of the project. Unfortunately, graduate programs are closely tied to research programs so, as a result, the distortion in funding becomes self-perpetuating in the numbers of scientists available in the various disciplines.

The real tragedy of the separately and locally funded centers for biotechnology is that they have been justified to the donors on the basis of greatly exaggerated claims for benefits of biotechnology in production systems. When these claims are not realized (and they will not be realized because of the imbalance of scientific effort discussed earlier), the funds will be curtailed or withdrawn, leaving the whole of biological research in disarray because of the failure of one branch of research to fulfill its promises. Thus, biotechnology, which has real benefits to offer mankind, will be discredited and mistrusted in the minds of the public, not because the promises were not real but because of the failures to pursue them with a rational effort by all segments of biology. In the pursuit of the spectacular, the chance to achieve reality will have been missed. Eager biotechnologists, like the child who is too eager to wait for the rose to bloom, will have dismembered the bud and left a disfigured stump instead of a beautiful flower.

Notes

1. David Nash, "Biotechnology, Past and Prospects," *New Trail* (Summer 1986, University of Alberta, Edmonton), 9.

2. Barbara McClintock, quoted in Evelyn Foxkeller, *A Feeling for The Organism* (San Francisco: Freeman, 1983) 198.

3. "Biotechnology, Past and Prospects," 8.

4. *Federal Register* 49, no. 30, 5571.

6

University-Industry Relationships in Biotechnology: Convergence and Divergence in Goals and Expectations

**WILLIAM F. WOODMAN, BRIAN J. REICHEL,
and MACK C. SHELLEY II**

Introduction

Arrangements between modern universities and other organizations are hardly new. As early as 1952, Lyle Lanier was compelled by the increasingly common phenomena of contract research to attempt to dispel the common assumptions that such research was basically incompatible with academic institutions.[1] During that same year, Ralph Morgan argued that engineering education benefits by contract research.[2] By 1978, MacCordy could write that not only was the federal research and development (R&D) budget topping $5 billion, but that two-thirds of that expenditure went to only fifty universities.[3] MacCordy's point was not that federally funded research was becoming concentrated at few sites, but that industry needed to exert greater rather than less influence on the research endeavor. The year 1982 saw the *New England Journal of Medicine* publishing an article on "Taking University Research into the Marketplace."[4] By 1986 things had changed to such an extent that MacCordy was writing to an eager audience about the minimum elements a university-industry agreement should contain in order to protect all parties.[5]

It is clear that the rules of the game in externally sponsored university research have changed radically over the past thirty years. Not so clear, however, is the shape of the evolving relationship between our traditional institutions of higher education and private industry. The best place to see this relationship may be through the examination of the expectations held by both parties as such arrangements are created. The authors were presented such an opportunity by events in the state of Iowa.

The state of Iowa has pledged some $17 million in state lottery revenues over the period of four years to support the accelerated development of biotechnology applications, principally at Iowa State University (ISU), with economic development as the explicitly stated end goal. Almost as an afterthought, the sum of $50,000 (later supplemented by private grant money) was allocated for the examination of the ethical, social, and institutional dimensions of biotechnology. One subcommittee of the Bioethics Committee formed as a result of that thrust was charged with understanding the effects of the accelerated expenditures on the university and affected client groups. It was out of such a series of surveys that the research reported here was generated.

Methods of Data Gathering

The following groups were studied: ISU faculty (both those involved in biotechnology and those not involved); graduate students; ISU administrators (all at or above the level of departmental executive officer); and biotechnology companies (obtained from a listing which appeared in *Genetic Engineering News,* 1986).[6] Clearly, faculty and graduate students represented the client group most likely to benefit from UIR agreements, particularly those involved in biotechnology, while ISU administrators represented the institutional vantage point on the contract arrangements, and biotechnology companies represented the industry view on UIRs.

The specific samples were obtained. Faculty were targeted first by their involvement in biotechnology (relying on a published listing, from which 137 were selected). Then, the balance of the general faculty was sampled, from which came the balance of the 264 ISU faculty, representing 59 percent of the total 450 questionnaires sent out. Graduate students were selected by random listing of on-campus students only (94 or 38 percent were returned from 250 mailed); for university administrators, some 115 (71 percent) were returned from 161 mailed; and for biotechnology companies 130 (38 percent) were returned from 342 mailed. University-generated mailing labels were used to assure accurate initial contact with respondents. There were no follow-ups carried out with the exception of a single series of telephone follow-ups for the graduate student sample.

The authors feel that the samples drawn are representative and reflect no systematic bias.

Analysis

Areas of Consensus

There are a number of areas in which the various groups surveyed expressed substantial agreement as to both goals and means. Among these were the following areas addressed by survey questions.

With regard to whether biotechnology represents an appropriate area emphasis for ISU Figure 6.1 demonstrates that there was enormous consensus as to the appropriateness of the direction. Indeed, only a mere 3.3 percent of graduate students demurred, and they represented the largest such source of disagreement. At the same time, the extent of agreement between administrators and industrial people was striking (93.9 percent and 97.5 percent respectively). Note that graduate students represented the least

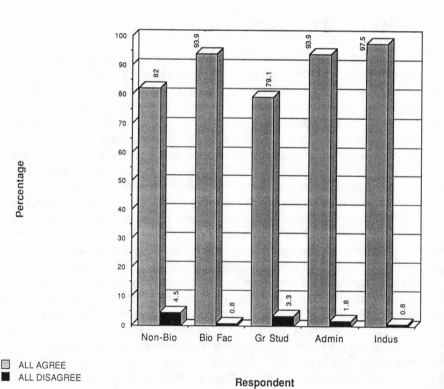

Fig. 6.1. Biotechnology is an appropriate ISU emphasis.

agreement (a "mere" 79.1 percent), with faculty tending to closely reflect the responses of the administrators.

Very similar responses were collected from an item asking whether Iowa State University should work more closely with industry.

In Figure 6.2 it can be seen that in the policy area concerning whether more public (in this case, state) money should be spent on biotechnology, there again existed a strong consensus that such represents a prudent use of state funds.

Note that the percentage of agreement varied only to a minor degree with the exception of graduate students, among whom a mere 54.5 percent agreed that such expenditures made sense. However, among faculty, administrators, and industry respondents there was a mere 3.6 percent range of variation. Among those who disagreed, there also existed an amazing congruency, as less than 3 percent separated all categories of respondents

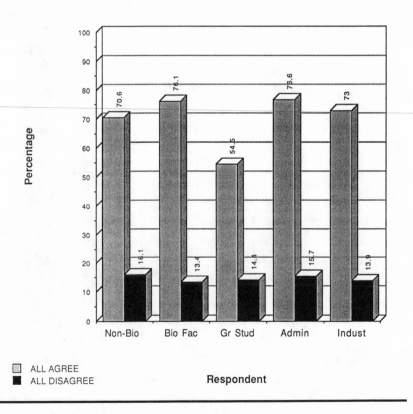

Fig. 6.2. More public funds should be used for biotechnology.

disagreeing with the statement. It is interesting to speculate as to the response differences demonstrated by graduate students on almost every item. Clearly it could be argued that they are incipient faculty in many cases and should thus be likely to mirror faculty interests, although the small numbers of students entering academe relative to the demand suggests that the differences may actually reflect a mixing of conflicting perceived interests (i.e., between those expecting to enter the academic versus the nonacademic marketplace).

Figure 6.3 shows responses to the question of whether biotechnology research holds the promise of a more successful future for agriculture. From the percentages it can be seen that some differences surface in this area, for fully 97.5 percent of industry respondents espoused the expected answer and agreed with the statement. A high percentage of administrators, 66.9 percent, also agreed, with a mere 3.5 percent of administrators dis-

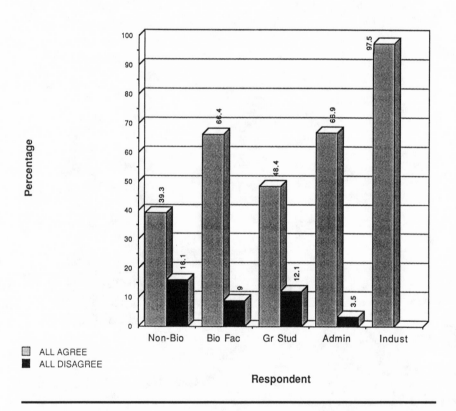

ALL AGREE
ALL DISAGREE

Fig. 6.3. Biotechnology will lead to a more successful future for agriculture.

agreeing. Biotechnology faculty were slightly less likely to agree, with 66.4 percent agreeing while 9 percent disagreed. Of faculty not involved in biotechnology research, a mere 39.3 percent agreed, and 16.1 percent disagreed.

Once again, graduate students were less likely to agree (only 48.4 percent), with the highest percentage of disagreement (12.1 percent) to be found among graduate students. In these instances, there appear to have surfaced some differences of opinion as to outcomes to be anticipated from biotechnology research in Iowa.

As shown in Figure 6.4, even the area of economic market ends, often thought of as exclusively the province of business, seems to have become an area of agreement for respondents from both academic and biotechnology areas. In response to the question, "Should universities strive to perform significantly more work oriented toward industry and market need?" 56.1

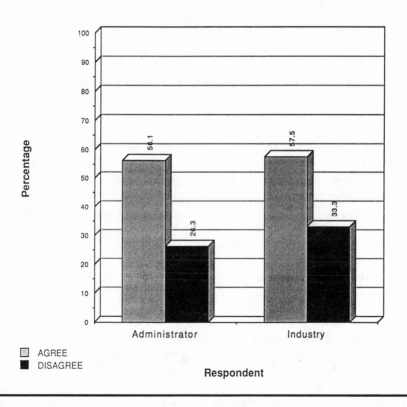

Fig. 6.4. Universities should work more toward market needs.

percent of administrators and 57.5 percent of industry CEOs agreed that universities should be oriented toward meeting the needs of the economic marketplace, with even fewer administrators (26.3 percent) than CEOs (33.3 percent) disagreeing.

This graph highlights two surprising facts. First, there exists a strong consensus between industry and the particular university under study as to the need to orient university activities toward the marketplace; and second, a hefty minority of both administrators and industry CEOs appear to have some misgivings about developments in this direction.

Thus, while there seems to be considerable consensus about certain aspects of economic development goals which biotechnology is presumed to support, there are some nagging questions extant as to how they are to be accomplished and the specific nature of the outcomes to be expected. Some of these differences appear in the following section.

Areas of Dissent

One of the first areas of disagreement as to the expectations for biotechnology relates to some of the solutions it will offer. Figure 6.5, shows that respondents indicated the extent to which they agreed that biotechnology will help solve the problem of farm surpluses by finding new uses for crops and livestock.

As can be seen, slightly over half (50.7 percent) of the faculty members involved in biotechnology envision the field holding such a promise, while only CEOs came close to agreeing in such numbers (44.5 percent). At the lower end of agreement, graduate students (at 32 percent) and nonbiotechnology faculty (34.5 percent) demonstrated little support for the proposition with administrators falling between the two extremes (at 37.7 percent). It is interesting to note that the undecided category constituted a substantial part of the results (ranging from 24.3 percent of biotechnology faculty to 41.9 percent of graduate students).

Thus, these results suggest substantial ambivalence or uncertainty as to the potential of biotechnology to deal with the problems of commodity surpluses.

In Figure 6.6, the issue of the likely beneficiaries of biotechnology is examined. Specifically, the respondents were asked to indicate their agreement with the statement that "Advances in biotechnology will probably benefit persons with large farm operations more than persons on middle sized or small farms." As can be seen, with some moderate differences in the ranges of 40.9 percent to 56.2 percent, university personnel (from graduate students to administrators) saw it as rather obvious that large farms would be more likely than small farms to benefit from biotechnology.

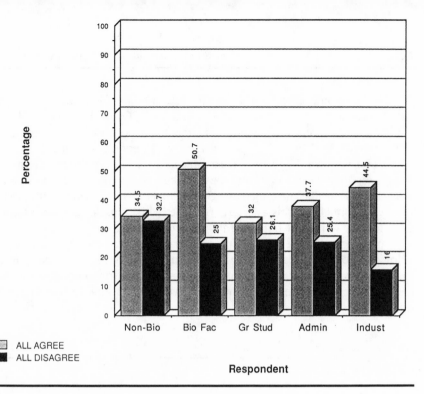

Fig. 6.5. Biotechnology will solve farm surpluses.

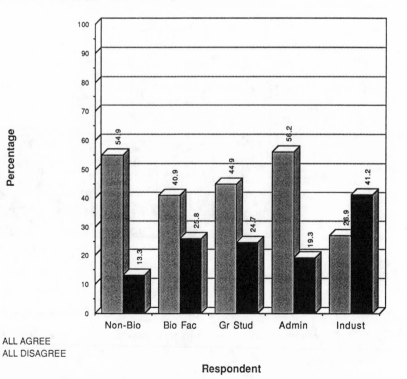

Fig. 6.6. Biotechnology will likely benefit large farms more than small farms.

Note that nonbiotechnology faculty and administrators demonstrated almost identical percentage responses. Among CEOs in the biotechnology industry, 41.2 percent indicated disagreement, the only group to do so, and fully one in four CEOs (26.9 percent) expressed agreement with the statement. This item was important to examine because it indicated the extent to which respondents saw biotechnology as likely to accelerate current trends in agriculture which have increasingly endangered small scale farms. University personnel saw this as clearly the case, while those in the biotechnology industry were split but generally tended to disagree.

In Figure 6.7, the same issue was examined through the statement, "Biotechnology will lead farmers to become more dependent upon large corporations for many of their inputs such as seeds, growth hormones, and feed additives." The unanimity among university personnel was significant (ranging only from 57.5 percent to 63.2 percent across four categories of

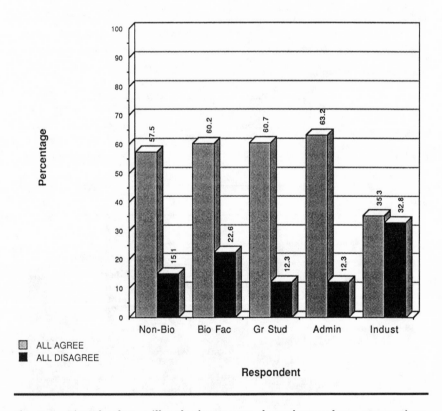

Fig. 6.7. Biotechnology will make farms more dependent on large corporations.

respondents), while CEOs split almost evenly with 35.3 percent indicating agreement and 32.8 percent indicating disagreement. Note that no value connotation was built into the format as asked; thus, there exists no means of knowing whether respondents saw increasing dependency as a positive or negative outcome.

It should be noted too that here again there was not only a split between university and industry views, but there also existed substantial ambivalence on the part of all parties as evidenced by the "no opinion" or "no answer" option (from a high among university staff of 27 percent for graduate students and 31.9 percent among CEOs). In sum then, almost one in four graduate students and one in three CEOs were unable or unwilling to express an opinion on this issue.

Figure 6.8 reveals findings derived from questions asked only of university administrators and CEOs. In this case, they were asked to agree or

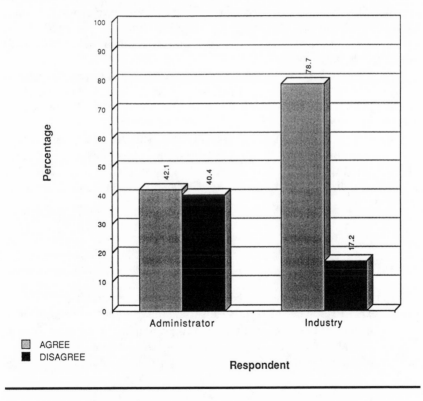

Fig. 6.8. Universities should withhold data until patent is obtained.

disagree with the question, "Should universities agree with industrial sponsors to withhold research results from publication until patent protection can be obtained?" There seemed to exist a clear break on these kinds of issues relating to what could be called the "nuts and bolts" of research marketing, for while administrators almost equally split on the issue (42.1 percent approval to 40.4 percent disapproval), biotechnology industry CEOs approved of the idea by a whopping 78.7 percent (although it was interesting that almost one of five, or 17.2 percent, of CEOs felt that universities should have no obligation to withhold publications until patents are obtained). In these less clear areas, it seems that the conflict between the traditional role of the university as a knowledge-producer as contrasted with that of an equal partner in entrepreneurship is an issue about which both industry and university people exhibit considerable differences of opinion.

Finally, as shown in Figure 6.9, the question of direct market involvement was engaged when respondents in university administration and biotechnology companies were asked to agree or disagree with the statement, "New discoveries by university scientists should be patented by the university and sold to the highest bidder who would then make these products commercially available."

As Figure 6.9 shows, while university administrators were strongly in favor of the proposal (66.4 percent), fewer than half of the biotechnology industry CEOs approved (43.4 percent), and almost one in three CEOs (31.7 percent) did not see this as an appropriate role for universities. It may well be that these individuals, as well as those who declined to express an opinion, saw this as an unwarranted movement of a state-sponsored entity into the marketplace. If this is the case, then the surprising figure is not the number who disagreed among industry CEOs, but the 43.4 percent approving of the idea.

Conclusions

Certain parallels between the goals of universities and industries obviously exist. Even more obvious seems to be the existence of serious questions as to logical and acceptable means to accomplish some of the goals. The most productive areas of inquiry now appear to relate to the kinds of accommodations and compromises necessary to reach these goals.

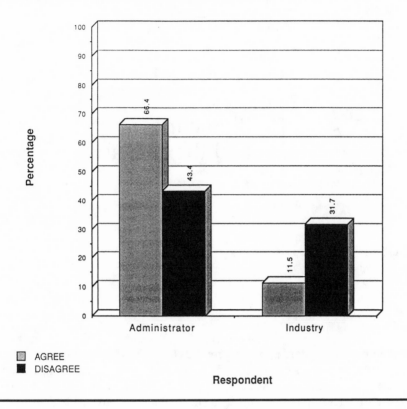

Fig. 6.9. Universities should sell marketing rights to products.

Notes

1. Lyle H. Lanier, "Contract Research as a Problem in the Sociology of Science," *American Psychologist* 7 (December 1952):707–9.

2. Ralph A. Morgan, "The Role of Sponsored Research in Training Personnel," *Journal of Engineering Education* 43 (November 1952):145–48.

3. Edward MacCordy, "The Untapped University Resource," *Les Nouvelles* 13 (September 1978):178–85.

4. Gilbert S. Omenn, "Taking University Research into the Marketplace," *New England Journal of Medicine* 307 (September 1982):694–700.

5. Edward MacCordy, "Industry-University Research Agreements," *Chemtech* 16 (October 1986):612–14.

6. "Fifth Annual GEN Guide to Biotechnology Companies," *Genetic Engineering News* 6 (November/December 1986):13–48.

7 Biotechnology and the Military

SUSAN WRIGHT

In November 1969 at a hearing on the U.S. chemical and biological warfare policy before the House Foreign Affairs Committee, molecular biologist Joshua Lederberg pointed to the rapid advances being made in his field and warned of their potential military use and the possibility of "the development of agents against which no reasonable defense can be mounted." And, he commented, "[W]hatever pride I might eventually wish to take in the eventual human benefits that may arise from my own research is turned into ashes by the application of this kind of scientific insight for the engineering of biological warfare agents. In this respect, we are in somewhat the same position as the nuclear physicists who foresaw the development of the atomic weapon."[1]

Lederberg, whose own work on the molecular genetics of bacteria had contributed to the dramatic advances of molecular biology in the 1950s and 1960s, was in a good position to make this assessment. His concern was echoed by others as the new fields of genetic engineering and hybridoma technology developed in the 1970s. Nevertheless, over the first decade of development of the new biogenetic technologies, this problem was largely set aside. In the 1980s, however, it has been reactivated by news that the Department of Defense and other military establishments are pursuing extensive programs of research and development in these areas, by the ambiguous nature of some of this work, by a widespread perception that legal barriers to biological warfare have been eroded, and by some voices within the Pentagon that claim that a new generation of biological weapons is within reach.

This chapter examines the evolution of perceptions of the military potential of biotechnology in relation to actual developments in genetic engineering as well as to changes in the military, political, and legal contexts of this technology. Biogenetic technologies have obviously evolved dramatically in the last eighteen years—from the first stumbling efforts of the late 1960s, of which Lederberg was aware, to the precise manipulations that are feasible today. However, the central argument of this chapter is that the recent surge of concern has been influenced not so much by technical advances as by changes in the social contexts in which biogenetic technology is being pursued. Stated in another way, international tensions and hostilities, increasing military interest in biotechnology, and a new perception of the fragility of the international legal regime banning biological warfare have all combined in the 1980s to place the power of genetic engineering, hybridoma technology, and other biogenetic technologies in a different light. What was the subject of exuberant enthusiasm in the 1970s and early 1980s is now qualified as a field that definitely promises "dual use." The product of science and technology is not a thing in itself; it can be seen only through the lens of social, legal, and military arrangements.

Perceptions of the Military Potential of Molecular Biology in the Late 1960s

In the late 1960s, the field of molecular biology had reached maturity. For almost two decades, the DNA model proposed by Watson and Crick in 1953 had provided the basis for a series of dramatic theoretical and technical achievements. Given the instrumental orientation of the field inherent in the description of the living cell as an information-processing machine as well as the utilitarian emphasis of U.S. biomedical research policy, it was hardly surprising that American scientists began to entertain the possibility of genetic engineering—of active intervention in and manipulation of biogenetic processes. Many experiments aimed at introducing foreign genes into living cells were undertaken at this time. These early efforts were unsuccessful, but there was little doubt in the minds of many of those involved at this time that the goal would eventually be achieved.[2]

At this time, the United States was supporting an active chemical and biological warfare effort. In the 1950s, the U.S. chemical and biological warfare (CBW) program had grown substantially under the influences and impetus of the Cold War. Funding was stepped up in the late 1950s. (Support for the program grew from about $10 million in the 1950s to over $300 million in the late 1960s.)[3] Legal restraints on the development of the program were weak. The United States had signed but had not ratified the Geneva Protocol banning the use of chemical and biological weapons. During the Cold War, an earlier policy of no first use of chemical weapons was abandoned and replaced with one of "preparedness for use at the discretion of the President."[4] A new arsenal of chemical and biological agents was developed: nerve agents, herbicides, incapacitating agents, psychotropic drugs, and a wide range of lethal biological agents. Although these weapons were not broadly assimilated into military practice, the extensive use of herbicides in Vietnam demonstrated that they could be and that moral scruples could be overcome by arguments of military expediency. Biological agents remained the least assimilated.[5] Their unreliability and delayed action were seen as important disadvantages. However, the development of new techniques of delivery and dissemination during and after World War II gave them new potential for use as tactical weapons as well as the capability of inflicting immense damage, not only on civilian populations but also on animals and plants.[6]

The advances being made in molecular biology did not go unnoticed. As early as 1962, a Defense Department spokesman told members of Congress that "it is not unlikely that . . . major contributions to biological weaponry and defense will result from research and a better understanding of the sciences of genetics." And by the late 1960s, the department was anticipating that "within the next 5 to 10 years, it would probably be possible to make a new infective microorganism which could differ in certain important aspects from any known disease-causing organisms. Most important of these is that it might be refractory to the immunological and therapeutic processes upon which we depend to maintain our relative freedom from infectious disease."[7] At the same time, the National Academy of Sciences-National Research Council was approached about initiating an exploratory program of research in this area.[8] From a military perspective, molecular biology offered a potential for overcoming the disadvantages of biological weaponry that had previously constituted barriers to assimilation.

This was also a period of deepened sensitivity to the destructive power of modern science and technology. As the antiwar movement gained strength, the use of the electronic battlefield, napalm, herbicides, and antipersonnel gases in Vietnam was increasingly seen as a reflection of the complicity of modern science in warfare. The United States CBW program became the target of strong criticism both at home and abroad. Within the United States, criticism concentrated on the use of chemicals in Indochina, the open-air testing of chemical weapons, and the use of medicine and biology for the development of weapons of mass destruction. Thousands of American scientists signed petitions calling on the government to declare its commitment to refrain from initiating use of chemical and biological warfare.[9] In the United Nations, repeated complaints were made against the U.S. use of chemicals in Vietnam. A 1969 UN resolution affirmed the position of the majority of nations (from which the United States dissented) that the Geneva Protocol prohibited "any chemical agents of warfare . . . which might be employed because of their direct toxic effects on man, animals or plants."[10]

In large part as a result of the broad international concern about growing use of chemical warfare, a number of initiatives were taken in the late 1960s to place the question of chemical and biological disarmament on the agenda of the UN General Assembly and the Committee on Disarmament. In 1968, it was agreed that this issue would be taken up by the Committee on Disarmament. In July 1969, a report issued by the Secretary General of the United Nations warned of the growing destructive capacity of CB weaponry developed since World War II and called on all nations to halt development, production, and stockpiling of CB weapons and to achieve their elimination from military arsenals.[11] In July 1969, the United Kingdom submitted a proposal for a treaty aimed at comprehensive biological disarmament to the Committee. At that point, President Nixon ordered a review of U.S. chemical and biological warfare policy.

It was in this context of increasing assimilation of chemical and biological weaponry by the military and emerging military interest in molecular genetics on the one hand, and of the efforts towards chemical and biological disarmament abroad and policy reassessment at home on the other, that scientists close to the field of molecular biology began to share their concern about future military use of the field. The tangible model of assimilation and use of biological research by the military provided by the use of herbicides in Vietnam added special urgency to their unease.

An important event that illustrates biologists' apprehensions about the future military use of their field was a meeting in Boston in the summer of 1969 arranged to address policy issues associated with chemical and biological warfare. The meeting was sponsored by the American Academy of Arts and Sciences and the Salk Institute, and the participants included several leading molecular biologists and others close to the field, as well as specialists in international law, political science, and military strategy. Two representatives of the Nixon administration were also present. The discussion ranged widely over possible routes to strengthening the Geneva Protocol's ban on the use of chemical and biological weaponry. Comprehensive chemical and biological disarmament, biological without chemical disarmament, persuading the administration to include herbicides and incapacitating agents in its definition of chemical weapons, and U.S. ratification of the Geneva Protocol were among the issues discussed.

In the background of this discussion was a concern about future military use of advances in molecular biology. This concern was not specific; there was no concrete hypothesis about how molecular biology would be used — as, for example, physicists had anticipated the use of nuclear fission for the creation of nuclear weapons.[12] Rather, it derived from a general argument that military exploration of a powerful new field of science would eventually produce discoveries that could be turned to military use. Biochemist Paul Doty expressed this concern as follows: "The ultimate consequences of developing successive generations of biological weapons probably go far beyond what we can expect out of the chemical side because the diversity [of the organisms] with which one can work is much greater."[13] This view was reinforced by Nobel laureate James Watson, who proposed that it might be better to seek a limited ban on biological weapons rather than struggle with the "enormous vested interests" built up by the military in chemicals.[14] The idea was also supported by strategist Herbert Scoville, who proposed that the group submit a paper to the president on "the security aspects of giving up biological warfare."[15] Thus there was a sense emerging among at least some of the participants at this meeting that the way to deal with military interests in new advances in molecular biology was to stop them in advance.[16]

Similar concerns were aired by biologists in more public settings. Joshua Lederberg's testimony to the House Foreign Affairs Committee in November 1969 was cited earlier. In testimony to the UN Committee of the Conference on Disarmament in August 1970, Lederberg elaborated his position, warning that the future application of molecular biology was double-edged, depending entirely on the manner of exploitation:

> From the very beginning, it was inescapable to me that these new approaches for the understanding and manipulation of living organisms had potential implications for human progress of very great significance. On the one hand molecular biology could increase man's knowledge about himself and lead to revolutionary change in medicine in such fields as cancer, aging, congenital disease, and virus infections. It might also play a vital role in industry and agriculture. On the other side it might be exploited for military purposes and eventuate in a biological weapons race whose aim could well become the most efficient means for removing man from the planet. As a student of evolution, and having studied it in the microcosmos with bacterial cultures, I knew man had no guaranteed place on our earth. He has faced and continues to face natural disasters like the infestations that have wiped out the American chestnut and the European grapevine. To these long-standing threats would now be added new ones, potentially of our own invention.[17]

To summarize, this first phase in the development of perceptions of the military potential of molecular biology was marked by considerable apprehension about the possibility of future military use of the field. Within the community of molecular biologists who understood both the practical significance of advances in their field as well as the potential for military involvement, there was a shared understanding that further advances were likely to provide techniques for overcoming the disadvantages of biological weapons. As one scientist put it, their fear was that biological weapons would not remain forever "in the category of useless weapons."[18]

Perceptions of the Military Potential of Molecular Biology: Late 1960s–Early 1980s

The second phase in the development of perceptions of the military potential of molecular biology opened on a more optimistic note. In November 1969, President Nixon announced two major changes in the U.S. chemical and biological warfare policy: first, an unconditional renunciation of the production and stockpiling of biological weapons, and second, his intention to submit the Geneva Protocol to the Senate for ratification. Henceforth, Nixon affirmed, U.S. interests in biological warfare would be confined to research for *defensive* purposes, and stockpiles of biological weapons would be destroyed.[19] In February 1970, Nixon further announced that the U.S. renunciation of biological weapons would be extended to toxin weapons.

It is difficult to assess the reasons for Nixon's decision. It is clear in a general way that it came in response to growing public and international criticism of U.S. CBW policy. In some respects, the decision may be seen as a compromise, going part way to satisfy the demands of critics of U.S. policy by renouncing those weapons which had the least military utility and at the same time preserving the U.S. option to use herbicides and tear gas in Vietnam. It is possible, also, that the decision was in part a response to the concerns documented here about the future military use of advances in biology. Evidence for this view comes from the testimony of Fred Ikle, director of the Arms Control and Disarmament Agency during the Nixon administration. Ikle stated that "with such a prohibition, new developments in the biological sciences might give rise to concern because they could be abused for weapons purposes. Such anxieties could foster secretive military competition in a field of science that would otherwise remain open to international cooperation and be used solely for the benefit of mankind." A third and possibly decisive reason for supporting renunciation of biological weapons was that advances in biological weaponry might encourage a change in the balance of power in favor of smaller countries. As Ikle also stated, "[W]idespread adherence to the Convention can help discourage some misguided competition in biological weapons."[20]

Following Nixon's announcement (and in part because of it), rapid progress toward a treaty banning biological weapons was made by the twenty-six–nation Conference of the Committee on Disarmament. The Biological Weapons Convention (BWC) banning the development, production, and stockpiling of biological weapons and toxins was completed and opened for signature in 1972. The treaty was subsequently ratified by the Soviet Union, the United States, and most other major states.

The treaty was (and is) a major achievement in the history of disarmament. Indeed, it is the only arms control agreement in modern times in which parties are required to forego *possession* as well as *use* of a category of weapons. As one observer states, "The illegal status of biological weapons is more widely and significantly established than for any other category of dubious weaponry or tactic, including chemical and nuclear."[21]

Nevertheless, the treaty is not entirely unambiguous with respect to the limits it places on construction of biological weapons. For example, Article I of the treaty allows possession of biological agents in quantities that are not specified for "protective, prophylactic and other peaceful purposes." That qualification allows certain types of activity that do not violate the letter of the convention but might undermine its overall purpose. Note particularly that the construction of novel pathogens via genetic engineering is not excluded.

In addition, the convention is silent on the question of research. And in fact, the U.S. policy developed at the time of Nixon's renunciation of biological weapons allowed research for "defensive" purposes and construed the term "defensive" broadly. The limits of such research were defined in a National Security Council memorandum issued by Nixon's National Security Advisor, Henry Kissinger, on November 25, 1969. The memorandum indicated that research into "offensive" aspects of biological agents would be allowed as long as it was pursued for "defensive" reasons. In other words, the operative criterion for permissible biological defense research was not the *product* of the research but the *motive* guiding it.[22] Thus the Kissinger guideline allowed for the possibility of research in a gray area where defensive and offensive activities could not be easily distinguished.

At the same time that the limits of permissible biological activities were being established by international law and domestic policy, the first experiments in genetic engineering, demonstrating the possibility of deliberately linking genes of different species and introducing these hybrid genes into new cellular environments, were being accomplished. Almost immediately, this achievement was seen as double-edged. On the one hand, molecular biologists immediately recognized the scientific and practical potential of the techniques. On the other hand, it was also appreciated that there might be costs which could be generated inadvertently through accidental release of organisms or deliberately, through social misuse.

As is well known, these concerns about the potential hazards and possible misuse of genetic engineering — which were widely shared by molecular biologists in this period — led to the initiation of official and semiofficial policy processes designed to assess the extent of the problem and produce recommendations for policies for the promotion and control of recombinant DNA technology. In the United States, the National Academy of Sciences appointed a committee of molecular biologists, chaired by Paul Berg of Stanford University, which proposed in 1974 three main actions: a partial suspension of work (the so-called genetic engineering moratorium which captured headlines in 1974); an international meeting of scientists, which was held in California at Asilomar in 1975; and finally, the establishment of an advisory committee within the National Institutes of Health to develop future policy. In Britain, the Advisory Board for the Research Councils (the body responsible for advising the secretary of state for education and science on research policy) established a working party under Lord Ashby, a former chairman of the Royal Commission on the Environment, to assess the technology and make policy recommendations.

Many of those who participated in these decisions were aware of the concerns about the social misuse of advances in molecular biology raised in the 1960s. Of the members of the Berg Committee, molecular biologist David Baltimore had been active in the antiwar movement and Nobel Prize winner James Watson had been a member of the Presidential Chemical and Biological Warfare Advisory Board and had attended the American Academy meeting in 1969. Of the participants at the Asilomar Conference, in addition to Baltimore and Watson, Joshua Lederberg, as we have seen, had warned eloquently of the dangers of military application of his field, and plasmid biologist Richard Novick had been deeply involved in struggles to persuade professional organizations to oppose military use of the biological sciences.

Not surprisingly, therefore, the issue of weapons applications of genetic engineering was raised on several occasions during this crucial period of policy formation. In the United Kingdom, Sydney Brenner of the Cambridge University Laboratory for Molecular Biology warned the Ashby Committee of the problem of controlling "institutions which can and often do practice secrecy in their activities, such as defense research laboratories."[23] In the United States, the issue was raised on the Berg Committee by Richard Roblin, an assistant professor of microbiology at Harvard Medical School. In a draft for the statement issued by the Berg Committee, Roblin inserted a paragraph warning that "these new technological capabilities could potentially be used to create new sophisticated weapons of biological warfare" and urging citizens, scientists and government officials to "take appropriate action to prevent such applications."[24] Finally, at the Asilomar Conference, an expert working group on the use of plasmid and phage in genetic engineering, chaired by Richard Novick, stated in its report to the conference: "We believe that perhaps the greatest potential for biohazard involving the genetic alteration of microorganisms relates to possible military applications. We believe strongly that construction of genetically altered microorganisms for any military purpose should be expressly prohibited by international treaty, and we urge that such prohibition be agreed upon as expeditiously as possible."[25]

Despite these clearly articulated misgivings about military exploitation of genetic engineering, the pattern of response to the concerns raised directly on the public and private bodies responsible for the early stages of policy for genetic engineering was entirely uniform. Repeatedly, the issue was set aside.

The Roblin insert in the Berg letter was taken out of a later draft so that the final version of the letter, which received international press coverage, made no mention of the possibility and problem of military use. When the problem was raised by reporters at a press conference on the Berg letter in July 1974, the terse response from David Baltimore was that military applications were "a challenge we're going have to meet when we know it exists."[26] In his opening remarks at the Asilomar Conference, David Baltimore acknowledged that use of genetic engineering in biological warfare could be a serious issue, but he went on to dismiss it as "peripheral" to the concerns of the meeting.[27] And the prohibition on use of genetically engineered organisms for military purposes was quietly dropped from the final report of the meeting.[28] Similarly in the United Kingdom, the Ashby committee dismissed the problem as beyond the scope of the charge to the committee.[29] Generally, these early stages of policy making on the genetic engineering issue were dominated by what I have called elsewhere a "technological paradigm" in which the agenda of decision making remained focused on technical dimensions of the issue and social dimensions were set aside.[30] This pattern of nondecision making continued into the early 1980s.

Why was the biological warfare issue ignored during this period when biotechnology was making such dramatic advances? Three general types of explanation may be entertained, resting respectively on technical, legal, and social conditions. First, it might be argued that there were technical reasons why, once advances such as recombinant DNA and hybridoma technology were achieved, scientists then realized that the earlier vague apprehension concerning military use of advances in molecular biology was unfounded because these techniques were incapable of enhancing the military utility of biological weaponry. Indeed, that argument was made in this period.[31] However, there appears to have been little consensus on this argument at this point, for the obvious reason that there was no sustained scientific debate about it. Other scientists held that exploratory military research might well reveal novel properties that would be tempting targets for military exploitation. It should also be remembered that in the 1970s, claims about the potential civilian application of genetic engineering were expansive. Scientists and heads of genetic engineering firms projected a vast array of novel products and organisms resulting from civilian development of genetic engineering. The claim that military uses could never be realized seemed to some scientists to be an unfortunate example of wishful thinking. Furthermore, there is evidence that military exploration of genetic engineering was under consideration within the Pentagon. A Department of Defense report to Congress in 1980 stated that "recombinant DNA technology could make it possible for a potential enemy to implant virulence factors, or toxin-producing genetic information into common, easily-transmitted bacteria such as *E. coli.*"[32] In 1981, a report commissioned by the Department of Defense concluded that "toxins could probably be manufactured by newly created bacterial strains under controlled laboratory conditions."[33] It is difficult to see how scientists could have concluded that biogenetic technology would be completely immune from military exploitation.

Second, it might be argued that scientists believed that the BWC was sufficient to relieve policy makers from any need to address the problems of military use. (This argument was made in 1982 at a meeting of the National Institutes of Health [NIH] Recombinant DNA Advisory Committee in response to a proposal for the inclusion of a prohibition on research aimed at the construction of biological weapons in the NIH guidelines.)[34] However, as shown earlier, the BWC does not address research, and some scientists had issued warnings in the 1960s that weapons research would be destabilizing. Furthermore, with the advent of genetic engineering, the ambiguities of the BWC noted earlier become particularly relevant to a discussion of its effectiveness in restraining use of this technology. As long as military establishments confined their research to work with known pathogens and toxins, it was unlikely that they would produce anything that might be judged an improvement of the combat potential of biological weapons. But the capacity of biogenetics to construct *novel* organisms and substances meant that discoveries made in the course of military research could be escalatory. Evidence from a 1977 Congressional hearing on the implications of genetic engineering suggests that scientists were aware of this problem.[35]

Neither technical nor legal explanations for nondecision making on the

BWC issue in this period seem satisfactory. A comprehensive explanation needs also to take into account the social conditions affecting biomedical research in this period. In the late 1960s and early 1970s, scientists lost much of their earlier political influence and their autonomy in deciding on the direction of research. Funds for research no longer flowed so readily from federal coffers, and scientists came under increasing pressure to justify their need for support. This change in the federal climate for research funding was particularly pronounced in the biomedical fields. The turning point came in 1966 when President Lyndon Johnson summoned the directors of the NIH to the White House and demanded of his shocked audience whether "too much energy was being spent on basic research and not enough in translating laboratory findings into tangible benefits for the American people." That event was symptomatic of a larger political change in which the utilitarian dimension of biomedical research support assumed a new prominence and federal funding entered what a former director of the NIH called a "new era of selective growth."[36]

This was also a period of heightened public sensitivity to the impacts of science and technology. The environmental and consumer safety movements peaked in the early 1970s and Congress was still engaged in enacting reformist legislation aimed at protecting the environment, the workplace, and consumers from harmful effects of technology. At the same time, leaders of the scientific and business communities, sensitive to the impact of continued public criticism of science on business and federal support for research, had begun to organize to contain the environmental and other protective movements and to cultivate a more supportive ideological climate. As Philip Handler, President of the National Academy of Sciences, stated in 1970, "Until the terms 'science' and 'education' again evoke warm, positive associations in the public mind, we cannot expect significant growth of the fundamental endeavors."[37] In other words, science had to be *seen* to be harnessed more purposefully to human needs.

Genetic engineering made its appearance in this social context of public criticism of science and politicians' demands for accountability and safety on the one hand, and the counterreaction of scientific and industrial leaders aimed at containing public discontent on the other hand. The field of molecular biology, despite a long record of brilliant scientific advances in the 1950s and 1960s, had produced little of practical significance. Leaders of science wished to argue that here at last were the practical benefits of two decades of federally funded research in molecular biology, a demonstration that basic research did pay off and that these benefits could be pursued safely with no danger to the public. These claims were strongly reinforced by those who moved to invest in the industrial applications of genetic engineering in the late 1970s and early 1980s. However, the question of the safety of genetic engineering became the subject of intense public debate and controversy in the late 1970s. To have acknowledged at that point that genetic engineering might be used to construct novel biological weapons, more effective than any that had appeared to that point, would have removed the entire foundation of the case for a safe, productive, profitable form of advanced technology. As Richard Novick, one biologist who was present at Asilomar and active in subsequent policy decisions, recalls,

> [In 1975], the whole genetic engineering enterprise was already teetering in the public mind. All one had to do was to warn that biological weapons might emerge, and that would cause a legislative ban and so forth. [The organizers of the Asilomar Conference] were afraid of that. . . . Because of that fear, the biological weapons prohibition was never included. The excuse was that the outcome of Asilomar was only concerned with laboratory safety.[38]

This concern that a focus on biological warfare might block the genetic engineering enterprise continued into the 1980s. When the question of issuing a prohibition on the use of genetic engineering for the construction of biological weapons was brought before the NIH Recombinant DNA Advisory Committee in 1982, committee members were reminded by David Baltimore that "it is extremely important to make sure that we are not doing anything to hamper basic research."[39]

As long as the military was not involved in biotechnology and as long as the BWC appeared to be working satisfactorily, the social pressures on molecular biologists to demonstrate that genetic engineering was useful, profitable, and safe made it tempting to ignore the question of military use. The easiest course of action was to believe that military application would never become a problem.

Perceptions of the Military Potential of Genetic Engineering in the 1980s

In the 1980s, military use of biotechnology *did* become a problem, at least in the minds of a growing number of biologists. Since 1982, the political, legal, and military contexts in which biogenetic technologies were advancing changed significantly. And these changes induced a new wave of concern about military development.

In the first place, there was an important reversal of attitude in official circles toward chemical and biological disarmament. A principal cause of this change was the general deterioration of international relations, particularly East-West relations, in the 1980s. As Cold War hostilities were renewed and the arms race accelerated, the U.S. commitment to arms control and to international law generally weakened. These trends were accompanied by and reflected in an accusatory posture that tended to view the behavior of adversary states, particularly the Soviet Union, with deep suspicion and distrust.

Although there was no *formal* change in the U.S. biological warfare program since 1980 (the United States remained formally committed to its obligations under the Geneva Protocol and the BWC), this accusatory posture transformed official behavior on matters related to biological warfare. Since 1981, for example, the Reagan administration repeatedly accused the Soviet Union of producing toxin weapons ("yellow rain") for use by the Vietnamese in Southeast Asia, of using chemical warfare in Afghanistan, and of maintaining a biological weapons program — all major violations of the BWC and the Geneva Protocol. Especially prominent in the mid-1980s was the claim that the Soviet Union was pursuing a biological weapons program and using genetic engineering for that purpose, a charge that has received dramatic elaboration in the conservative press. William Kucewicz, in an eight-part feature in *The Wall Street Journal,* alleged that the Soviet biological warfare operation was aimed at a first-strike weapon — a flu virus armed with the gene for cobra venom which would be released around the world to eliminate everyone except immunized Russians.[40] Jack Anderson made similar claims. Both writers were given access to classified material by the Department of Defense.

In each instance, however, the evidence for these charges proved to be highly questionable — either flimsy or technically inconsistent.[41] For example, the "yellow rain" charges collapsed under the weight of contrary evidence, much of it generated by the U.S. government itself.[42] And, Defense Department officials, when pressed to substantiate claims of a Soviet biological weapons program, acknowledged that they had little more than a "working hypothesis."[43]

Despite the tenuousness of the Reagan administration's charges, they had a cumulative and damaging effect on the chemical and biological warfare legal regime. The difficulty of producing a definitive resolution of these charges raised doubts as much about the effectiveness of the BWC as about the commitment of the Soviet Union. Indeed, some conservative analysts used the evidence as grounds for suggesting the withdrawal of the United States from the treaty.[44]

Furthermore, as superpower relations improved in the late 1980s and concerns about the Soviet CBW "threat" faded away, confidence in the CBW again began to erode in the late 1980s, this time a result of what was widely regarded by experts as premature claims of "proliferation" of chemical and biological weapons, particularly in the Middle East. Although the only conclusively demonstrated case of possession and use of chemical weapons was that of Iraq, charges of the spread of chemical and biological weapons to other nations began to circulate in the mid-1980s, escalating toward the end of the decade.[45]

Secondly, the chemical warfare and biological defense programs, which entered a long decline in the 1970s, experienced a rapid resurgence. In the first seven years of the 1980s, support for these programs rose in real terms by 55 percent, to $1.44 billion.[46] A major and controversial item on the Reagan administration's program of expansion was renewal of support for production of a new form of chemical nerve gas—"binary" nerve gas. Production of chemical weapons had ceased in 1969 with President Nixon's announcement of his intention to ratify the Geneva Protocol. In the 1970s, Defense Department interests in renewing production of these weapons were resisted both by the president and Congress on the grounds that breaking the moratorium on production of chemical weapons would jeopardize the continuing negotiations for a treaty. Thus, no production funds for chemical weapons were requested for fiscal years 1977 to 1980. By contrast, in the 1980s the Reagan administration repeatedly pushed Congress to appropriate funds for production of new chemical weapons. (In 1986, Congress finally succumbed to the pressure and appropriated funds, but with several important conditions attached.)

Congress was also persuaded to support a renewed emphasis on programs of research and development for chemical and biological warfare. Support for these programs was projected at $370.5 million for fiscal year 1987, a real increase of 286 percent over fiscal year 1980.[47] In real terms, spending on research and development for chemical and biological warfare was close to the highest levels of research and development spending during the Cold War.

One of the functions of the charges of Soviet noncompliance was to justify this expansion. Year after year, Reagan administration officials testifying in Congress for the need to expand the chemical warfare and biological defense programs, bolstered their case by reference to nefarious Soviet intentions as evidenced by "yellow rain," Soviet biological warfare (BW) programs, and other claimed violations of the CBW legal regime. The reiteration of these allegations in congressional hearings meant that, despite their fragility, they hardened into "fact." The willingness of members of Congress to listen despite the growing case against the administration's claims reflected the return of Cold War postures.

In the third place, many of these new research dollars were used to support an expanding program of research in biogenetic technologies, particularly genetic engineering and hybridoma technology. It is difficult to obtain a precise picture of military support for research in these fields, but a rough measure is provided by the fact that support for research in the life sciences and in medical defense under the chemical warfare and biological defense programs increased by about 400 percent to $176 million in 1980–85.[48] This work was aimed at a wide range of projects, including vaccines against organisms deemed to be of military importance; the study of the gene for acetylcholinesterase, the neurotransmitter interrupted by nerve agents; the development of enzymes that degrade nerve agents; antidotes to bacterial and fungal toxins; and basic research in protein structure and gene control.[49]

In response to concerns about the purpose of this work, Defense Department officials have asserted that all research in progress is purely defensive in the sense that its results—vaccines, detection systems, and better understanding of the action of toxins and nerve agents—could not be applied directly to the development of weapons.[50] Yet these assurances cannot be unequivocal. In the absence of a chemical weapons treaty, there can be no assurance that biological technologies will not be used for developing new chemical weapons. In addition, for the reasons discussed earlier, the line between "defensive" and "offensive" biological warfare activities cannot be drawn unambiguously.

Moreover, in 1984 evidence emerged of Pentagon plans to move into the far more ambiguous area of exploration of weapons potentials where defensive and offensive purposes overlap. For example, the Defense Department has revealed plans to build a high-containment facility for testing aerosols of lethal pathogens and toxins at its chemical and biological warfare test site at Dugway Proving Ground, Utah.[51] The plans indicated that the facility would be used initially to test defenses against "conventional" biological warfare agents such as anthrax, Q-fever, and tularemia. However, in addition, the Army acknowledged that genetically altered pathogens might also be tested. (Indeed, this is the rationale for requiring the highest level of containment.)[52] Thus, types of work that were forbidden at the Asilomar conference and for many years after by the guidelines issued by the NIH were anticipated.

Concern about military use of biotechnology intensified in the late 1980s as a result of claims emanating from the Pentagon that biotechnology now presented an immediate military threat, making biological warfare easier, cheaper, and far more effective than in the past. This view was aired with particular force in a Defense Department report to Congress in 1986.[53] The scope of the Pentagon's biological vision was reminiscent of the exuberant speculation in the 1970s about civilian application of biotechnology, except that now the emphasis was on the dark side of the technology. The possibility of using genetic engineering both for the construction of microorganisms with novel properties and for their use as "factories" for the production of novel substances, previously portrayed in glowing terms, was now projected as a biological warfare threat of menacing proportion. The report concluded that "biological warfare is not new, but it has a new face."

While these claims are dismissed by many well-informed scientists, both the new military attention to biogenetics and plans to move into "gray areas" of research have generated a new concern that focuses not on an immediate biological warfare threat but rather on the *dynamics* of future military assimilation of the field. The real focus of concern, these scientists argue, is not that a novel biological weapon could be unleashed by any nation in the immediate future. (It is pointed out that the Soviet Union, portrayed in many Pentagon documents as the most immediate threat, is far behind the United States in the theoretical knowledge and techniques necessary even to embark on such a development.) Rather, it is that a concerted effort to explore military dimensions of biogenetics on the part of the United States will stimulate similar work by the Soviet Union and that the subsequent interaction and escalation of exploratory activity will eventually produce some development with perceived military utility that will undermine legal restraints and initiate a new biological dimension to the arms race.

We have come full circle, back to the concerns of scientists before the advent of genetic engineering. As Joshua Lederberg stated in 1970 at the Committee of the Conference of Disarmament,

> My gravest concern is that scientific breakthroughs of a rather predictable kind will be made and their potential military significance exploited, so as to result in a transformation of current doctrine about "unreliable" biological weapons. We are all familiar with the process of mutual escalation in which defensive efforts of one side inevitably contribute to further technical developments on the other hand and vice versa. The mere existence of such a contest produces a mutual stimulation of effort. . . . The potential undoubtedly exists for the design and development of infective agents against which no credible defense is possible, through the genetic and chemical manipulation of these agents. It is thus clear to me that if we do not do something about this possibility, work will go forward and my fears will become realities.[54]

Conclusions

The possibility of a spiral of interaction and escalation between the two superpowers leading to exploration of the weapons potential of biogenetics, assimilation, and ultimately to a biological arms race has become a renewed focus of concern for many scientists. For example, scientists at the Cold Spring Harbor Phage Meeting in 1987 circulated a statement opposing the use of biological research for military purposes and pledging not to engage knowingly in research and teaching that would further the development of chemical and biological warfare agents.

It is becoming clearer that crucial choices concerning the militarization or nonmilitarization of biotechnology are likely to be made in the near future. What can be done at this point to ensure that this field is not used for the development of novel weapons? The most fundamental need is to strengthen the BWC. Despite the ambiguities of this treaty, its basic prohibition on developing, testing, producing, and stockpiling biological weapons remains a most significant barrier to the assimilation of these weapons into existing military theory and practice. (It is conceivable that nuclear weapons would never have been produced had their development been preempted by a similar legal regime.)

However, strengthening an international treaty is not simply, or primarily, a matter of amending its text to remove loopholes and ambiguities. Perhaps even more important, particularly at present, are actions taken by parties to the legal agreement to signal reassurance and commitment to the basic treaty obligation. How the effectiveness of the treaty is perceived is to a great extent a function of the political behavior of states in relation to the treaty requirements.[55]

At the present time, the United States has an important opportunity to reverse the trends that threaten to undermine the treaty. At home, the United States could, by revoking or withholding funding for ambiguous and provocative research and development projects such as the Dugway expansion, proclaim its intention to avoid research and development in "gray" areas of the BWC where defensive and offensive interests cannot be easily distinguished. It should also renounce all secret biological warfare activities. Secrecy of military research (even if these activities are entirely legitimate) always tends to exacerbate fears on the part of other nations and plays into the scenarios of worst-case planners in military establishments. Openness, on the other hand, tends to defuse suspicion and reduce military interest in gaining advantage through technological surprise.

In addition, the United States should be more careful in relation to allegations of noncompliance. And it should seek to resolve any serious concerns through the formal procedures contained in the treaty itself and developed further in the past year. Finally, Congress should seize a new opportunity to reinforce our commitment to the BWC by passing legislation to make the provisions of the treaty binding on individuals and institutions within the boundaries of the United States.

Abroad, the United States should both participate actively in efforts to articulate and extend the CBW legal regime and encourage other nations to do likewise. Advances in this direction were made in the second half of the 1980s. Negotiations for the Chemical Weapons Convention made significant progress, particularly with respect to the previously intractable issues associated with verification of compliance. And the Second Review Conference on the BWC in September 1986 took some steps toward rebuilding confidence in the treaty by establishing mechanisms for exchanges on information about high-containment facilities for research involving pathogens and unusual outbreaks of disease. But much more remains to be accomplished. The Chemical Weapons Convention needs to be completed as an urgent priority. And the BWC needs to be strengthened further both through expanding confidence-building measures as well as, eventually, through negotiating a new protocol requiring verification of compliance with the treaty.[56]

Finally, the international community of biologists can make a special contribution because of its role in producing the theory and technology that is in danger of becoming militarized. As they did in the 1960s, biologists can again call on their professional organizations to censure research aimed at biological weapons, and they can seek, with their colleagues around the world, to use their authority to reinforce cultural and moral barriers to biological warfare.

In the early years of the controversy surrounding genetic engineering, the philosopher Hans Jonas warned that the root problem of modern technology was its capacity to turn the world itself into a laboratory and nature into the object of an experiment of global proportion.[57] Depletion of the earth's ozone layer by chlorofluorocarbons, the global warming produced by chemical pollutants, and the fallout from the Chernobyl accident all confirm the imminence of the danger. In the future, a major challenge will be to prevent military establishments from pursuing a similar global experiment with biotechnology.[58]

Notes

1. U.S. Congress, House, Committee on Foreign Affairs, Subcommittee on National Security Policy and Scientific Developments, *Hearings on Chemical-Biological Warfare: U.S. Policies and International Effects,* 91st Cong., 1st sess., 1969, 87.

2. For analysis of the orientation of molecular genetics toward manipulation of the biogenetic processes, see Susan Wright, "Recombinant DNA and Its Social Transformation, 1972–1982," *Osiris* 2 (1986):303–60.

3. U.S. Congress, House, Committee on Foreign Affairs, *Hearings on Chemical-Biological Warfare: U.S. Policies and International Effects,* 465.

4. U.S. Congress, Senate, Committee on Human Resources, *Hearings on Chemical-Biological Warfare: U.S. Policies and International Effects,* 95th Cong., 1st sess., 1977, 40–44.

5. Julian Robinson, "Chemical Biological and Radiological Warfare: Futures from the Past" (draft submission to the Independent Commission on Disarmament and Security Issues), September 1981.

6. See *Proceedings of the Conference on Chemical and Biological Warfare,* sponsored by the American Academy of Arts and Sciences and the Salk Institute, July 25, 1969, in Committee on Foreign Affairs, *Hearings on Chemical-Biological Warfare: U.S. Policies and International Effects,* 454, 467.

7. Quoted in Stockholm International Peace Research Institute, *The Problem of Chemical and Biological Warfare,* vol. 2 (Stockholm: Almqvist and Wiksell, 1973), 316.

8. Ibid., 314.

9. J. V. Reistrup, "5000 Scientists Ask Ban on Gas in Vietnam," *Washington Post,* February 15, 1967.

10. UN General Assembly, Resolution 2603A, December 16, 1967.

11. UN Secretary General, *Chemical and Bacteriological (Biological) Weapons and the Effects of Their Possible Use* (New York: United Nations, 1969).

12. On physicists' perceptions of nuclear fission, see Daniel J. Kevles, *The Physicists: The History of a Scientific Community in Modern America* (New York: Vintage, 1979), 324.

13. *Proceedings of the Conference on Chemical and Biological Warfare,* 492.

14. Ibid., 494.

15. Ibid., 496.

16. Ibid. See, for example, comments of Henry Wiggins, p. 493.

17. UN Committee on the Conference on Disarmament (CCD)/312, August 27, 1970: Remarks by Dr. Joshua Lederberg at informal meeting of CCD, August 5, 1970.

18. Bernard Feld, *Proceedings of the Conference on Chemical and Biological Warfare,* 494.

19. Statement of President Richard Nixon, November 25, 1969, reprinted in Committee on Foreign Affairs, *Hearings on Chemical-Biological Warfare: U.S. Policies and International Effects,* 83–84.

20. U.S. Congress, Senate, Committee on Foreign Relations, *Hearing on Prohibition of Chemical and Biological Weapons,* 93d Cong., 2d sess., December 10, 1974, 15–16.

21. Richard Falk, "Inhibiting Reliance on Biological Weaponry: The Role and Relevance of International Law," in *Preventing a Biological Arms Race,* ed. Susan Wright (Cambridge, Mass.: MIT Press, 1990).

22. Henry Kissinger, National Security Decision Memorandum 35 (November 25, 1969). For further analysis, see Susan Wright and Robert L. Sinsheimer, "Recombinant DNA and Biological Warfare," *Bulletin of the Atomic Scientists* 39 (November 1983):24.

23. Sydney Brenner, "Evidence for the Ashby Working Party." Paper submitted to the Ashby Working Party on the experimental manipulation of the genetic composition of microorganisms, September 26, 1974. Recombinant DNA Controversy Oral History Collection, Institute Archives and Special Collections, M.I.T. (hereafter cited as RDNA Collection, MIT).

24. Richard Roblin, draft of NAS Committee Statement, April 25, 1974 (RDNA Collection, MIT).

25. Working Party on Potential Hazards Associated with Experimentation involving Genetically Altered Microorganisms, with Special Reference to Bacterial Plasmids and Phages, "Proposed Guidelines on Potential Hazards with Experiments involving Genetically Altered Microorganisms," February 24, 1975 (RDNA Collection, MIT).

26. David Baltimore, transcripts of news conference, National Academy of Sciences, July 18, 1974 (RDNA Collection, MIT).

27. Quoted in Michael Rogers, *Biohazard* (New York: Knopf, 1977), 52.

28. Richard Novick, interview with author, October 1987.

29. *Report of the Working Party on the Experimental Manipulation of the Genetic Composition of Micro-organisms,* Cmnd. 5880 (1975), 9.

30. Susan Wright, "Molecular Politics in Great Britain and the United States: The Development of Policy for Recombinant DNA Technology," *Southern California Law Review* 51 (September 1978), 1383–1434.

31. Report of the Preparatory Committee for the Review Conference of the Parties to the UN Convention on the Prohibition of the Development, Production and Stockpiling of Bacteriological (Biological) and Toxin Weapons and on Their Destruction (February 8, 1980), UN Document BWC/CONF.I/5.

32. Defense Department, Annual Report on Chemical Warfare and Biological Defense Research Programs for fiscal year 1980 (December 15, 1980), sec. 2, p. 4.

33. Frank Armstrong, A. Paul Adams, and William H. Rose, "Recombinant DNA and the Biological Warfare Threat," unclassified version of a classified report commissioned by the U.S. Army Dugway Proving Ground, TECOM No. 8-CO-513-FBT-021 (May 1981).

34. National Institutes of Health, Transcript of Proceedings, Meeting of the Recombinant DNA Advisory Committee, June 28, 1982.

35. For example, U.S. Congress, House, Committee on Science and Technology, Subcommittee on Science, Research and Technology, *Hearings on Science Policy Implications of DNA Recombinant Molecule Research,* 95th Cong., 1st sess., March–September 1977, 168–72.

36. Robert B. Semple, "President Orders a Medical Review," *New York Times,* June 28, 1966, p. 35; Donald Frederickson, "Health and the Search for New Knowledge," *Daedalus* 106, no. 1 (Winter 1977): 165.

37. Philip Handler, "Science's Continuing Role," *BioScience* 20 (October 15, 1970): 1101–06.

38. Richard Novick, interview with author, October 1987.

39. Proceedings, NIH Recombinant DNA Advisory Committee, June 28, 1982, p. 38.

40. William Kucewicz, "Beyond Yellow Rain: The Threat of Soviet Genetic Engineering," *The Wall Street Journal,* April 23, 25, 27; May 1, 3, 8, 10, 18; 1984.

41. See analysis of Elisa Harris, "Sverdlovsk and Yellow Rain: Two Cases of Soviet Noncompliance?" *International Security* 11 (Spring 1987): 41–95.

42. Julian Robinson, Jean Guillemin, and Matthew Meselson. "Yellow Rain: The Story Collapses." *Foreign Policy* 68 (Fall 1987): 100–117.

43. Jonathan Tucker, "Gene Wars," *Foreign Policy* 57 (Winter 1984–85).

44. Robert L. Bartley and William P. Kucewicz, "Yellow Rain and the Future of Arms Agreements," *Foreign Affairs* 61 (Spring 1983): 817–20.

45. For detailed analysis of charges of "proliferation" of chemical and biological weapons in the late 1980s, see Susan Wright, "The Evolution of U.S. Biological Warfare Policy, 1945–1990," in *Preventing a Biological Arms Race* (Cambridge, Mass.: MIT Press, 1991).

46. Office of the Secretary of Defense.

47. Office of the Secretary of Defense; Department of Defense, Annual Report on Chemical Warfare and Biological Research Programs, FY 1980.

48. Department of Defense, Annual Reports on Chemical Warfare and Biological Research Programs, FY 1980, FY 1985.

49. This description is based on an analysis of roughly one hundred abstracts of projects involving recombinant DNA and hybridoma technology sponsored by the Department of Defense in 1985.

50. See, for example, statements of Defense Department officials quoted in Susan Wright, "The Military and the New Biology," *Bulletin of the Atomic Scientists* 41 (May 1985):14–15.

51. Susan Wright, "New Designs for Biological Weapons," *Bulletin of the Atomic Scientists* 43, no. 1 (January/February 1987):43–46.

52. *Foundation on Economic Trends* v. *Caspar W. Weinberger,* Civil Action No. 84–3542, Memorandum Opinion and Order, May 31, 1985.

53. U.S. Department of Defense Biological Defense Program, "Report to the Committee on Appropriations, House of Representatives," May 1986.

54. Joshua Lederberg, CCD/312. See note 17.

55. For perceptive analysis of this dimension of international law, see Richard Falk, "Strengthening the Biological Weapons Convention of 1972," *Biological and Toxin Weapons Today,* ed. Erhard Geissler (Oxford: Oxford University Press, 1986), ch. 8.

56. For detailed proposals for strengthening the BW legal regime, see Richard Falk and Susan Wright, "Preventing a Biological Arms Race: New Initiatives," in *Preventing a Biological Arms Race,* ed. Susan Wright (Cambridge, Mass.: MIT Press, 1990).

57. Hans Jonas, "Freedom of Scientific Inquiry and the Public Interest," *Hastings Center Report* 6 (August 1976): 15–17.

58. I wish to gratefully acknowledge the National Science Foundation grant (No. SES-8511131) that supported research for this chapter.

PART

III

Public
Perceptions

8

The Dark Side of Agricultural Biotechnology: Farmers' Appraisals of the Benefits and Costs of Technological Innovation

GORDON BULTENA and PAUL LASLEY

Introduction

Technological innovation has long been a driving force in the social and economic evolution of U.S. agriculture. The mechanical revolution early in this century and the petro-chemical revolution following World War II transformed farming from a system of small, labor-dependent units to one dominated by large, capital-intensive operations. The new technologies provided many benefits, including higher crop yields, greater production efficiencies, and reduced food prices. But these benefits were partly offset by other, more adverse impacts, including a massive social and economic restructuring of farming, displacement of many farm operators, expansion of corporate control over farm inputs and marketing, and the onset of economic malaise in many agriculturally dependent communities. *degredation*

Now, in the waning years of the twentieth century, U.S. agriculture is poised on the brink of a third technological revolution, one spurred by developments in biotechnology. This revolution is expected to bring significant gains in productivity; better quality foodstuffs; less reliance on farm chemicals; and increased resistance of crops to drought, heat, frost, and pests. But there is also a dark side to the new biotechnologies. Among other things, they are projected to bring (1) a substantial decline in farm numbers by the year 2000 (down 50 percent), (2) increased capital and managerial requirements for farm operators, (3) intensified concentration of food and fiber production and financial wealth, (4) more vertical integration and absentee ownership, and (5) shifts in the production locales of commodities.[1] Beyond the farmstead, biotechnology will have far-reaching, and potentially disruptive, consequences for agribusiness, rural communities, land grant universities, and the Third World.[2]

Several pending biotechnologies have already generated public concern and controversy. Bovine somatotropin (BST), while not yet released, has been actively opposed by farm groups and others concerned about social, economic, and political consequences of a continued overproduction of milk. Many Wisconsin dairy producers acknowledged in a recent survey that BST is a touchy subject in their communities, despite its contribution to improved production efficiencies.[3] Opposition to biotechnology has also arisen among potato growers in California who are concerned with potential consequences of a biotech-produced frost retardant[4] and among farm groups that have joined with nonfarm interests in opposing the patenting of new life forms.[5]

Public protests against agricultural biotechnology are symptomatic of a generally growing politicization of agricultural research.[6] Social and economic impacts of new agricultural technologies have caused increased concern. Historical evidence indicates that many new farming technologies are beneficial to some operators but harmful to others.

But despite scattered opposition, agricultural biotechnology is generally welcomed in both the farm and nonfarm populations. Iowa farmers, for example, largely endorse the genetic manipulation of plants and animals for the purpose of improved farm production.[7] Also, despite sporadic protests against it, BST is expected to be rapidly adopted by dairy farmers when it is released in the next few years.[8] Apart from farmers, the American public holds generally positive views of biotechnology, feeling that its likely benefits outweigh any potential costs or dangers.[9]

As with the earlier mechanical and petro-chemical revolutions, agricultural biotechnology is an enigma in that it offers decidedly attractive benefits to farmers and consumers (e.g., greater production efficiencies and higher quality, less expensive food) while at the same time extracting significant social and economic costs. Especially troublesome to many is the likelihood that benefits of the new technologies will accrue primarily to larger farm operators. Despite the publicized "scale-neutrality" of many of the impending biotechnological innovations, it is anticipated that they, not unlike the technologies that preceded them, will be adopted first on the large, intensively managed farms.[10] It is such early adopters who stand to reap the initial financial gains of the technology because of lowered production costs. But more importantly, because of greater production efficiencies, these adopters also may be the most immune from price declines that will likely occur as the new technologies stimulate higher aggregate production levels.

Theoretical Perspective

This study examines farm operators' perceptions of biotechnology—specifically, the favorableness accorded some anticipated impacts of biotechnology on food production and farm structure. Farmers' receptivity to biotechnologies depends, in part, upon the perceived benefits and costs of this technology. They should welcome some effects of biotechnology, such as its contribution to improved production efficiencies, lessened labor demands, better quality farm products, and, most importantly, improved profits. But collectively they can be expected to oppose changes that are disruptive of established social and economic patterns and that threaten their personal survival in farming.

Study has shown that some demographic and farm characteristics (e.g., age, education, socioeconomic status, scale of operation) are associated with the speed with which farmers adopt new technologies.[11] These variables should also be important to farmers' receptivity to new, untested technologies. The study described below posited that younger (H1) and better-educated (H2) farmers would be the most supportive of agricultural biotechnology, and that this support would be strongest among those with the shorter tenure (H3), operating the largest units (H4), and receiving the highest gross farm incomes (H5).

Sample and Methodology

Sample

These respondents are Iowa farm operators. They were randomly drawn from a list of all farmers in the state compiled by the Iowa Department of Agriculture and Land Stewardship, Division of Statistics. Questionnaires were sent in 1987 to 3,527 farmers, of whom 1,943 (55 percent) provided information. Comparison of the respondents' personal and farm characteristics with the 1982 Census of Agriculture shows that the sample somewhat overrepresented farmers on the large operations (average farm size of respondents was 446 acres as compared with a statewide average of 283 acres).

Measurements of Variables

The respondents were queried about their attitudes toward some projected impacts of biotechnology. Two types of impacts were assessed—changes in the efficiency and levels of agricultural production and alterations in farm structure.

At this early stage of biotechnology development, it is difficult to anticipate how the pending innovations will affect farming. But some likely effects have been identified. Drawing upon a recent assessment,[12] fourteen statements were prepared about different types of biotechnology impacts, including greater production efficiencies, higher levels of aggregate production, enhanced ability to meet world food needs, and the development of new crops and farm products. Included in these statements were anticipated structural changes in agriculture, including a projected drop in the number of farms, greater control by large corporations over farm inputs, and more concentration of agricultural production (Table 8.1).

The respondents rated the desirability of each impact. The possible responses (on a fivefold Likert scale) were "very desirable," "somewhat desirable," "uncertain," "somewhat undesirable," and "very undesirable."

Information was also obtained about the respondents' personal and farm characteristics, including age, educational attainment, family income, tenure, farm size, and gross farm income. Measurement of these background characteristics is described in Table 8.2.

Findings

Impact Evaluations

The attractiveness of agricultural biotechnology to farmers is a function of the type of impact being considered. As shown in Table 8.1, there were substantial differences in how the fourteen impact statements were evaluated. Most appealing is the possibility that biotechnology will help farmers become less dependent upon agricultural chemicals (80 percent rated this desirable); that scientists will engineer new crop varieties (79 percent); and that corn will be developed that fixes its own nitrogen from the atmosphere, thus reducing the need for commercial fertilizers (79 percent). Also rated desirable by a substantial majority of respondents was the possibility that biotechnology will permit new uses for crops and livestock (61 percent) and will bring greater efficiency of feed-conversion in livestock feeding programs (61 percent). Notable exceptions to this pattern of support for production impacts are the largely negative appraisals of the two items about aggregate increases in production (items 1 and 13).

Statements of likely structural impacts from biotechnology drew predominately negative review. Three-fourths of the respondents rated the statement that biotechnology will contribute to a further decline in farm numbers as undesirable. A majority were also concerned that biotechnology will bring a continued concentration of farm production on fewer units

Impact[b]	Very desirable (5)	Somewhat desirable (4)	Uncertain (3)	Somewhat undesirable (2)	Very undesirable (1)	No response	Total[c]
			Percent				
+ (1) Through embryo transfers, gene inserts, and growth hormones, milk production from individual cows will be doubled in the next fifteen years	4	24	25	26	19	2	100
+ (2) The number of farms in the United States will decline from the present 2.2 million to 1.2 million by the year 2000	2	6	16	26	48	2	100
○ (3) Biotechnology will help solve the problem of farm surpluses by finding new uses for crops and livestock	34	27	16	4	5	3	100
○ (4) Biotechnology will help meet the growing worldwide demand for food products	21	30	33	7	6	3	100
+ (5) By the year 2000, it is estimated that 50,000 farms (about 4 percent of all farms) will produce 75 percent of the nation's foodstuffs	2	8	21	27	40	2	100
○ (6) Through genetic changes, new varieties of corn will be able to fix their own nitrogen from the atmosphere, thus reducing the need for commercial fertilizer	48	31	13	3	2	2	100
○ (7) Through biotechnology, scientists will be able to engineer new crop varieties	36	43	14	3	2	3	100
+ (8) Advances in biotechnology will probably benefit persons with large farm operations more than persons on middle-size and small farms	3	7	25	29	33	2	100
○ (9) Biotechnology will enable farmers to become less dependent upon agricultural chemicals	53	27	14	2	1	2	100
(10) Through biotechnology, scientists will be able to develop new species of animals	6	17	43	15	16	2	100
○ (11) Research in biotechnology will increase the efficiency of feed conversion in livestock production	27	44	18	6	3	2	100
+ (12) Biotechnology will lead farmers to become more dependent upon large corporations for many of their inputs, such as seeds, growth hormones, and feed additives	2	7	19	34	35	2	100
+ (13) Increased production of foodstuffs—i.e., greater quantities of crops and livestock products will be available for sale and export	5	19	30	26	17	2	100
○ (14) Biotechnology will bring improved levels of living for many farm families	31	29	32	3	3	3	100

[a]N = 1943.
[b]○ Item loaded on Factor I, "Production-Related Impacts."
   + Item loaded on Factor II, "Structural Impacts."
[c]Totals approximate 100 although slight deviations may occur due to rounding.

Table 8.2. Measurement of variables

Variable	Measurement Procedure
Dependent variables	
Production-related impacts	Scale score for seven attitudinal items that formed a common factor; Cronbach's Alpha was .85 (see Table 8.1)
Structural impacts	Scale score for six attitudinal items that formed a common factor; Cronbach's Alpha was .78 (see Table 8.1)
Independent variables	
Personal/family characteristics	
Age	Present chronological age (X = 52)
Education	Number of years of schooling completed (X = 12.6)
Farm-firm characteristics	
Years farmed	Number of years in farming (X = 29)
Farm size	Total acres owned and rented (X = 447)
Gross farm sales	Total farm sales for 1987 (nine categories ranging from less than $2,500 to $200,000 or more; X = $61,000)

(74 percent rated this undesirable), will foster increased dependency on farmers on large corporations for their production inputs (69 percent), and will prove more beneficial for large than for small farmers (62 percent; Table 8.1).

Factor Analysis of Impacts

It was felt that appraisals of biotechnology would be sensitive to the type of impact being considered, namely, that farmers would endorse developments that offer greater production efficiencies and would oppose those that alter the structure of farming. This was confirmed in responses to the individual impact statements, as described in the preceding section. A more definitive test of the argument was made by determining statistically if responses to the impact items tapped distinct attitudinal domains. For this purpose, the fourteen items were factor analyzed (principal component analysis with varimax rotation). Two strong underlying attitudinal domains were identified. All but one of the impact items loaded on these two domains, the exception being that biotechnology would permit scientists to develop new species of animals. The first factor, with an eigenvalue of 4.2, contained seven items and the second factor (eigenvalue of 2.9) six items. Each of the items loaded strongly on their respective factors (Table 8.3).

The contents of items in each factor substantiate the argument that farmers are divided in their reactions to biotechnology, their views being dependent upon whether the focus is on production-related or structure-related impacts. Factor I contains items that pertain to improved production efficiencies, development of alternative crops, reduced dependency on chemicals and fertilizer, enhanced ability to meet food needs, and improved levels of living for farm families. We have termed this first factor *production-related impacts.*

Factor II contains items about structural impacts (i.e., reduced number of farms, increased concentration of production, greater dependency on large agribusiness corporations, and disproportionate benefits to large farmers) and items about increases in food production (i.e., increased milk, crops, and livestock). That biotechnology would produce these effects was viewed as undesirable by most respondents. Also, although they wanted to become more efficient in their production techniques (Factor I), the respondents largely opposed increased aggregate output. We have called this second factor *structural impacts.*

Scale reliabilities of the two factors (Cronbach's Alpha) were .85 and .78, respectively (Table 8.3). A scale score was calculated for each factor by cumulating response scores for the constituent items. The response scores ranged from 1 ("very undesirable") to 5 ("very desirable"). The possible range of scale scores on Factor I was from 7 to 35, and was from 6 to 30 on Factor II. The respondents' scale scores show that the impact items in Factor I were generally favorably evaluated as a set (mean scale score was

Table 8.3. Factor loadings and scale scores

Items That Loaded on Factor		Loading
Factor I	*Production-Related Impacts*	
Item 3	New markets for crops and livestock	.72
4	Meet world food demands	.70
6	Nitrogen fixation for corn	.70
7	Development of new crop varieties	.77
9	Less dependency on agricultural chemicals	.69
11	Increased efficiency of feed conversion	.69
14	Improved levels of living for farm families	.73
Factor II	*Structural Impacts*	
Item 1	Increased milk production	.57
2	Decline in number of farms	.74
5	Concentration of agricultural production	.76
8	Large-scale operators benefit most	.72
12	Increased dependency on large corporations	.71
13	Increased production of foodstuffs	.55

Characteristics of Factors

Factor I	*Production-Related Impacts*
	Range = 7 (undesirable) to 35 (desirable)
	Average Score = 27.8
	Cronbach's Alpha = .85
Factor II	*Structural Impacts*
	Range = 6 (undesirable) to 30 (desirable)
	Average Score = 13.4
	Cronbach's Alpha = .78

27.8), whereas those in Factor II tended to be negatively evaluated (mean score was 13.4). The average item score for Factor I was 3.9 and for Factor II was 2.2 (Table 8.3).

Correlates of the Impact Assessments

Farmers' appraisals of biotechnology were predicted to be correlated with their socioeconomic statuses. Specifically, it was posited that favorable views of biotechnology would be predominately held by persons who were younger, better educated, and in more secure farming situations (i.e., had large farming operations and high gross farm sales).

To test the hypothesized relationships, personal and farm-firm characteristics of the respondents were correlated with their scale scores on each of the two factors. As expected, younger respondents (H1) most often endorsed the production-related impacts of biotechnology (Factor I) but, unexpectedly, displayed somewhat more negative attitudes than older persons toward potential structural impacts of biotechnology (Factor II). The argument was partly confirmed that the better-educated farmers would be the most supportive of biotechnology (H2). They were more positive in their evaluations of production-level impacts than were those with lesser education but did not differ significantly from their peers in assessments of structural impacts (Table 8.4).

Several farm-firm characteristics were associated with the respondents' appraisals of biotechnology. As predicted, persons with the shorter tenure in farming were the most accepting of production-related impacts (H3), but were also the most critical of structural impacts. Farmers on the large and more productive operations, as measured by farm size (H7) and gross farm sales (H8), expressed the most positive views about production-related benefits of biotechnology (Table 8.5).

To better pinpoint the locus of farm-based support for biotechnology, comparisons were made on selected personal and farm-firm characteristics

Table 8.4. Characteristics of "proponents" and "opponents" of biotechnology[a]

Proponents[b]	Opponents[c]
Younger (X = 49)	Older (X = 55)
Better educated (13.7)	Less educated (12.0)
Fewer years farmed (24.8)	More years farmed (31.5)
Larger farm units (591)	Smaller farm units (446)
Higher gross farm sales ($76,000)	Lower gross farm sales ($55,000)

[a]Averages are reported for the profiled characterstics. Differences in characteristics were statistically significant at the .05 level.
[b]Proponents were in the top quartile of scale scores on each of the two indices.
[c]Opponents were in the bottom quartile of scale scores on each of the two indices.

Table 8.5. Correlations between selected social and economic statuses of respondents and their attitudes toward biotechnology

Status	Factor I "Production Impacts"	Factor II "Structural Impact"
Demographic characteristics		
Age	−.23[b]	+.07[b]
Educational attainment	+.20[b]	+.02
Farm-firm characteristics		
Years farmed	−.23[b]	+.04[a]
Farm size	+.10[b]	+.09[b]
Gross farm sales	+.15[b]	+.05[a]

[a]Statistically significant at the .05 level.
[b]Statistically significant at the .01 level.

between persons expressing supportive and negative views toward all aspects of biotechnology. "Proponents" of biotechnology were those scoring in the upper one-fourth of all respondents on each of the two impact scales (i.e., were supportive of biotechnology, regardless of its effects). "Opponents" were those who expressed largely negative views of biotechnology, that is, were among one-fourth of all respondents who scored lowest on the two scales.

Proponents and opponents of biotechnology differ in that those expressing across-the-board support tend to be younger, better educated, newer to agriculture, and operating the largest units (i.e., farming more acres and having higher gross farm sales; Table 8.4). These farm-related characteristics suggest that proponents of biotechnology are among the larger and more prosperous farm operators in Iowa.

Summary and Conclusion

The pending commercial release of new agricultural biotechnologies is expected to have profound, far-reaching consequences for American agriculture. On the one hand, these technologies promise to bring important benefits in food and fiber production, including greater economic efficiencies and higher quality farm products. They should also aid the struggle against world hunger and perhaps lessen environmental degradation. On the other hand, the new biotechnologies present complex and controversial issues, especially as to their likely impacts on farm structure and, ultimately, on agriculturally dependent communities.

As shown in these data, Iowa farmers are divided in their assessments of biotechnology. For the most part, they enthusiastically endorse developments that are expected to improve production efficiencies. But they generally are critical of aggregate increases in food production made possible by biotechnology, as well as likely effects of biotechnology upon farm structure.

Consistent with previous research, persons on the larger, more capital-intensive units were found to be the most favorable toward biotechnology. But the gap between farm subgroups in acceptance of the new technologies is not large. Small farmers, in sizeable numbers, also saw substantial value in this technology.

It is likely that farmer acceptance of biotechnology may become more problematic as additional information becomes available about its potential structural effects. Especially troublesome will be disclosures about inequities in the distribution of benefits and costs of the technologies. To date, public views of biotechnology have been largely shaped by proponents of technological innovation. As Ruttan has observed, agricultural scientists today are among a group of "reluctant revolutionaries", that is, persons who "have wanted to revolutionize technology but have preferred to neglect the revolutionary impact of technology on society."[13] Until recently, the "dark side" of biotechnology, or its potential for altering the structure of agriculture and farm communities, has not received wide attention. Especially lacking has been scrutiny of who stands to gain and lose from the new technologies.

Because of its recency and complexity, many farmers admittedly remain uninformed about biotechnology. In fact, few of the study's respondents, less than 5 percent, said they were "very well informed" about the likely benefits and problems of biotechnology; conversely, a majority (55 percent) were "relatively uninformed" or "not at all informed." The preponderant lack of knowledge suggests the possibility of considerable volatility in farmers' attitudes as various costs and benefits of the impending biotechnologies become more evident. It has been common in the past for public support for new technologies to wane with the onset of social controversy.[14]

Farmers' appraisals of biotechnology probably will be increasingly politicized as the effects of these technologies on agriculture become more evident. As shown in the data collected, farmers' appraisals of the desirability of biotechnology are cast along class lines, with large-scale operators expressing support whereas smaller, less productive farmers tend to be unenthusiastic or even opposed. But to the extent that biotechnology comes to be perceived as weakening commodity prices or leading to inequitable benefits for certain types of farms, many larger farmers may also become disenchanted with technological innovations and join in protest against the development of new technologies and/or the scientific community.

The extent to which farmers embrace the biotechnological revolution will undoubtedly be shaped by many factors. Persons on small operations may come to perceive the major benefits of the new technologies as accruing primarily to large operations and, thus, as constituting a threat to their own economic well-being. Farmers with large operations will probably display lesser concern with equity issues than with the effects of the new technologies on the profitability of their own enterprises. In the past, technological change has permitted innovative farmers to reap windfall profits. But, given an expected rapid diffusion of these technologies, the time period for receiving windfall profits could be greatly diminished over historical patterns. Inability to obtain profits will likely bring lessened enthusiasm for technological change among progressive farmers and, ironically, could lead to a concerted call from a broad spectrum of farmers for a moratorium on future public financial support of biotechnology research and development.

Notes

1. U.S. Congress, Office of Technology Assessment, "Technology, Public Policy, and the Changing Structure of American Agriculture." (OTA-F-285. Washington, D.C.: U.S. Government Printing Office, 1986); Robert J. Kalter, "The New Biotech Agriculture: Unforeseen Economic Consequences," *Issues in Science and Technology* (Fall 1985); R. D. Smith and D. E. Bauman, "Bovine Somatotropin," *Animal Health and Nutrition* (December 1986:20–25); Lewellyn S. Mix, "Potential Impact of the Growth Hormone and Other Technology on the United States Dairy Industry by the Year 2000," *Journal of Dairy Science* 70 (1987):487–97; Dick Russell, "Rush to Market: Biotechnology and Agriculture," *The Amicus Journal* 9, no. 1 (1987):16–37; Jack Doyle, *Altered Harvest: Agriculture, Genetics, and the Fate of the World's Food Supply* (New York: Viking, 1985); Jack Kloppenburg, Jr., "The Social Impacts of Biogenetic Technology in Agriculture: Past and Future," in *The Social Consequences of New Agricultural Technologies,* ed. Gigi M. Berardi and Charles Geisler (Boulder, Colorado: Westview Press, 1984); 291–321; Lowell Hill, Jerry Stockdale, Harold Breimyer, and Gerald Klonglan, *Economic and Social Consequences of Biological Nitrogen Fixation in Corn Production* (AE-460B. Urbana-Champaign, Illinois: Department of Agricultural Economics, University of Illinois, 1986); Peter Nowak and Roy Barnes, "Potential Social Impacts of Bovine Somatotropin for Wisconsin Agriculture: A Survey of 270 Wisconsin Dairy Producers," paper presented at the meeting of the Rural Sociological Society, Madison, Wisconsin, 1987; Charles C. Geisler and Rex R. Campbell, "Cumulative SIA and Biotechnology: A Case of bGH," paper presented at the meeting of the Rural Sociological Society, Madison, Wisconsin, 1987.

2. Sheldon Krimsky, "Biotechnology and Unnatural Selection: The Social Control of Genes," in *Technology and Social Change in Rural Areas: A Festschrift for Eugene Wilkening,* ed. Gene F. Summers (Boulder, Colorado: Westview Press, 1983), 51–70; Frederick H. Buttel, "Biotechnology and Genetic Information: Implications for Rural People and the Institutions That Serve Them," *The Rural Sociologist* 5, no. 4 (1985):68–78; William D. Cole, William Lacy, and Lawrence Busch, "An Alternative under Attack? The Evolution of the Role of Agricultural Cooperatives," paper presented at the annual meeting of the Rural Sociological

Society, Salt Lake City, Utah, 1986; Martin Kenney, Frederick Buttel, Tadlock Cowan, and Jack Kloppenburg, Jr., "Genetic Engineering and Agriculture: Exploring the Impacts of Biotechnology on Industrial Structure, Industry-University Relationships, and the Social Organization of U.S. Agriculture," Bulletin #125 (Ithaca, New York: Cornell University Department of Rural Sociology, 1982); Martin Kenney, "Agriculture and Biotechnology," in *Biotechnology: The University-Industrial Complex* (New Haven, Connecticut: Yale University Press, 1986).

3. Peter Nowak and Roy Barnes, "Potential Social Impacts of Bovine Somatotropin for Wisconsin Agriculture: A Survey of 270 Wisconsin Dairy Producers," paper presented at the meeting of the Rural Sociological Society, Madison, Wisconsin, 1987.

4. Stephen S. Hall, "One Potato Patch That Is Making Genetic History," Smithsonian 18, no. 5 (1987):125–36.

5. Mark Crawford, "Religious Groups Join Animal Patent Battle," *Science* 237 (July 1987):480–81.

6. Frederick H. Buttel, "Agricultural Research and Farm Structural Change: Bovine Growth Hormone and Beyond," *Agriculture and Human Values* 3, no. 4 (1986):88–98.

7. Paul Lasley and Gordon Bultena, "Farmers' Opinions About Third-Wave Technologies," *American Journal of Alternative Agriculture* 1, no. 3 (1986):122–26.

8. W. Lesser, W. Magrath and Robert Kalter, "Projecting Adoption Rates: Application of an Ex-Ante Procedure to Biotechnology Projects," Report 85–23 (Ithaca, New York: Cornell University, 1985).

9. U.S. Congress, Office of Technology Assessment, *New Developments in Biotechnology: Public Perceptions of Biotechnology* (OTA-BP-BA-45. Washington, D.C.: U.S. Government Printing Office, 1987).

10. Frederick H. Buttel, "Biotechnology and Genetic Information: Implications for Rural People and the Institutions That Serve Them," *The Rural Sociologist* 5, no. 4 (1985):68–78; Burt W. Sundquist, "Impacts of Emerging Technologies on U.S. and Third World Agriculture: Discussion," *American Journal of Agricultural Economics* 67, no. 5 (1985):1176–77.

11. Everett M. Rogers, *Diffusion of Innovations* (New York: The Free Press, 1983).

12. U.S. Congress, Office of Technology Assessment, "Technology, Public Policy, and the Changing Structure of American Agriculture." (OTA-F-285. Washington, D.C.: U.S. Government Printing Office, 1986).

13. Vernon W. Ruttan, "Agricultural Scientists as Reluctant Revolutionaries," *Choices* 2, no. 2 (1987):3.

14. Allan Mazur, "Public Protests against Technological Innovations," in *Technology and Social Change in Rural Areas: A Festschrift for Eugene Wilkening*, ed. Gene F. Summers (Boulder, Colorado: Westview Press, 1983).

9 | Bovine Growth Hormone: Who Wins? Who Loses? What's at Stake?

MATTHEW H. SHULMAN

Introduction

Not many years ago, endocrinologists studying human hypopituitary dwarfism discovered that recombinant DNA techniques could create a synthetic hormone to fill the biological role of the missing natural growth hormone. Replacement of the missing hormone, somatotropin, has been a major breakthrough in overcoming this condition.

The same technology has been adapted for dairy cows. Synthetically produced bovine growth hormone (bGH) is not designed for hormone-deficient cows but to boost production in healthy, high-producing animals. And the strange thing is that, aside from some researchers and the pharmaceutical companies that will manufacture and market the product, no one seems to want it.

The present dairy environment is already marked by a chronic commercial surplus. As the federal government underwrites 63 percent of the five-year, $1.827 billion dairy buyout program to cut surplus production by 8.7 percent, bGH promises to increase total production by 15–20 percent. Taxpayers are understandably concerned about the costs of dealing with the projected surpluses.

Dairymen are not looking forward to bGH because studies suggest that up to half of them will be forced out of farming due to falling prices following increased milk surpluses. Rural communities will not benefit because the loss of large numbers of farmers threatens their employment picture and tax bases. Consumers are wary for they have been taught since childhood that milk is an unadulterated natural substance, the *real* symbol of health and growth.

Few argue technology's role in fostering America's productive capacity in agriculture. Yet, technology does not exist in a vacuum. It interacts with contemporary social, economic, and political realities and personal perceptions. The introduction of bGH may be the ultimate high-stakes, zero-sum game. How it came to be just one administrative step away from commercial authorization and what its broad effects may be are matters that affect us all.

What Is bGH?

Bovine somatotropin, a naturally occurring protein which stimulates cows to direct food energy into milk production, is found in tiny amounts in cows' pituitary glands. Biotechnology permits synthetic production in large amounts at relatively low costs. When lactating dairy cows receive bGH by injection or via implants, milk production increases dramatically.

During Cornell University–Monsanto experimental trials under optimal management conditions, production increased by as much as 41 percent during the final two-thirds of the cows' lactation cycle. Other Monsanto trials in Missouri and Mississippi and Cyanamid trials in Pennsylvania had lower production increases, but overall results suggest that bGH will yield net increases of 15–20 percent under field conditions following commercial introduction.

The hormone works, according to Cornell animal scientist Dale E. Bauman, by coordinating cows' physiological processes so more nutrients from feed are channeled directly into milk production. Cows compensate for the extra energy and nutrients they use by eating more. Production efficiency is enhanced because the energy needed to maintain non–milk-producing bodily functions is not increased. With proper management, fewer cows can produce the same amount of milk with less feed than is required for a larger number of untreated cows. The corollary is that the same number of cows will produce significantly more milk.

No one can predict the exact dimensions of the effects of bGH-enhanced production after commercial introduction, but at least three results are likely for the agricultural community. First, increased aggregate production will depress the farm price paid for milk. Second, unless the cost of other farm inputs (including debt service) falls along with the farm price of milk, already economically stressed farmers will be forced into bankruptcy. Third, until such time as "enough" farmers are eliminated to restore a balance between supply and demand, the government will either greatly increase expenditures for surplus dairy product purchases through the Commodity Credit Corporation or see the price support program eliminated or significantly modified due to the cost to taxpayers.

With this scenario, a balance between supply and demand will eventually be restored. However, it is unclear whether any long-term benefits would accrue to consumers, taxpayers, or the altered rural economy. Will a "lean and mean" high-tech dairy sector achieve the prosperous stability that has eluded the industry these past four decades? Will more efficient production make the dairy industry more competitive with other beverage sectors for consumers' dollars? Even if the answers here are yes, how does one make a cost-benefit analysis of such gains against the changed rural social and economic environment?

Who wins, who loses, and what's at stake form the three questions that fuel what may be not only the greatest challenge to face the dairy industry for the balance of the century, but also an economic watershed affecting the structure of rural communities across the nation. As in all landmark changes, divergent philosophies are entwined in arguments pro and con.

Initial Reactions

When the results of Cornell's bGH trials were released in 1985, responses were widespread and varied. At one end of the spectrum, Cornell economist Robert Kalter called bGH an efficient production tool that will lower retail prices, spurring consumer purchases "essential to the survival of the dairy industry." According to Kalter, the root cause of the surplus problem lies in the government price-support program. "Surpluses would disappear and milk consumption would increase," he contended, "if we let market forces operate more efficiently."

At the other end of the spectrum, antibiotechnology activist Jeremy Rifkin has said that bGH-generated surpluses will force out nearly half of the nation's dairy farmers within five years. "I don't think anybody in the dairy industry wants this product, except some scientists at the universities and the chemical companies." The small farmer doesn't want bGH and the large farmer can live without it, said Rifkin, arguing that bGH will result "in the single most devastating dislocation in U.S. agricultural history."

While not predicting such dire consequences, most other researchers do anticipate significant and lasting structural changes in the dairy industry if bGH receives commercial authorization. Cornell researchers William Magrath and Lauren Tauer say that bGH will drastically increase New York per-cow milk production for dairymen who must already "deal annually with actual surpluses or the likelihood of them."

The 1985 New York State Senate report, "When Hard Work Was Not Enough," concluded that not only will smaller dairies suffer from bGH but called it "the technological development most likely to have large scale ramifications for the dairy industry."

Noting bGH's potential for 10–30 percent per cow production increases, a 1985 New York Cooperative Extension report said, "current and near future developments have the potential for massive displacement of farming resources including both people and capital." Citing a New York farm finance survey, it projected over 7,200 (or 28.9 percent) of New York's farms out of business by 1990 – even without factoring in the full impact of bGH production increases on farmers' exit rates.

Agway's recently retired Director of Research and Development Lewellyn Mix reported that bGH will push per-cow production levels to over 20,000 pounds annually nationally and to over 25,000 pounds in the Pacific region, that there will be 51 percent fewer dairy farms and 195,000 fewer dairy farm employees by the year 2000, and that "farm milk prices are likely to decline."

Finally, a 1985 Congressional Office of Technology Assessment (OTA) report cited bGH as likely to increase production greatly. It said that this increased production will "trigger price support reductions and increased failure rates among dairy farms." It concluded that in the Northeast "over the next ten years, over half of the small-to-moderate dairies may be forced to leave agriculture." Similar data were presented for dairy regions in the upper Midwest.

One expects controversy with the introduction of any technology. What complicates the situation is that federal approval of bGH will, one way or another, profoundly affect not only the future structure of the dairy industry but also of America's rural and agricultural economic and social landscape. And, just how one thinks agriculture should be structured within the economic, social, and political fabric of American life determines the filters through which one analyzes bGH's costs and benefits.

These filters are apparent in the assumptions upon which Kalter's and Rifkin's statements are based. While both men's statements contain elements of truth, there is absolute truth in neither. Clearly, something other than a simple analysis of bGH technology is going on.

Conflicting Values

The introduction of biotechnology (in general) and bGH (in particular) into the mainstream of American agriculture involves several disputed values. These include (1) the relationship between technological advancement and societal progress, (2) the role of the traditional family farm in American society, and (3) the "proper" role of government intervention regarding technology innovation.

Because philosophical attitudes are not objective criteria from which public policy decisions are made, open discussion of value considerations is rare. Nevertheless, debate over these values influences the political process leading to the approval or denial of bGH for commercial use. Each value, by itself, injects considerable philosophical grist into the supposedly objective policy-making decision process. When all three values intersect, as is the case with bGH, even defining (let alone discussing and evaluating) the issues becomes difficult.

Technology and Progress

As a nation, we tend to equate technological advancement with progress. What's more, progress is generally perceived as an unmitigated "good." Those challenging emerging technologies are labeled as obstructionist Luddites—those early nineteenth-century English workers who smashed factory machines that threatened (and subsequently extinguished) their home-based cottage industries.

Clearly, our country would not have chosen to forego the benefits of automotive transportation to avoid the displacement of blacksmiths and wheelwrights. Yet questions of economic equality are important and need some consideration—especially if they transcend particular job classifications and have the potential of negatively affecting the economic and social viability of a sector as broadly defined as American rural communities.

Beyond the issue of economic equity, however, the last thirty years have seen technological advance increasingly questioned on the dimension of public health and safety. From thalidomide to acid rain, DES to Three Mile Island, Love Canal to toxic shock syndrome, the expanding number of unintended negative outcomes associated with technological development have resulted in greater public wariness. One does not have to resort to conspiracy theories to recognize that the limits of human knowledge can not only beckon us to find solutions but come back to haunt us when solutions turn out to have been founded on inadequate information, however much we may have been impressed by our own abilities at the time a given technology was advanced.

This tension between those who focus on the benefits of new technologies and those concerned with the negative economic, social, and public safety outcomes that may accompany a technology's introduction is very much at play in the bGH controversy.

The Family Farm

Few images speak more intensely to the American psyche than the archetypical God-fearing, self-reliant, pioneering, independent, and entrepreneurial farm family. While the family farm can be defended as the keystone to rural prosperity on economic criteria alone, nonfarm public sentiment (and considerable infusions from the national treasury) have been generated on social grounds as well.

Notwithstanding a stated public policy of preserving family farms, the structure of farm ownership continues to shift toward a bipolarization of agriculture. Five percent of our farms, with annual sales exceeding $200,000, now account for half of the nation's agricultural output. Another 71 percent (the majority of which do not rely upon agriculture as their principal source of income) have under $20,000 in annual sales representing just 12 percent of America's farm output. Both groups are growing.

Squeezed in the middle are the 24 percent of our traditional full-time family farms with annual sales of $40,000–$200,000 accounting for 38 percent of production. The treasury is hemorrhaging tens of billions of dollars to offset low farm prices, but nearly two-thirds of 1985's deficiency payments went to a small number of larger farms while tens of thousands of family farms collapsed.

In New York, former Agriculture Commissioner Joseph Gerace estimated that only 10 percent of the state's farms are operating solidly in the black. Another 25–40 percent, he said at the 1985 New York State Grange annual meeting, are "barely hanging on for survival." It is this latter group of moderate-sized, full-time family farms that is under greatest economic distress and perceives itself threatened with almost irresistible pressure to leave the land should commercial introduction of bGH be authorized.

Government Intervention in Technology

"Before a new animal drug such as bGH may be marketed commercially," said U.S. Commissioner of Food and Drugs Frank E. Young, "it must be found safe and effective in the target animal and safe from the standpoint of human food consumption." Despite acceptance of the government's role to intervene to protect public health, safety, and environment, there is considerable debate over how narrowly or broadly these interests are to be defined. FDA health and safety statutes require that decisions be made on a scientific basis *without consideration* of economic or social impacts being taken into account, according to Dr. Young.

There are those who believe that questions of economic and social impacts of new technologies, such as bGH, are solely matters of economic efficiency and should be resolved by the marketplace, unencumbered by government intervention. Government interference, they claim, preserves inefficient producers to the detriment of the entire agricultural sector and society as a whole. Moreover, future scientific advancement, as well as knowledge creation and utilization, may be compromised by the threat of additional regulatory reviews.

Others believe that the economic and social health of agriculture and America's rural sector represent the foundation of America's prosperity. They argue that these issues are an integral and paramount portion of the nation's health and safety and, as such, deserve consideration every bit as much as do the technical issues of a drug's physiological impacts and biological consequences.

In a discretely different vein, the FDA is caught in a dilemma of having to protect the confidentiality of the private sector's proprietary rights and simultaneously having to encourage adequate public input so full disclosure of any scientific risk may be heard, learned, and considered. As Dr. Young explained, the government is "prohibited by law from disclosing information submitted to us by these companies concerning their research . . . unless the companies publicly disclose or acknowledge the information themselves."

At best, the paucity of proprietary research data in the public domain makes it difficult to know if health issues are being adequately addressed and answered by private sector research and/or duly considered by the FDA. At worst, this process results in continuing claims that companies are selectively restricting FDA submissions to data that are favorable to their cause or that the FDA is not being aggressive enough in ferreting out and releasing data supporting or refuting claims of the technology's negative side effects.

Now that some of the issues underlying the debate over bGH introduction have been discussed, it may be useful to understand the evolution of the current dairy environment into which commercial bGH utilization may be inserted. Following this, specific reservations about the technology will be evaluated and policy options presented for consideration.

The National Dairy Environment

Since World War II, U.S. agricultural policy has allowed dairy farmers to sell as much milk as they could produce. The government established a floor on market prices with its dairy price-support program. Unlimited quantities of storable dairy products (cheese, butter, and nonfat dry milk) are purchased by the government whenever commercial markets fail to absorb supplies at prices above the support level.

For the thirty years preceding 1980, the program successfully balanced supply with demand without huge surpluses of dairy products. From 1950 to 1979, the program cost taxpayers an average of $250 million a year. Since 1980, commercial milk surpluses and, hence, program costs, began to soar. The difference between total production and commercial consumption averaged 12.3 billion pounds of milk annually. The treasury's cost to buy and store this surplus averaged $2 billion per year.

Put another way, the $11.9 billion spent purchasing dairy surpluses in the last six years was about one and one-half times the $7.5 billion cost for the preceding thirty years. The price tag has created tremendous pressure to do something to cut back on taxpayer-financed expenses.

Over the last four years, several voluntary programs have offered dairymen incentives to cut production. A rebate program in late 1983 was followed by the 1984–1985 diversion program. The nation is now in the final phase of the whole-herd buyout program designed to cut dairy production by paying farmers to stop farming. The hope is that it will ultimately prove less expensive to buy out farmers than to indefinitely pay for surpluses through unlimited government purchases.

At the same time that the government has been seeking to reduce total dairy production, advances in herd nutrition, management, and breeding (as well as biotechnological advances such as embryo transfer) have steadily increased milk production per cow — from 5,314 pounds annually, for example, in 1950 to 13,400 pounds in 1986. Even without the added impact of bGH, annual production increases in the range of 2–4 percent are forecast for the balance of the century.

Ohio State agricultural economist Robert Jacobson told dairy industry leaders last November in Syracuse that for every 1 percent increase in milk production, about 107,000 cows must be removed from the national herd to keep total output at its current level. With an average herd size of sixty-three milkers per farm, seventeen hundred American dairy farms must be pruned for each percentage increase in production efficiency.

The balance between supply and demand has gotten so far out of equilibrium that the nation's essentially conservative and fiercely independent farmers are again considering mandatory production controls to manage the runaway supply side of the equation. Whether one supports or opposes supply control is not important here except inasmuch as its inclusion as a serious policy option reflects the fragile status of the dairy environment.

Thus, whether one feels that commercial use of bGH will result in net national production increases of 25, 20, 15, or "only" 10 percent in the national milk supply, the scale of its impact on the structure of American agriculture, rural social infrastructure, and national treasury becomes less obscure.

Reservations about bGH

At its 1986 annual meeting, the New York State Grange requested a delay in granting commercial authorization of bGH pending study of seven specific reservations about the technology and its potential impacts on agriculture and rural society. These were (1) human health, (2) long-term herd health, (3) long-term animal fertility, (4) the impact of added milk production on farm prices, (5) the impact of added production on taxpayer-financed product removals, (6) the possible skewing of economic benefits to larger producers at the expense of smaller farms, and (7) concern about the state, federal, and private sector projections of massive numbers of bGH-related farm failures on the structure of agriculture and the health of rural communities.

A major source of concern has been an absence of public information about both the clinical and economic effects of bGH utilization over a wide range of environmental and management conditions and over extended periods of time. Because of the proprietary nature of the technology and the fact that several companies are competing to be the first to market bGH to gain market share, only limited data are available to the public scrutiny.

With data from all pharmaceutical company experiments being held confidential by the FDA, most known data about bGH come from limited experiments at agricultural land-grant schools. Of these, the Bauman study at Cornell is generally recognized as the most significant. Although carried out under exacting standards and scrupulously reported, the researchers recognized methodological constraints in their protocols limiting the application of their findings.

Long-Term Herd Health

Cornell researchers stated that no health problems presented themselves during the experiment, but data covered only the final two-thirds of a single lactation (milk production cycle). They noted this limitation in their findings, saying that "long-term multi-lactation studies will be required to confirm the safety of [bGH] treatment." No such multi-lactation studies are currently available for public scrutiny.

Despite this, pharmaceutical company press releases routinely cite the Cornell study without mentioning the scientists' expressed reservations. In the absence of data on multi-lactation studies, readers of these releases may erroneously conclude that the findings of the one-time, single lactation study can be extrapolated across time, differing environmental conditions, and varying levels of managerial ability without any such evidence ever having been presented.

Last April, a joint petition was filed by the Foundation on Economic Trends, the Humane Society of the United States, the Wisconsin Family Farm Defense Fund, and the Secretary of State of Wisconsin asking the FDA to conduct a study to produce an environmental impact statement on bGH. Among the cited justifications for the study were claims that animals injected with bGH would exhibit increased physiological stress and be subject to a host of diseases.

FDA Regulatory Affairs Commissioner John M. Taylor denied the petition in September 1985. The FDA declined to indicate whether it would prepare an environmental impact statement because confidentiality regulations prohibit "disclosing the existence or nonexistence of NADAs (new animal drug applications)" unless an applicant has publicly disclosed or acknowledged the existence of an application.

Taylor indicated that environmental assessments had been submitted to the FDA with investigational new animal drug (INAD) applications pending before the agency and concluded that "the investigational use of [bGH] in dairy cows does not result in significant impacts on the human environment."

An October 1985 Freedom of Information Act petition seeking the environmental assessments was denied by the FDA under the "trade secret and confidential information" exemption. Rifkin's foundation appealed this decision in February 1987 and announced its intention to sue the FDA to force public disclosure of the environmental assessments if the appeal is denied.

Whether these allegations prove to be accurate or mistaken will not be known until the FDA finds a way to legitimately transcend the limits imposed on it to preserve the confidentiality of proprietary secrets. Some suggest that the FDA release a summary statement that provides a wider data base for public scrutiny without compromising individual companies' interests. Until this occurs, however, questions will remain about FDA's role in the maintenance of the public interest.

University of Pennsylvania scientist David S. Kronfeld has recently presented findings suggesting negative outcomes with bGH use. Kronfeld, who has studied the effects of growth hormone since the early 1950s, told the 1986 California Animal Nutrition Conference that bGH-related shifts in metabolism are not limited to increased milk production.

His research demonstrates that high levels of growth hormone result in subclinical hypermetabolic ketosis, a condition associated with reduced reproductive efficiency, mastitis, decreased immune function, and "the full gamut of other diseases typical of early lactation." Noting that health problems were not limited to his research but have also been reported by the University of Missouri, Kronfeld concluded that bGH "should be expected to influence health and reproduction as well as milk production."

Kronfeld calculated that bGH-induced hypermetabolic ketosis might incur economic costs due to mastitis of $1.55 billion annually. He suggested that specific research be conducted on the health effects of bGH, replacing ancillary data from production trials as the basis of health findings.

When asked why other experiments did not show the disease incidence his research would suggest, Kronfeld said that one had only to look at the ratio of cows recruited to the number completing the trials. Sick cows were sometimes removed from trials and replaced on the assumption that they had picked up illnesses prior to joining the experiment. While acknowledging that this may legitimately occur, Kronfeld believes that this practice may well have masked bGH-related disease incidence reports.

Long-Term Animal Fertility

Administration of low-level doses of bGH in sponsored university experimental trials has not generated data suggesting any resultant reproductive inefficiency. Significant inefficiencies, however, have appeared at higher dosage levels. Initial data come from small experimental populations; results from larger studies are not yet available.

During the single Cornell-Monsanto partial-lactation study, reproductive performance of bGH-treated animals was not compromised. "Cows became pregnant and delivered normal, healthy calves at term," stated the Bauman report. Without questioning the Cornell findings, it remains undetermined if the same results would occur after an animal had gone through previous bGH-enhanced lactations. At this point, assertions about the benign nature of bGH given the associated physiological processes related to extra feed consumption and milk management remain conjecture.

During the University of Pennsylvania-Cyanamid study, reproduction rates were mixed. In control and lower bGH daily dosages, pregnancy rates ranged from 75–100 percent. However, only 14 percent of the cows receiving the higher bGH dosage rate successfully conceived. Research leader William Chalupa explained that the 50 mg/day dose yielding the problematic results is higher than the anticipated recommended rate for commercial application. Yet, none of the experiments to date has data to evaluate the potential for aggregate reproductive impairment after several lactations' exposure to bGH at any level.

Complicating the evaluation of health and fertility data is the fact that published university results are based on tiny samples. The Cyanamid experiment in Pennsylvania, for example, had just thirty animals divided into one control and three trial groups. While one may measure differences between trial groups, the total experiment numbers are too small to permit reliable statistical projections. Whether results from larger studies still in progress will be available or treated as proprietary information remains an uncertainty.

Also at question is the effect of commercialized farm-level management capabilities and varying environmental conditions on herd health and fertility at research farms other than those equipped and managed by universities and the private sector. As animal nutritionist Marhall McCullough pointed out in *Hoard's Dairyman,* "Many of these [questions] will not be answered in routine Food and Drug Administration clearances."

Human Health

In order to be approved for commercial utilization, bGH must be cleared by the FDA as being safe for human consumption. At the same time bGH's biologically active qualities are being analyzed and evaluated by the FDA, a far less scientific jury of consumer opinion will render its verdict on its psychologically defined perception of health risks.

Research going back several decades suggests that bGH has no negative effects on human beings. First, because somatotropin (like insulin) is a protein, direct ingestion results in its digestion. Also, experiments using animal somatotropin to correct hypopituitary human dwarfism established that the hormone is species-specific.

American Cyanamid reports that milk from cows receiving daily doses of bGH has less than 2.5 parts per billion of somatotropin in the milk. Noting that low levels of somatotropin are degradable by enzymes in the digestive tract, Cyanamid adds that "pasturization [sic] and other manufacturing processes may inactivate the low levels of somatotropin usually found in milk."

Resolution of bGH safety issues requires experimentation with large numbers of animals. Use of milk from bGH-treated experimental animals "is an important cost savings activity for researchers," explained University of Pennsylvania scientist Chalupa. That's one reason why the FDA responded to pharmaceutical firms' data submissions on human safety by granting permission to allow milk from a limited number of bGH-treated cows undergoing controlled experiments to enter the commercial food chain.

If one accepts that current scientific information is unanimous in commending bGH's safety, what reservations could the public possibly express? The first reservation is the recognition of human fallibility and the shifting limits of human knowledge. One need only look at recent events to realize that honorable, well-trained scientists can make claims of safety based only on their current thresholds of knowledge.

Second, while claims regarding the safety of bGH are based on studies going back several decades, these studies used naturally occurring (and therefore only minutely available amounts of) the growth hormone. If any study has been conducted to determine if there are cumulative uptake effects from extended consumption of bGH-enhanced food production, its reports are not currently available to the public.

The third reservation is consumer chariness towards any synthetic additive to milk—*the* food of choice for young children and lactating mothers. Sales of this ubiquitously consumed beverage are likely to be compromised as long as any trace of public concern about health and safety over bGH remains. Given this environment, it may be more prudent to conduct additional laboratory and clinical trials prior to commercial authorization of bGH.

Impact on Farm Prices

There are legitimate arguments about the scale of added production that will occur depending on the rate of commercial bGH adoption. However, unlike previous technological innovations which were primarily extended to the agricultural community by county agents, it is expected that pharmaceutical companies will carry out their own aggressive program of marketing targeted at larger dairy farms. Thus, instead of taking many decades for bGH adoption to percolate to farmers, adoption rates of over 80 percent are forecast by the year 2000.

It is likely that the strategy will have "carrot and stick" elements. The "carrot" will be the lure of significantly increased production (hence, increased income) before the aggregate increased milk supply leverages down the per unit price paid to farmers. The "stick" will be the threat of finding oneself producing traditional levels of milk after the bGH-induced production increase has caused the per unit farm price to plummet.

"It's not the percentage increase in production that's crucial," said science consultant George H. Kidd. "It's the increase in gross revenues that the farmer sees." Whether the increased volume of production would offset the lower per unit price paid farmers is problematic. Producers with smaller herds would be relatively disadvantaged.

Even those who enthusiastically support commercial use of bGH acknowledge that its introduction will be accompanied by an indeterminate period of significantly depressed farm prices. Professor Kalter, who believes that price-support programs cause the surplus and would prefer to see market efficiencies determine prices, concedes that "as aggregate milk production increases through the use of [bGH], milk prices will fall and the number of dairy cows and dairy farms will have to decline substantially."

Whether one accepts Rifkin's estimate of a 50 percent farm failure rate or Kalter's more conservative 25–30 percent projection, the scale of the impact on the nation's dairy-producing families is clear.

Impact on Price Supports

Falling farm prices that accompany bGH-generated production increases will result in added stress to the federal dairy program. Congress will feel pressure to review the current policy of making unlimited purchases of surplus production. Whether such a move gets off the ground is a matter of speculation, but if the cost to taxpayers mounts at the scale of bGH-related production, as projections suggest, several policy options will probably be considered.

One option is to entirely eliminate the price-support system in favor of a free market situation. Despite the desire to get government out of the farming business, this does not seem likely. The fact that every industrialized as well as developing nation (of every possible political persuasion) involves itself in farming indicates that agriculture around the world is consistently viewed as an unstable economic sector. Our government's involvement acts as a hedge against the low rate of return to producers to ensure ample supplies of relatively low-cost food. In some form, this policy is likely to continue.

Another option might be to establish additional voluntary supply management programs, such as more farmer buy-out cycles, to ease the transition out of farming for those forced towards bankruptcy.

A third policy option might be to establish mandatory supply management quotas. While such quotas would have to address differences in regional product scarcity or oversupply, their aim would be to bring supply in line with demand by allocating production amongst all producers.

No solution is cost free. Society, through the political process, will have to weigh the long-term economic and social price of losing a significant portion of the rural sector against the immediate costs of underwriting less than optimally efficient production.

Skewing Benefits to Larger Farms

Questions of equity are inevitably raised when any potentially revolutionary technology presents itself. Will bGH be introduced on a "level playing field," or will the process structurally favor one class of farmers over another?

Proponents of bGH argue that the technology is "size neutral." By this they mean that efficient farmers can use bGH to their advantage regardless of the size of their operation. "This technology is not capital intensive," stated Kalter, "hence it can be used by every farmer whether he has ten cows or one thousand."

This is accurate, as far as it goes, because of bGH's low dosage cost. However, successful long-term use of bGH requires more feed with more nutrients (a "hotter," more expensive feed) as well as significant capital investment in high-tech computer and automated feed management systems to customize feeding programs for individual animals.

Good herd management often comes down to large farms' ability to achieve economies of scale in purchasing and operating the ancillary tools necessary for successful bGH use. Farms with thirty to sixty milkers will have a much harder time getting a return on such investments than farms with several hundred, or over a thousand, animals.

Unless smaller farms, as a class, can find a way to pool capital investments through a cooperative mechanism, they will find themselves outflanked by the resources and economies of scale enjoyed by large farms. Small farms may have equal access to bGH without ever realizing the production yields that come only when bGH is incorporated in a complete herd management program.

Yet, it is not even clear that small farms will have equal early access to the bGH technology. The leading research on likely adoption patterns for bGH relies on a "diffusion of innovations" model. After wide use by the United States and multilateral development efforts overseas, this model has been largely discredited because it kept producing negative, unintended outcomes—namely, fostering adoption by larger farms to the relative disadvantage of smaller farms.

Add to this the previously mentioned proprietary strategy of marketing directly to larger farms to amass market share, and one can see how rapid adoption of bGH technology will hasten the polarization of agriculture and further diminish the role of moderate-size family farms in America.

Computer projections by both Cornell and congressional OTA studies confirm the likelihood of these outcomes with bGH. A Cornell study of one thousand New York dairy farmers indicated that larger operators will most likely be the first to adopt bGH. The OTA study concluded that well-financed farmers are best positioned to take advantage of bGH. "In the main," it said, "such farmers will tend to be those with relatively large operations," and they will adopt bGH at the expense of small and moderate-size operators.

Structure of Rural America

After all is said and done, will bGH really alter the structure of agriculture and the fabric of America's rural countryside? The answer is both yes and no.

Certainly the trend towards larger and fewer farms predates and would continue without bGH. However, commercial introduction of bGH will significantly accelerate the process. And it may be the accelerated pace of change, more than the change itself, that outruns rural America's adaptive capacity.

As farm numbers diminish, local agricultural support services wither. Farm suppliers, implement dealers, marketing firms, and agricultural lenders feel the change first. In the absence of adequate off-farm employment to absorb former farm families and farm laborer families, the impact spreads to the service and retail sectors. The empty Main Streets of many small midwestern farming communities already bear witness to this phenomenon.

While some of the former farm units will be incorporated into larger farms, this consolidation will not be uniform. Soil quality, microclimatic variation, topography, market proximity, and alternative viable agricultural options are some of the factors that will determine whether farmlands will be recycled. Where significant acreage is diverted from agriculture and is not converted to an equal or greater value of land use, rural tax bases and tax-supported town, country, and school district services will be impacted.

The Senate committee report, "Governing The Heartland: Can Rural Communities Survive the Farm Crisis?" notes that the loss of revenue for local governments will be greatest on those regions which rely most heavily on agricultural income to sustain businesses and tax bases, such as the Upper Midwest milkshed. However, even in nominally insulated states such as New York (where the presence of population centers from Buffalo to Albany tends to support land values in surrounding agricultural communities), more isolated regions of the state are likely to feel significant economic consequences from the major projected decreases in dairy farm numbers.

The commercial introduction of bGH may yield efficiencies per unit of production for those who can successfully adopt the technology, but it carries another harvest in its wake.

Who Wins? Who Loses?

There are five categories of players in the bGH equation: input suppliers, farmers, consumers, taxpayers, and universities. How each fares individually, and how the nation benefits or suffers as a consequence, are of concern far beyond the commodity being addressed.

The biggest winners will be the input suppliers—specifically, the two or three pharmaceutical companies that are positioned to benefit from the production and marketing of bGH. According to University of Wisconsin bGH researchers Dave Combs and Larry Satter, a single dose may cost ten cents to manufacture and sell for a half dollar. Dairy cows would receive one dose per day during the final two-thirds of each animal's lactation cycle. "Manufacturers have mentioned $2 returned per dollar invested," said Satter.

By the year 2000, Dr. Chalupa expects bGH-related production increases to contribute to a reduction in the national dairy herd to 8.8 million cows. According to adoption projections, 80 percent of these animals will each receive about 205 doses of bGH annually. Monsanto alone ultimately expects to produce one hundred metric tons of bGH per year, according to Senior Vice President Howard Schneiderman. Vice President Lee A. Miller estimates a total worldwide market for bGH topping $1 billion annually.

The concentration of commercial bGH production in the hands of three or four manufacturers has already raised the specter of monopoly pricing, which could transfer a sizeable share of any production efficiency gains from farmers to input manufacturers. In testimony before Congress in 1985, Cornell's Robert Kalter noted that "monopoly power could develop with the resultant effect being bGH market prices substantially higher than production, marketing and delivery costs." Even if questions of monopoly pricing never become an issue, commercial authorization of bGH presents these firms with a sizeable continuing market.

Some farmers who use bGH in the early days of the adoption cycle will see positive economic returns. All will suffer from lowered prices after adoption has become widespread but before the mass exits begin. Up to half the dairymen in the nation, mainly operators of traditional moderate-size family farms, are expected to exit farming by the year 2000. It is not clear what standard of living or level of security will be achieved by those who remain.

Consumers may see significant price reductions in the short run as widespread adoption of bGH brings a glut of milk to the marketplace. How much of the price cut will be absorbed by intermediate dairy processors, handlers, and distributors before the retail prices are set is, of course, unknown. At some point the mass exits of dairy farmers will bring supplies in line with demand. Eventual retail price levels from this point cannot be predicted.

We can't forget that consumers are also taxpayers. What they receive in temporary lower retail prices at the checkout counter may well be more than offset by the less visible cost (namely, taxes) of dealing with more prolonged surpluses. The long-range implications for taxpayers will further be clouded by the uncertainty of other indirect local, state, and federal costs associated with helping those forced from the land.

Thus far, university research on biotechnology has largely focused on physiological and economic efficiency aspects of a variety of new technologies—the first of which to capture widespread public notice is bGH. This narrow research focus is caused, in part, by the presence of proprietary research grants directed towards bringing products to market and the absence of public funds to look at the wider social and economic implications of such technologies to society. How the public ultimately views the outcome of privately financed research at publicly funded institutions in the case of bGH may well influence the scope of future biotechnological research at public institutions.

Farmers' Responses

Speaking with dairy farmers across the country, one senses an attitude that is one part apprehension, one part distrust, and two parts resignation. Producers have been told for over two years that the bGH technology is coming and that, like it or not, they must adapt and adopt or get out of farming.

While this may be true, not all farmers are passively awaiting bGH's authorization and its consequences. At least one cooperative has decided not to accept any member's milk produced with bGH. Several others are considering the marketing of bGH-free milk to consumers who wish, for any reason, to avoid the hormone-enhanced product. Farmer and consumer activists met in Wisconsin in October 1988 to create the framework for consumer boycott of dairy products not labeled "bGH-Free."

In Washington, Representative Steve Gunderson (R-Wis.) last year introduced legislation to deny approval of commercial use of bGH until federal impact statements completely cover the environment impacts of bGH, the impact of bGH on the dairy price-support program, and the impact of bGH on the structure of the national dairy sector.

At its national session held last November in Wisconsin, the National Grange requested that commercial introduction of bGH be delayed until "additional data is acquired over several lactations to provide adequate data-sets" on long-term herd stress, health, and fertility.

Though the USDA has begun to respond to this pressure by initiating one study on the impact of bGH on the price-support program and another on bGH's impact on the structure of dairying (with particular emphasis on moderate-size family farms), at present the only legal obstacle holding up the commercial introduction of bGH is FDA approval.

Whether farmers and nonfarming Americans who value the role of the moderate-size family farm have any lasting influence on the bGH debate may depend upon their ability to convince Congress and the administration that a wider definition of public health and safety needs to be considered, that the economic and social benefits of having a strong and diverse rural agricultural sector transcend production efficiencies per unit of output.[1]

Notes

1. M. Shulman's paper is less formal than many other papers in this volume. It was originally prepared to stimulate discussion within the farm community. — the editors

10 | Agricultural Biotechnology: Public Perception of Risk

AMMERTTE C. DEIBERT

Introduction

Biotechnology, or the ability to alter hereditary materials of biological systems, carries with it the potential to alter the planet in ways equal to or greater than those accompanying the development of the ax and hoe, the domestication of animals, the creation of the internal combustion engine, the splitting of the atom, or the development of the digital computer. In short, the potential impacts are revolutionary. Clearly, the practical applications of agricultural biotechnology hold much promise for a better world. The existing limitations on crop production because of disease, pests, drought, length of growing seasons, soil nutrients, and so forth may be greatly reduced as a result of advances of agricultural biotechnology. Similarly, there are possibilities for new varieties of livestock with improved disease resistance, nutrient efficiencies, and lean/fat ratios, which will yield products with desirable tastes and textures. Such advances have major implications to help resolve problems of world hunger and to help the United States reclaim lost prestige as a leader in development of knowledge and technology. With such promises, an obvious question is, "Why all the 'hullabaloo' over this emerging technology?" The answer is that all technologies, including biotechnology, involve risk.

This chapter does not delve into the scientific complexities and intricacies of agricultural biotechnology itself. Rather, it attempts to provide a context within which to view the risks of agricultural biotechnology and to understand conflicting positions that are emerging as the promises and realities of biotechnology become known. Special consideration is given in this chapter to the value orientation the United States places on technology and how that has impacted on our society's response to new technologies, as well as what past risk perception studies tell us about technologies and how this relates to agricultural biotechnology. Finally, a study on public perceptions of biotechnology is examined to suggest prospects for the future.

Risk is viewed as the probability of the occurrence of some negative future event.[1] Risk, that is, involves the chance of loss or the probability of ruin as well as the alternative chance of some payoff that exceeds the investment in scarce resources.[2] Risk perception refers to the definitions and interpretations members of the public attach to a risk object or topic.

American Value System

Robin Williams cites four components in defining values: (1) they are abstractions drawn from the flux of an individual's immediate experiences; (2) they are emotionally charged and represent actual or potential mobilization to action; (3) they are the criteria by which goals are chosen, not the concrete goal of action; and (4) they are important ideas, not trivial or of slight concern.[3]

In describing the main pattern of values in the United States and how it relates to technology, Williams systematically studied the American culture and postulated major cultural themes. His work describes the American value system as stressing achievement (value accomplishments) and success (rewards), with our heroes having an eager interest in knowledge. The business ideal permeates society. Goodness is assumed in the American love for bigness. Williams sees Americans also valuing activity and work. Within this competitive atmosphere they seek to "demonstrate the world of nature to subdue and exploit the physical world."[4] This orientation developed out of a "religious tradition, frontier experience, ceaseless change, vast opportunity and fluid social structure."[5] High values are placed on team work, managerial roles, and technology. Valuing efficiency and practicality, Americans have long emphasized adaptability, technological innovation, economic expansion, up-to-dateness, and expediency in getting things done. The urge for material comfort makes Americans ready to accept most gadgets. Pressure has been exerted on individuals and society to search for better methods in technology, including those which give people control over nature and mastery over the environment. As a consequence, high honor and prestige are assigned to the sciences. Since Americans value progress, they tend to emphasize the future rather than the past or present.

There is a high propensity to change, a faith in progress, an ever-present thinking that the best is yet to be, and that nothing is impossible. Clearly, the value of progress defends the course of technological innovations. In such a value system, technologies have negative consequences to individuals such as small-time merchants, farmers, or urban workers. Such impact is considered regrettable but necessary in the greater goal of progress.

When this value is interfaced with developments in agricultural biotechnology, the initial assessment of America's response would seem to be one summed up in a modern day colloquialism—Go for it! Indeed, a recent Louis Harris poll of adult residents of the United States found 67 percent felt genetic engineering would make the quality of life better.[6] In another recent study, 1,273 telephone interviews that asked the questions, "How much have you heard or read about genetic engineering?" and, "Based on what you know or have heard, what is meant by *genetic engineering?*" found only moderate awareness of genetic engineering.[7] Even though the American public has only this moderate awareness, it thinks its quality of life would be improved by genetic engineering. When Americans were asked if the quality of life for people such as themselves would be greatly improved by genetic engineering, 66 percent said their lives would be much or somewhat better. The reason for this could be that 80 percent of the respondents said "no" to the question, "Have you heard about any potential dangers from genetically engineered products?" For the 19 percent who have heard of potential dangers, the problem of containment— the difficulty of controlling the product's spread—is most often cited (16 percent). This information from the public about the perceptions of biotechnology seems to reinforce theory about the acceptance of technology. But then a contradiction arises: when asked the question, "From what you have heard and read, how likely do you think that the threat of genetically engineered products will represent a serious danger to people or the environment?" over 50 percent responded either that it was very likely or somewhat likely. Could this be due to the fact that Americans are willing to accept the technology even if there is a danger to the environment or to people, as long as they see a potential future benefit?

There seems to be more awareness when discussing the issue of agricultural biotechnology. When asked the question, "Have you heard about using gene splicing or recombinant DNA to produce hybrid plants and animals by direct genetic manipulation?" 41 percent responded with a "yes." Only 24 percent of the respondents felt that it was morally wrong to create hybrid plants and animals through direct genetic manipulation of DNA. Of that 24 percent, 35 percent objected to this genetic manipulation because of tampering with nature and 31 percent objected because of their religious beliefs. Over 60 percent of all of the respondents felt if there were no direct risk to humans, they would approve of genetic manipulation for the following: (1) new treatment for cancer (96 percent), (2) new vaccines (91 percent), (3) cures for human genetic diseases (87 percent), (4) disease-resistant crops (87 percent), (5) frost-resistant crops (85 percent), (6) more productive farm animals (74 percent), and (7) larger game fish (66 percent).

The public seems to be saying, "Go for it!" when asked the question, "From what you have heard or read, how likely do you think it is that the use of genetically engineered organisms in the environment will very likely or somewhat likely (1) create antibiotic-resistant diseases (61 percent), (2) produce birth defects in humans (57 percent), (3) create herbicide-resistant weeds (56 percent), and (4) endanger the food supply (52 percent)?" When asked the question, "If new plants or animals produced by direct genetic manipulation can reproduce, how likely do you think it is that this will pose a danger to the environment?" the respondents also were very likely to somewhat likely (total 68 percent) to think that bacteria created by direct genetic manipulation will pose a danger to the environment.

Most respondents would approve the environmental use of genetically engineered organisms designed to produce disease-resistant crops, bacteria to clean oil spills, frost-resistant crops, more effective pesticides, and larger game fish.

Most of these findings can be interpreted to support Williams's outline of the U.S. value system and a better world through technology.

However, in spite of the value placed on science and technology as postulated by Williams and in spite of the support found in recent opinion polls, there are also growing indications of restlessness and concern (founded or unfounded) by some groups and individuals about the potential impacts of biotechnology on society. Some of these impacts are (1) passing on a risk to the environment or to human health;[8] (2) intentional release of genetically engineered organisms;[9] (3) organisms being created in the laboratory which have never existed in nature before;[10] (4) fewer and fewer farmers and corporations controlling most of what is produced, processed, and exported;[11] (5) new biotechnology allowing scientists to consciously shape the genetic destiny of the human race in ways that prove dangerous or destructive;[12] (6) biotechnology accidents with escaping organisms;[13] (7) implications and consequences of university-industry cooperation;[14] (8) property ownership;[15] (9) continuation of eugenics;[16] and (10) social impacts of advances in genetic research.[17] Some of the individuals and groups concerned with these potential impacts are (1) Nobel laureate James Watson,[18] (2) socially concerned scientists in Cambridge, Massachusetts,[19] (3) Congress,[20] (4) American biologists,[21] (5) Gordon Conference attendees and presenters,[22] (6) Asilomar Conference attendees and presenters,[23] (7) Jeremy Rifkin,[24] (8) Stanford President Donald Kennedy,[25] (9) the National Institutes of Health,[26] and (10) religious leaders.[27]

Some of this restlessness and concern can be better understood by looking at past technologies that have threatened environmental quality and long-term protection of quality of life. But, for the purpose of this chapter, a study of risk perception will be utilized to help understand this concern.

Risk Perception

Although the general literature on risk perception is not well developed, there are several principles that provide insights into understanding risk perception as it relates to agricultural biotechnology.

As used in this chapter, *risk perception* refers to the definitions and interpretations that members of the public attach to a risk object or topic — in this case the risks the public attaches to agricultural biotechnology.[28] The perceived risks may or may not be justified based on expert judgment or empirical verifications. Risk definitions between risk objects or topics by the public (for example, perceptions of impacts of biotechnology versus impacts of tobacco use) may at times appear inconsistent. Indeed, certain of these apparent incongruities prompted development of risk perception as a field of inquiry.[29]

Today, the caution and controversy surrounding agricultural biotechnology baffle some molecular biologists. Similarly, nuclear physicists were surprised a generation ago at the extent of public opposition to nuclear power plants. Some clarification and understanding of risk perception about agricultural biotechnology emerge when factors in the formation of risk perceptions are considered alongside characteristics of the technology itself.

Voluntary versus Involuntary Risk

Both theoretical reasoning and empirical documentation suggest a tendency for risks voluntarily encountered to be more acceptable than risks involuntarily encountered. Thus, higher levels of risk may be acceptable for a sport activity voluntarily engaged in than for a chemical residue emitted into the atmosphere by industry. This tendency for less acceptability of involuntary risk was recognized by Starr in 1968. He postulated the margin of safety or acceptability of technologies having an involuntary risk dimension may need to be as much as one thousand times greater than for technologies involving wholly voluntary exposure to risk.[30]

For the most part, the public perceives risk associated with agriculture biotechnology as tending to be more involuntary than voluntary. At least this would be the case for some of the more sensational fears, such as creating and releasing into the environment an altered organism having negative and perhaps uncontrollable consequences.

Active Control versus Passive Acceptance

There is unequivocal documentation that the risk of death is greater for automobile travel than for commercial airline travel. Yet, perceptions and behaviors of a significant portion of the population suggest a reverse interpretation. Among the characteristics accounting for differences is degree or perceived degree of control. Automobile travel offers the individual some opportunity to beat the odds (i.e., "I'm a safe driver," or, "I don't ride with people who are careless drivers"), whereas the airplane passenger is at the mercy of others. Covello found risks are perceived to be higher if the activity is perceived to be not personally controllable.[31]

The average person has little opportunity to have active control over the agricultural biotechnology process — the technology is developed in laboratories far away from the realm of most people.

Highly Complex versus Simplistic

Research suggests that highly complex issues also have higher risk perceptions. Tobacco smoking is a major and well-understood risk in the United States. The link is made between smoking and cancer and heart attacks, and that link is understood. This may be contrasted to an issue which is highly complex, such as the environmental impacts of nuclear energy. Because knowledge of the mechanisms involved in the many impacts of pollutants on man and his environment is still lacking, higher risk is associated with nuclear energy. Risks were perceived to be higher among highly complex issues than less complex issues in a study of risk perception by Covello.[32]

Agricultural biotechnology is a highly complex area of research and application. The links between agricultural biotechnology, the environment, and society are not clear. This topic will most likely have higher risk perceptions than will a topic that is less complex and better understood.

Familiarity versus Unfamiliarity

Familiarity seems to breed confidence. Farmers who live on flood plains, smokers, pedestrians, drivers, and people living near nuclear plants — all tend to perceive their activities as low-risk. One of the findings in risk perception is that individuals have a strong but unjustified sense of subjective immunity. In other words, in very familiar activities, there is a tendency to minimize the probability of bad outcomes.[33]

Most of the public is unfamiliar with agricultural biotechnology; this is reflected in their perception of the risk. The people who are familiar with agricultural biotechnology (researchers, farmers, and other people associated with the agricultural biotechnology process) will most likely have a lower risk perception.

Credible Probabilities versus Conflicting Messages

When experts give out conflicting messages, the perception of the risk increases. Ground water quality is an example of this. Some experts say there is absolutely no connection between ground water quality and serious health problems such as cancer. Other experts say there are links. In addition, some experts have known for years that chemicals are seeping into the water, but until recently it was not brought to the forefront for public scrutiny. Janet Fitchen found that experts "who disagree among themselves or change their mind lose credibility" and that "delays in providing information undermine confidence in the risk management."[34]

It is a well-known fact that for years agricultural biotechnology experts have been in disagreement over the risks and social impacts of this research. These conflicting messages may exacerbate the perception of risk. Whether agricultural biotechnology will bring credible probabilities about the risks to the environment and social impacts to the forefront as each issue arises is yet to be seen. This will be an important aspect in understanding the public's perception of this topic.

Severity of Consequences

Risk perceptions are affected by catastrophic potential,[35] but uneventful experiences reduce the perceived risk.[36] The nuclear industry has the potential of catastrophic disasters; this is one reason the public continues to perceive this industry as high risk. Without incidence, there was a possibility of lowering this perception of risk; but with recent catastrophes in the industry, this possibility has been negated.

Although the public's risk perceptions of agricultural biotechnology could be based on catastrophic potential because of media coverage and conflicting expert opinions, most likely the repeated uneventful experience will lower the perceived risk.

Whether you are trying to understand the perception of risk from the public standpoint or the researcher's, the study of risk perception gives insight into this very complicated topic. Risk perception studies can also give insight into the prospects for the future. The Office of Technology Assessment reports the findings of a nationwide survey of public knowledge about issues concerning genetic engineering and biotechnology.[37] This will be discussed in the context of future prospects of biotechnology.

Prospects for the Future

This section deals with the knowledge of how the public looks at the future of biotechnology, using data which was accumulated in the recent report by the Office of Technology Assessment.[38] With this knowledge there are ethical dilemmas that can challenge us, as well as insights into the American public and how it feels about this area of research.

When asked, "Would it be better if we did not know how to genetically alter cells at all?" the survey respondents disagreed by 65 percent. In a similar question, "Do you think that research into genetic engineering should be continued or should be stopped?" 82 percent of the 1,273 respondents felt the research should be continued. Even though they want the research to continue, the respondents (43 percent) thought the government funding should remain the same with 10 percent wanting decreases and 40 percent wanting increases.

This report also found that most (82 percent) respondents thought that environmental applications of genetically altered organisms to increase agricultural productivity or clean up environmental pollutants should be permitted on a small-scale, experimental basis. Even when asked if their "community was selected as the site to test a genetically altered organism — where there was no direct risk to humans and a very remote potential risk to local environment," 53 percent said they were strongly or somewhat in favor of this idea. When asked, "Who should be responsible for deciding whether or not commercial firms should be permitted to apply genetically altered organisms on a large-scale basis, the company that developed the product, an external scientific body, a government agency, an industrial trade association, or other group?" 37 percent of the respondents felt government agencies should be the responsible agents. Although this shows some strong credibility in the government, when asked, "Suppose a federal agency reported that the use of a genetically altered organism did not pose a significant risk to your community, but a national environmental group said it did pose a significant risk, would you tend to believe the federal agency or the national environmental group?" 63 percent of the respondents said they would believe the environmental group.

One should not, of course, assume that the present studies of perceptions of biotechnology will foretell future attitudes. They do tell us, however, about the public perceptions of this risk and how the public has confidence in the guardians of this technology. This confidence in turn could influence the future of agricultural biotechnology.

Notes

1. James F. Short, Jr., "The Social Fabric at Risk: Toward the Social Transformation of Risk Analysis," *American Sociological Review* 49 (1984):711–25.

2. David Hertz, Howard Thomas, and John Wiley, *Risk Analysis and Its Applications* (Chichester: John Wiley and Sons, 1983).

3. Robin M. Williams Jr., *American Society* (New York: Alfred A. Knopf, Inc., 1960).

4. Ibid., 235.

5. Ibid., 239.

6. American Enterprise Institute, "Opinion Roundup," *Public Opinion* (February/March 1986):21–29.

7. U.S. Congress, Office of Technology Assessment, *New Developments in Biotechnology* OTA-BP-BA-45 (Washington, D.C.: U.S. Government Printing Office, 1987).

8. Marvin Rogul, "Risk and Regulation in Biotechnology: Context for the Seminar Series," *Biotechnology and the Environment* (Washington, D.C.: American Advancement of Science, 1985); Joseph Fisksel and Vincent T. Covello, *Biotechnology Risk Assessment* (New York: Pergamon Books, Inc., 1986).

9. Rogul, "Risk and Regulation in Biotechnology."

10. Martin Alexander, "Spread of Organisms with Novel Genotypes," *Biotechnology and the Environment* (Washington, D.C.: American Advancement of Science, 1985).

11. Jack Doyle, *Altered Harvest* (New York: Viking Penguin, Inc., 1985).

12. Zsolt Harasanyi, "Biotechnology and the Environment: An Overview," *Biotechnology and the Environment* (Washington, D.C: American Advancement of Science, 1985).

13. Doyle, *Altered Harvest.*

14. Rogul, "Risk and Regulation in Biotechnology."

15. Ibid.

16. Ted Howard and Jeremy Rifkin, *Who Should Play God?* (New York: Delacorte Press, 1977).

17. Rogul, "Risk and Regulation in Biotechnology."

18. Ibid.

19. Sheldon Krimsky, "Regulation of Biotechnologies: State and Local Roles and Initiatives," *Biotechnology and the Environment* (Washington, D.C.: American Advancement of Science, 1985).

20. Doyle, *Altered Harvest.*

21. Ibid.

22. Rogul, "Risk and Regulation in Biotechnology."

23. Ibid.

24. Harasanyi, "Biotechnology and the Environment: An Overview;" Howard and Rifkin, *Who Should Play God?;* Jeremy Rifkin, *Algeny* (New York: Viking Press, 1983).

25. Rogul, "Risk and Regulation in Biotechnology."

26. Doyle, *Altered Harvest.*

27. Daniel J. Kevles, "Genetic Progress and Religious Authority: Historical Reflections," *Responsible Science* (San Francisco: Harper and Row, 1986).

28. Mary Douglas, *Risk Acceptability according to the Social Sciences* (New York: Russell Sage Foundation, 1985); Short, "The Social Fabric at Risk."

29. Chauncy Starr, "Social Benefits versus Technological Risks," *Science* 165 (1969):1232–38.

30. Ibid.

31. Vincent T. Covello, "The Perception of Technological Risks: A Literature Review," *Technological Forecasting and Social Change* 23 (1985):285–97.

32. Ibid.

33. Douglas, *Risk Acceptability according to the Social Sciences.*

34. Janet M. Fitchen, "The Importance of Community Context in Effective Risk Management," paper presented at the Society for Risk Analysis, Washington, D.C., 1985.

35. W. W. Lowrance, *Of Acceptable Risk: Science and the Determination of Safety* (Los Altos, California: William Kaufman, 1976); Paul Slovic, Baruch Fischoff, and Sarah Lichenstein, "Rating the Risks," *Environment* 21 (1979):3.

36. Alice C. Whittemore, "Facts and Values in Risk Analysis for Environmental Toxicants," *Risk Analysis* 3, no. 1 (1983):23–33.

37. U.S. Congress, Office of Technology Assessment, *New Developments in Biotechnology.*

38. Ibid.

PART

IV

Economic Prospects

11

Bovine Somatotropin's Scale Neutrality and Constraints to Adoption

ROY C. BARNES and PETER J. NOWAK

Introduction

The expected release of bovine Somatotropin (bST) by 1990 or earlier poses difficult questions to a number of groups. For example, the dairy farmer is confronted with the question of whether or not to adopt this productivity-enhancing innovation in a hostile climate of increasing supply and falling commodity prices. Furthermore, if our hypothetical farmer decides to use bST, a second critical question is how it will be used. Will it be used to maintain current levels of production with fewer cows, or will it be utilized to maximize production? To those interested in agricultural policy, questions surrounding the effect of bST on the dairy industry loom large. Assuming that a percentage of dairy operations across the country adopt bovine Somatotropin, how will such adoption patterns affect the price of dairy products and the returns to investment? For the sociologist, one concern involves the social impacts of the diffusion of this productivity-enhancing technology. Questions addressing who will benefit and who will lose from the release of bST, as well as the more basic question of who will adopt this technology, present themselves as research topics.

This chapter will touch on a few of these questions. To begin, a number of ongoing methodological problems with existing bST studies are highlighted. The intent of this discussion is not to downgrade this research but to advance analyses and increase the accuracy of predictions. In particular, the assumption of equal adoption of this innovation across all regions of the country and by dairy operations of all sizes at similar rates is criticized. Predicting socioeconomic consequences of bST diffusion on a national level requires recognition of inter- and intraregional variation in adoption rates. It is argued that the constraints to adopt bST are not equally distributed across either geographic regions or structures of agricultural production.

In order to refine ex-ante research efforts in anticipation of the consequences of bST diffusion, a number of constraints to adoption are identified. These fall under three major inputs to an economic organization: management, labor, and capital. Effective use of bST will involve a farm's relative ability to make changes in management skills, meet increased labor demands, or accommodate larger capital expenditures. After having delineated farms along these three dimensions, farmer's opinions of bovine Somatotropin, the characteristics of who will adopt, and anticipated production strategies are examined.

Previous Work

Several studies have greatly influenced the tone and direction of current research on the potential socioeconomic impacts of bovine Somatotropin. Central among these are studies conducted by Kalter and his colleagues.[1] Their studies have predicted that 80 to 90 percent of the dairy farms will adopt bST within three years of its release. It is anticipated that although feed requirements will increase, there will be a net return for some dairy operations in using bST. The overall market response to the increased production will create a transition period in which prices will drop, thus causing a necessary decrease in the number of farms. Their study suggests that farms with access to good land, operations in strong financial position, and operators with superior management skills will be better able to survive the transition. They have also suggested that operations with the above characteristics are beginning to increase herd size now in preparation for the commercial release of bST.

The 1985 study by Kalter and others used an innovative approach in predicting the adoption rate of bovine Somatotropin. In their study, simulated extension fact sheets and advertisements on bST were prepared and given to the respondents for examination. Respondents were then questioned as to whether or not they would use bST. There are, however, three major methodological problems with this particular study. (1) It was assumed that the respondents possessed no previous knowledge or opinions with respect to bST. By regarding the respondents as equivalents with respect to previous knowledge while ignoring other studies that have found structural biases relative to access to information and assistance, it is nearly impossible to ascertain the real effect of their experimental treatment. (2)

The second point concerns the nature of the treatment itself. Respondents were not presented with the potential negative consequences of bST diffusion — prospects of an accelerated decline in the numbers of dairy farms, the subsequent increased concentration of production in the dairy industry, and the possibility of increasing vertical integration of the dairy industry by a few multinational corporations.[2] Presenting only the positive side of bST likely inflated the adoption rate, especially by those who would be less likely to be influenced by potential negative impacts. Such a proinnovation bias also ignored the inevitable trade-offs associated with any agricultural innovation. (3) The sample response rate of 13 percent raises serious questions about generalizing the findings.

The second study which has greatly influenced the area of predicting socioeconomic impacts of biotechnologies is the report issued by the Office of Technology Assessment (OTA).[3] Though speaking generally about the effects of many different biotechnologies and information technologies, the OTA's analysis suggests that the most dramatic effects will be seen first in the dairy industry due partially to the commercialization of bST. The combined effects of economies of size, managerial bias, and differential finance lending practices are predicted to shift the comparative advantage of the lake and northeast states to the larger dairies in the West and Southwest.

Utilizing the pooled knowledge of OTA staff and agricultural experts, regional adoption rates were predicted and used in the analyses. A point of contention, however, concerns the level of analysis. By focusing exclusively on the macro process and cumulative impacts of biotechnology, the study precludes the micro processes which take place on each farm. Thus, although their interregional analyses may be valid, these results cannot be projected to intraregional settings without being subject to a fallacy in logic. Consequently this study does little to address the question of whether bST will be scale neutral.

The final study to be considered was conducted by the Economic Research Service of the USDA. Specifically addressing the consequences of the anticipated release of bST in conjuncture with varying governmental price-support strategies, this report generates four scenarios. There are two general findings that address the price-support–bST-release interaction and the question of regional shifts in the dairy industry. Not surprisingly, at current and slightly lower price supports, large dairy farms will experience the greatest increase in profitability while the smaller farms remain profitable. Nevertheless, at price supports of $8.60 per hundredweight (cwt), many small, inefficient dairy farms will go out of business if bST is released. As to the question of interregional shifts, the study predicts, irrespective of the price supports, no substantial change in the location of milk production above the current trends.

This study's adoption rates are based on the results of econometric models in which the distribution of economic characteristics of farms in various portions of the country is a modeled parameter. Unfortunately, this model nullifies any influences of behavioral imitation, stratification effects due to actual knowledge and knowledge generation, and the distribution of managerial capability. This is because the study assumed that all dairy operations, regardless of their characteristics, would adopt bST at equal rates. In effect, an underlying assumption of this study was that bST is scale neutral. However, it is maintained that the scale neutrality of this technology is a research question and should not be treated as an assumption.

Research Setting

All of the previously mentioned studies assumed that dairy operators will have equal access and ability to adopt bST. The validity of this assumption is the focus of this chapter. To address this assumption, a series of questions were added to a periodic survey of Wisconsin farmers. Researchers at the Department of Rural Sociology sought to address whether there were constraints to adopting bST and to map the distribution of these constraints.

The survey was administered for the *Wisconsin Agriculturalist* by the Miller Publishing Company, a Minnesota-based survey firm. Wishing to obtain 400 completed surveys from Wisconsin's 82,000 farmers, the interviewers were instructed to selectively sample 200 males and 200 females. The number of respondents selected throughout the state were proportional to the percentages of farms in five areas of the state: East, West, Central, North, and Southeast. The interviewers were instructed to travel in a specified direction from a given community within these areas and to interview at every active farm beginning with the third farm on their survey route. The interviews took place in the period between April 1 and April 10, 1987. Within the sample of 400, there were 270 dairy farmers on whose responses the following analyses are based. The general characteristics of the respondents have been presented in two previous papers.[4]

Constraints to Adoption

Far from the "magic pill" which promises to increase milk production without any strings attached, the successful administration of bovine Somatotropin could necessitate a number of changes in the farm operation.

For example, studies show that with bST, the feed intake of the cow increases, although in a smaller proportion than the increase of milk production.[5] To meet this increased feed requirement, a number of options are available. One could shift the current land use to accommodate more feed production or, if need be, purchase or rent more land. Other capital requirements could include barn modification, feed storage facilities, animal waste handling, and upgrading milking machines and bulk milk storage.

Not only will the herd's gross feed intake increase; the feed will also have to possess a more accurately balanced nutritional content. Careful monitoring of each cow's nutritional need will require a record-keeping system with analytical or problem-solving capabilities. Furthermore, knowledge required for nutritional testing or resources to have an outside party provide nutritional tests present themselves as additional baggage to the growing list of requirements effective bST adoption entails. Overall, the response to bST has been found to be directly related to the quality of the management surrounding the herd health program, milking practices, nutritional programs, and environmental conditions.

Although it is not clear what form the actual administration of bST will take, there is a possibility for increased labor. This labor demand may stem from two sources. First, if the technology is used in a daily injection form, there will be increased labor for administration. The second source of increased labor will present itself regardless of the form of the technology. This demand relates to the increased need for feed and management tasks. Thus, whether the cows are injected daily or whether they receive a time-released injection, it makes little difference in terms of the increased hours of record keeping, monitoring, testing, and feeding which will be required for efficient use.

Measuring Capital, Management, and Labor Constraints

To assess whether the respondents' current capital, management, and labor resources would be a constraint to bST adoption, a series of questions were asked about the problematic nature of a series of operational changes. These potential changes to the operation were derived from working with dairy scientists to determine what would be needed to use bST in a efficient manner. For each potential change, respondents were asked whether it would be a major problem, a minor problem, or no problem at all. Responses were scored *3, 2,* or *1,* respectively. Eleven of these potential changes were included in the analyses. (Table 11.1). A principal component factor analysis of these items was executed. The use of factor analysis is warranted when it is believed that there is more than one dimension to the concept that one is trying to measure. In this case, we hypothesized that there would be several factors in the broader concept of constraints to adoption. The rotated factor matrix is presented in Table 11.1.

Table 11.1. Oblique rotated factor matrix of eleven constraint items

	Factor Loading[a]		
Constraint Item	Factor 1: capital	Factor 2: management	Factor 3: labor
1. Purchase additional forage	.86037	. . .	. . .
2. Increase feed storage	.85808	. . .	. . .
3. Rent more land to grow more feed	.75659	. . .	. . .
4. Increase manure storage	.68628	. . .	. . .
5. Shift land use to grow more feed	.60529	. . .	.17615
6. Purchase additional concentrates	.59577	.21103	. . .
7. Have someone balance feed rations	. . .	.88144	. . .
8. Adopt forage testing program	. . .	.84965	. . .
9. Enroll herd in DHI program	. . .	.81339	.14611
10. More work from family	−.11176	. . .	.91521
11. Hire more non-family labor	.25387	. . .	.72754

[a]Factor loading 0.1 and less left blank to reduce clutter.

Three factors fall out very clearly. The first factor has been termed a *capital constraint factor,* given that the items clearly involve some increased financial burden. The second factor taps managerial capabilities and thus has been termed the *management constraint factor.* Here the balancing of feed rations, the adoption of a forage testing program, and the establishment of a record-keeping system are all included. The last factor, termed the *labor constraint factor,* has to do with the labor requirements of the dairy operation. In general, the two sources of labor included are family and nonfamily.

A few comments on the three factors are necessary for the interpretation of the later results. Having generated the above factor loadings, factor scores were created for each of the three factors. These scores will enable us to determine the extent to which changes in the dairy operation that involve managerial adjustments, for instance, would be problematic. It should be pointed out that these scores relate only to the data presently under analysis since factor analysis generates scores based on the data on hand. Thus, the scores themselves may vary from data set to data set. Nevertheless, it is hoped that similar factor patterns will be generated in subsequent replications.

The first factor, capital constraint, has six items. These items yield an alpha, an estimation of the reliability of these items in forming a scale, of 0.8492. The scores range from −2.904 to 2.755. A negative sign indicates that capital investments would not be problematic whereas a positive factor score indicates that it would be problematic. The second factor, managerial constraint, has an alpha of 0.8276 with a minimum of -2.028 and a maximum of 3.113. The directionality of the degree of constraint should be interpreted in the same manner as the first factor. The last factor, labor constraint, ranges from -3.123 to 2.216 and has an alpha of 0.6308. Although this is lower than is generally accepted as a reliable scale, its relation to the factor matrix and the importance of labor as a constraint justify its inclusion in the latter analyses. It is interpreted in the same fashion as the first two factors. These factor scores tap the respondents' relative ability to adopt bST with respect to capital, managerial, and labor requirements and can be used in an assessment of the scale neutrality issue.

The Scale Neutrality Assumption Reconsidered

The analyses that follow are divided into three sets for both practical and statistical reasons. The first set considers the breakdown of the three constraint factor scores by the categorical variables that indicate the farm size, the respondents' education, and opinion of private industry's research and development of bovine Somatotropin. The second set of analyses involves a series of correlations between eight indicators of farm scale and the three constraint-to-adoption scores. The last set of analyses examine the respondents' production plans under a bST scenario in relation to their constraints to adopt.

Constraints by Operation and Operator Characteristics

Technology is often conceptualized as an impersonal factor that is insensitive to the social context in which it is placed. This may seem to be the case when one concentrates on the technology; however, it may be quite different from the perspective of the dairy operator. In other words, although a technology potentially can be adopted by anyone, the more important question is whether those who decide to adopt are in any way preselected on dimensions not directly related to the technology itself. This section explores the possibility of differences among dairy operators in their anticipated ability to adopt bST.

In Table 11.2, the capital, managerial, and labor constraint scores are broken down by the six categories of gross income that were included in the questionnaire. There appears a pattern in the capital and managerial constraints in which the higher the gross income, the easier it will be for the

Table 11.2.   Breakdown of factor scores by farm size and operator characteristics

| | | Mean Factor Score | | |
Farm Scale Variable	N of cases	Capital constraint	Management constraint	Labor constraint
Gross farm income, 1986	247			
$0–$19,999	15	.0587 (.0785)	.6207 (.0714)	−.3241 (.4446)
$20,000–$59,999	68	.2274	.0310	−.0818
$60,000–$99,999	61	−.0439	.0863	.1478
$100,000–$149,999	50	.0624	−.2654	−.0265
$150,000–$299,999	36	−.3544	−.1870	.1534
$300,000 or more	17	−.4855	−.0495	.2454
Years of education	268			
8 years or less	27	−.0175 (.4463)	.4119 (.0021)	−.0308 (.5510)
9–12 years	34	.1096	.2449	−.0304
High school graduate	138	−.0722	.0627	.0571
Some college	50	.2171	−.4632	−.2105
College graduate	19	−.2202	−.1078	.1635
Opinion of private R & D	200			
Favor R & D	23	.1729 (.2052)	−.4081 (.1051)	−.2480 (.3827)
Neutral, undecided	61	−.1748	.1114	.0792
Oppose R & D	116	.0756	.0152	−.0208

farmers to adopt bST. For example, the mean factor score for the capital constraint is 0.0587 for the group with gross incomes in 1986 of less than $20,000. On the other hand, for the farms with sales between $150,000 and $300,000 and $300,000 and above, the mean factor scores are −0.3544 and −0.4855 respectively. In much the same pattern, the managerial constraint appears to be less of a problem for farmers in the upper income categories than for those indicating incomes of less than $100,000. It is noted that these breakdowns have an overall F-statistic with a probability of less than 0.08.

Relative to labor constraints, the results are the opposite of what was found on the other two constraint factors. More specifically, farms with low gross incomes seem to be less troubled by increases in the need for additional labor than do the farms with high gross incomes. An explanation may be that farms in the lower income range are predominantly farms in which family members supply much of the labor. However, farms in the upper ranges of gross income are most likely industrial-type and thus depend on hired labor.[6] One interpretation of this disunity between the capital and labor constraint is that the acquisition of more labor on the part of more industrialized farms is viewed as a problematic conflict rather than a trade-off between labor and free time for the family farm.[7]

The results of the mean factor scores for each of the education groups are not clear with one exception. There is a statistically significant relation between the managerial constraint and the education group. The more years of education, the less problematic the changes that involve better management skills. This is an expected finding since those who attended college or completed college tend to be more open to changes and perhaps more capable of implementing those changes.[8]

A dairy operator's opinion of the development of bST will obviously affect the operator's decision to adopt or not adopt this new technology. It may also reflect a more general attitude toward technological innovation. It is felt that views on the development of new technologies and on the research on bST in particular may not be evenly distributed across levels of constraints. In other words, do farmers who have little constraint on their abilities to adopt bST have more favorable opinions on the research and development of this technology? The data in Table 11.2 do not support the hypothesis of a negative relationship between a favorable opinion of bST research and the level of constraint to adopt bST. There does appear to be a slight pattern in the managerial constraint score in which those in favor of bST's development have very low managerial constraints. Yet the lack of a pattern in the remaining two constraint areas prohibits making any generalizations about the relation between opinion of bST and ability to adopt.

Constraints by Scale of Operation

There are many facets of a dairy operation that are indicative of its size. Although there is not a perfect correlation between size and type of operation (i.e., whether it is a family farm or an industrial-type farm, to use Rodefeld's typology), there is a certain degree of correspondence between size and the number of cows milked, the extent of hired labor, and the managerial skills possessed by the farmers themselves. In considering the adoption rates and potential impacts of bST diffusion, a fundamental question becomes whether the scale of a dairy operation is irrelevant with respect to the constraints to adoption. This question is addressed in Table 11.3.

Table 11.3. Correlation matrix of constraint factor scores and farm scale items

Farm Scale Variable	N of cases	Pearson Correlation Coefficient		
		Capital constraint	Management constraint	Labor constraint
Days of full-time family labor	270	$-.0226$ (.356)	$-.0075$ (.451)	$-.0866$ (.078)
Days of part-time family labor	270	.1110 (.034)	.0195 (.375)	$-.0965$ (.057)
Days of hired nonfamily labor	251	$-.0282$ (.328)	$-.1054$ (.048)	$-.0410$ (.259)
Number of cows on farm	265	$-.1170$ (.029)	$-.0752$ (.111)	.0362 (.279)
Pounds (cwt) of milk sold	190	$-.1621$ (.013)	$-.2216$ (.001)	.0716 (.163)
Computed herd average	141	$-.0651$ (.222)	$-.2609$ (.001)	.0295 (.364)
Age of respondent	263	$-.0449$ (.234)	.2067 (.000)	.0813 (.094)
Knowledge score	213	$-.0626$ (.182)	$-.2438$ (.000)	.1297 (.029)

Considering the relationships between the capital constraint factor and the various measures of the dairy operations scale, three relationships are found to be significant at the 0.05 level. As the number of days family members worked part-time increases, the capital constraint level also increases. This may indicate that dairy farms that are forced to increase the amount of labor the family contributes are in position of an ever-increasing cost-of-production bind.[9] This increase in the labor demands placed on the family can reasonably be seen to reflect a concurrent state of capital constraints. There is also a negative relation between herd size and the degree to which changes that require capital outlays are seen to be problematic. The larger the herd, the easier it will be to accommodate such capital changes as greater feed intake and larger waste-handling needs. Another scale indicator, the amount of milk sold annually, shares this same negative relation to capital constraints.

The relationship between the level of managerial constraint and the scale of the dairy operation demonstrate that larger farms and more knowledgeable farmers will be least constrained in their adoption of bST. Farms that hire more outside labor also scored lower on the managerial constraint factor. The pounds of milk sold and the pounds of milk produced per cow both demonstrate a strong negative association with the managerial constraint score. The higher the aggregate production and the higher the productivity per cow, the less constrained the operator will be in terms of using bST. There is a strong relationship between the age of the respondent and the level of constraint. This implies that older farmers will have a more difficult time incorporating the extra managerial skills into their dairy operation. Finally, there is a strong negative relation between the operator's level of knowledge of bST and managerial constraint. Those who know more about bST will have the least problems in effectively using this new product.

The labor constraint factor only highlights one significant relation between the level of knowledge and the degree to which increasing labor on the farm will be a problem. The positive correlation coefficient suggests that the less knowledgeable an operator is about bST, the less constrained this operator will tend to be in terms of meeting increasing labor demands. This finding is consistent with two of the previous findings. First, farms with low gross incomes were seen to be least constrained in terms of meeting extra labor demands. It was proposed that the smaller farms had an extra degree of labor flexibility. Yet, second, it was also found that smaller farms lack the managerial capacities often associated with being "in the know" in terms of the new technologies.

Constraints by Production Strategies

The effect of bST's adoption and diffusion is contingent on a number of factors, the most obvious of which is the actual extent of adoption. Another factor is whether the technology will be used to stabilize production by reducing the number of cows, or used as another tool to maximize producers' output vis-à-vis competitors. Will the adoption of bST be dominated by those who have followed a pattern of increasing production, or will it be used by status quo operators as the way to reduce their inputs relative to output? To complicate matters further, as the above analyses suggest, the ability to adopt bST may not be evenly distributed across the population of dairy producers and so these questions acquire an added element of contingency. The analyses that follow will address who is most able to adopt in relation to their use plans, production strategies, and production patterns.

The relation between anticipated time of adoption and constraints is portrayed in Table 11.4. Respondents were asked if they planned to use bST as soon as it became commercially available, wait and see what happens as others use it, or not use it under any circumstances. In general, there appears to be a relationship between the respondent's ability to adopt bST and their use plans. Although the strength of this relationship varies across the three factors, it is clear and logically consistent that those who indicate that they are least constrained plan to use bovine Somatotropin as soon as it becomes commercially available. This is particularly the case with respect to the labor and managerial constraint factors. The capital constraint factor shows that those who are planning to use bST right away average a very

Table 11.4. Breakdown of factor scores by production items

Production and Use Variable	N of cases	Mean Factor Score		
		Capital constraint	Management constraint	Labor constraint
Your use plans	239			
Use it right away	10	−.3480 (.5042)	−.3213 (.2092)	−.3834 (.0824)
Wait to see	143	−.0132	−.0750	−.1009
Do not plan to use it	86	.0828	.1552	.1833
Your production strategy	215			
Reduce herd, produce same	63	−.0337 (.8910)	.2353 (.0854)	.1532 (.3809)
Same herd, produce more	146	.0466	−.1376	−.0803
Increase both herd and production	6	−.0611	.1682	.0389
Production pattern	248			
Reduction mode	22	−.3119 (.0862)	.4774 (.0039)	.6213 (.0031)
Status quo mode	168	−.0477	.0229	.0236
Expansion mode	58	.2494	−.3579	−.2655

low constraint score, yet those who do not plan to use it under any circumstances are, interpretively, noncommittal.

Advocates of the widespread diffusion of bST often make the claim that bST diffusion need not lead to a glut of milk; rather, the dairy sector can continue to produce the same levels of milk with fewer cows. However, results from this survey suggest that Wisconsin dairy producers will not adopt the satisficing production strategy in which they reduce the number of cows in their operation while producing the same amount of milk with bST. The results indicate that approximately 68 percent of the respondents plan to produce more milk with the same number of cows.[10] The picture becomes more complicated if we add in the issue of constraints to adoption. The results stemming from the breakdown of the capital constraint yield no differences between the average level of constraint by the respondent's production strategy. However, this is not the case in terms of the managerial and the labor constraints. If we ignore the six cases who plan to increase both the herd and the overall production, the farmers who prefer a satisficing modus operandi are more constrained in terms of their managerial skills or their abilities to meet additional labor requirements. This is especially the case relative to the managerial constraint as those in the "reduce herd/produce the same" category have an average factor score of 0.2353, whereas those who plan to produce more with the same number of cows have an average constraint score of −0.1376.

The last variable to be considered in this section deals with the operation's production pattern. This is actually a composite measure designed to measure both past changes in the scale of the operation as well as planned future changes. The three categories developed are defined as follows: the expansion mode includes all operations that increased herd size in the past and that plan to increase in the future, the status quo mode needs no explanation, and the reduction mode includes the farms that decreased the size of their operation in the past and that are planning to reduce herd size further. The issue is whether bST will be adopted by those farmers who are currently on an expansion mode or a reduction mode. Impacts on the dairy industry will be deferentially affected by adoption rates occurring in either of the two groups. The overall increase in milk production will be greater, for example, if bST is adopted by farms expanding versus those decreasing their productive capabilities. Table 11.4 presents significant findings for the managerial and the labor constraint items. As can be seen, those operations that fall under the reduction mode are most constrained to adopt bST with respect to the increased managerial demands (mean factor score equals 0.4774) and labor requirements (mean factor score equals 0.6213). Conversely, those operations that are in the expansion mode exhibit quite low levels of constraint on these items. For example, the average managerial constraint score for those expanding is -0.3579 and the average labor constraint score for the same group is -0.2655. These findings strongly suggest that it is the expanding dairy operators who will be able to adopt bST. It appears that the past philosophy of "get big or get out" will be reincarnated under a bST future.

Discussion

The goals of the above analyses were threefold. First, the "structure" of constraints to adoption needed to be identified and quantified in a reliable fashion. The second phase, which was predicated on the first, was to determine the distribution of these various dimensions of constraint across dairy farmers in Wisconsin. The last facet of the research was to address specifically the question of bovine Somatotropin's scale neutrality. The factor analyses generated reliable, and logically consistent, measures of constraints to adoption, which in turn were found to be unequally distributed among the 270 dairy farmers in the survey.

In particular, it was found that dairy operations which scored low on the capital constraint factor were farms that had higher gross incomes, possessed a production pattern that was expanding, and milked a larger number of cows which produced more milk per cow. Dairy operations that had strong managerial capacities were the farms that had higher gross sales, were in an expanding production mode, hired more nonfamily labor, produced more milk, and did so more efficiently. In addition, these farms have operators who possess more years of education, know more about bST and are, on the average, younger. The last constraint item, the labor constraint, only differentiated between the dairy operations that were expanding versus reducing productive capabilities and between those operators that had high versus low levels of knowledge concerning bST.

The data presented above strongly support earlier findings by Kalter and others that the larger dairy operations with well-developed managerial capabilities will be in the best position to benefit from the release of bST given their lower levels of capital and managerial constraints. Further, it is no surprise that these operators are more favorable and knowledgeable with regards to bST. The upshot of these findings supports the notions that this new technological development will accelerate the existing trend of polarization within the dairy industry.

Some caution must be exercised in generalizing these results due to the limited sample size and restricted geographical area. Further, a more valid and reliable method of measuring the constraints to adoption is needed.

The current work is based on asking respondents about perceived constraints to a future technology that they may know little about. A better method would have been a farm visit where various capital, managerial, and labor processes could have been observed and measured.

Conclusions

The conclusions that are drawn from these data can be thought of as addressing the same issue on two different levels of analysis. For the farm, these data suggest that bovine Somatotropin is far from scale neutral. This, as will be discussed, has a number of implications for the adoption and diffusion of bST. The second level of analysis challenges the conclusions reached by the recent ERS study that there will be no interregional shifts in dairy production.[11] To the extent that the structural parameters that will influence adoption rates are unequally distributed among the production regions, the consequences of bST diffusion will also be unequally distributed.

Questions of the appropriateness of new technologies have been raised in response to a number of Third World development plans.[12] It is argued that the use of capital intensive technologies such as high-yielding varieties of seed and large-scale mechanical harvesting devices create inequities and distortions in the social and economic systems of Third World countries. Bovine Somatotropin, being developed with the anticipation of being the first agricultural biotechnology to be commercially available on a large scale, suggests similar concerns. Proponents of this technology argue that the technology is scale neutral because of the apparent lack of capital investments. This research found this not to be the case as capital expenditures may be required as well as additional labor and enhanced managerial inputs. Most importantly, the ability to meet these additional inputs is not equally distributed among the sampled farm population. Consequently, inequities and distortions in our social and economic systems, as witnessed in Third World settings, are also probable.

As to the question of regional shifts, it seems clear that the process of polarization will most likely include a relative increase in dairying in the South and Southwest and a relative decline of the role of dairy production in the Midwest and Northeast. This assertion is based on two sets of facts. The first is the qualitative and quantitative differences between dairy farms in California, as an example of farms in the South and West, and dairy farms in Wisconsin, as representative of farms in the Northeast and Midwest states.[13] This analysis indicates that farms in the West fall very much under the "industrial-type" of dairy farming. The scale of production coupled with the technological and managerial sophistication associated with these 1000 + herds in California is in stark comparison to the smaller, and perhaps less-efficient, farms of rural Wisconsin. Such a structure of dairy production establishes the parameters into which bST will be introduced. If certain regions are more advanced managerially and productively than other regions, and if the adoption of bST is heavily dependent on the operation's productive and managerial capacities, then it logically follows that certain regions will benefit more from the use of bST than will others. The second fact is that the Northeast and lake states currently export a significant amount of milk to other regions in order to meet demands. The diffusion of bST will allow these importing regions to better meet their own needs. These two factors have a number of consequences. Not only will the centers of aggregate output be altered, the actual livelihood of many smaller farmers in the Northeast and Midwest are placed in jeopardy. The larger farms in the South and West have the capabilities to effectively use bST while the smaller farmers, in their attempts to stay on the technological treadmill, will most likely be eliminated from the dairy industry.[14]

Elimination will occur as a result of two processes. First, market forces in response to increased supplies will decrease support prices. Second, a number of dairy operators who do not have adequate capital, labor, or managerial resources will adopt bST. The sophisticated managerial requirements for using bST in an efficient manner will also give rise to a classic example of an inappropriate technology for parts of the dairy industry. In the early 1990s a number of dairy operations will begin to use bST even though they do not possess the necessary managerial, labor, or capital resources. This decision will be made in response to market pressures as the national milk supply increases and in an effort to stay current on the technology treadmill. A double economic penalty will then accrue to these operations. First, they will pay for a product but not use it in a economically efficient manner, thereby losing money in the process. Second, due to insufficient management, any production gains will be offset by the depletion of the energy reserves of the animals. This stress can result in reproduction difficulties and other health problems. The end result will be to lower the overall viability of their herds. All this implies that bST may prove to be a "damned if you do and damned if you don't" situation for operators of small- to medium-size herds. On the one hand, they have neither the managerial, labor, or capital resources to use bST in an efficient manner. On the other hand, they will be penalized for not adopting when market prices decline in response to bST-induced increases in the milk supply.

Notes

1. Robert J. Kalter, "The New Biotech Agriculture: Unforeseen Economic Consequences," *Issues in Science and Technology* 2, no. 1 (Fall 1985):125–33; Robert J. Kalter et al., "Biotechnology and the Dairy Industry: Production Costs and Commercial Potential of Bovine Growth Hormone," *A.E. Report 84–22* (Department of Agricultural Economics, Cornell University, Ithaca, New York, 1985); Robert J. Kalter and L. Tauer, "Potential Economic Impacts of Agricultural Biotechnology," *American Journal of Agricultural Economics* 69 (1987):420–25.

2. Gordon Bultena and Paul Lasley, "The Dark Side of Agricultural Biotechnology: Farmers' Appraisals of the Benefits and Costs of Technological Innovation," paper presented at the annual meeting of the Rural Sociological Society, Madison, Wisconsin, August 12–15, 1987, 7.

3. Office of Technology Assessment (OTA), *Technology, Public Policy, and the Changing Structure of American Agriculture* (U.S. Congress, Report OTA-F-285, Washington, D.C.: U.S. Government Printing Office, 1986).

4. Peter J. Nowak and Roy Barnes, "Potential Social Impacts of Bovine Somatotropin for Wisconsin Agriculture: A Survey of 270 Wisconsin Dairy Producers," paper presented at the Wisconsin Dairy Leadership Conference on bovine Somatotropin, Madison, Wisconsin, July 8, 1987.

5. Kalter et al., "Biotechnology and the Dairy Industry."

6. Richard D. Rodefeld, "Trends in U.S. Farm Organization Structure and Type," in *Change in Rural America,* eds. Richard Rodefeld et al. (St. Louis: C. V. Mosby Company, 1978), 158–77.

7. Harriet Friedmann, "The Family Farm in Advanced Capitalism: Outline of a Theory of Simple Commodity Production," paper presented at the annual meeting of the American Sociological Association, Toronto, Canada, 1981.

8. Everett M. Rogers, *Diffusion of Innovations* (New York: The Free Press, 1983).

9. Friedmann, "The Family Farm in Advanced Capitalism."

10. Peter J. Nowak, Jack Kloppenburg, Jr., and Roy Barnes, "bGH: A Survey of Wisconsin Dairy Producers," *As You Sow,* no. 18 (Department of Rural Sociology, University of Wisconsin, Madison, July 1987).

11. Economic Research Service (ERS), *bST and the Dairy Industry: A National, Regional, and Farm-Level Analysis* (U.S. Department of Agriculture, Washington, D.C.: U.S. Government Printing Office, 1987.)

12. Frances Moore Lappe and Joseph Collins, *Food First* (New York: Ballantine Books, 1978); William C. Thiesenhusen, "Economic Effects of Technology in Agriculture in Less Developed Countries," in *Technology and Social Change in Rural Areas,* ed. Gene F. Summers (Boulder: Westview Press, 1983), 235–52.

13. Jess Gilbert and Raymond Akor, "Dairying in California and Wisconsin," *Wisconsin Academy Review* 33, no. 1 (December 1986):56–59.

14. Willard Cochrane, *The Development of American Agriculture* (Minneapolis: University of Minnesota Press, 1979).

12 Economic Issues and Analyses in Biotechnology

BRIAN BUHR, MARVIN HAYENGA, DENNIS DiPIETRE, and JAMES KLIEBENSTEIN

Introduction

Although the evolution of the food and agricultural system in developed countries seems to be progressing at an ever-increasing rate, in reality this trend has largely been realized as small bursts of progress followed by relatively dormant periods until another breakthrough is achieved. These breakthroughs occur in a variety of ways, including the scientific (such as the development of new hybrids and genetic lines, useful chemicals, etc.), the political (such as the opening of trade with China and other communist countries), and the behavioral (such as the shift toward leaner meats, more natural food products, and more highly processed heat-and-serve preferences among consumers).

Many of these changes have had profound effects on our way of life and in the products which are now available for purchase. Some of these include

1. the mechanization of agriculture with its resulting shift of population from rural to urban areas
2. antibiotics in animal health and nutrition with the resulting increases in resource productivity and the enabling of large-scale production units
3. hybridization of corn and other plant species resulting in dramatic increases in productivity both here and abroad (the Green Revolution in LDC's)

4. refrigeration and storage techniques, which have resulted in less
   waste and the availability of seasonally produced commodities year
   round
5. the development of chemical fertilizers and pest control, which
   have had the dual impacts of increasing yields while reducing the
   risks of production.

As we scan the present and look to the future, biotechnology with its
recombinant DNA procedures, gene transfers, embryo manipulation and
transfer, and so forth, appears to be the location on the broad frontier of
agricultural progress where the next burst of change is most likely to occur.
Some of these changes, born of technology, are already reaching the commercial marketplace in animal agriculture, with the major plant developments likely to come significantly later.

The major biotechnology innovations that have had some press coverage include the bovine growth hormone, which stimulates increased milk
production, and the ice-minus bacteria, which reduce strawberry plants'
susceptibility to freezing weather; but these are merely the tip of the biotech
iceberg that will gradually emerge over the next ten to twenty years. Likely
possibilities include significant improvements in feed efficiency; reduced fat
content of meats due to growth hormones; herbicide-resistant, pest-resist-
ant, and viral-resistant plants through gene transfers; and improved nutrition, taste, texture, or shelf life in, for example, fruits, vegetables, and
oilseeds. Generally these new developments can probably be classified as
production increasing (e.g., more production per animal or acre), *cost reducing* (e.g., less input required for some output), *risk reducing* (e.g., less
susceptibility to disease), *quality improving* (e.g., less saturated fats in oilseeds), or *new-product generating* (e.g., a new product clearly distinct from
existing products).

There is little doubt that these developments will have significant impacts, directly or indirectly, on a broad spectrum of participants in the food
and agricultural sector. Since we are all consumers of food, we are all
involved to some extent. Participants most likely to be dramatically affected in the private sector include farmers; farm supply companies such as
feed, seed, and farm chemical dealers; farm credit agencies including banks
and lending institutions; companies developing biotechnological products;
food processors; and merchandisers of food, fiber, and animal products.

Those involved in the public policy and regulatory process will play a
major role in responding to these developments while encouraging or restraining biotechniques through the political process and allocation of
funding. Economists will play a major role in this process by providing

input to these decisionmakers regarding the economic impacts of biotechnology.

In this chapter, we consider some of the more provocative economic issues regarding potential biotechnological developments and then briefly outline the analytical approaches which economists might use to address the economic and social trade-offs of innovation prior to their commercial realization.

The Ethical Dimension

Most economists, by training, conceive of themselves as dispassionate purveyors of value-free facts who provide information to policymakers. These policymakers, in turn, use these facts to make the kind of value-laden decisions concerning what ought to be in a society. This view of the role of economists in the life of a society is gradually giving way to a more realistic perception. The methods used by economists to model human behavior and predict the impacts of various potential developments (whether in biotechnology or other areas) have inescapable moral or ethical dimensions. Since all economic models and methods are essentially reductionist in nature, they require both implicit and explicit value judgments regarding the nature of the individual; the components of the biological and socioeconomic process that ought to be considered in the model and the analytical process; and the ultimate weighing or valuing, usually in monetary terms, of the measures of performance. Inasmuch as the results of these models are used to create reality by their implementation in the world, economists bear some of the responsibility and ethical burden for the reality they help to create.

In their recent pastoral letter, *Economic Justice for All,* the U.S. Catholic bishops suggest that the ethical notions of economic decisions can be judged in light of three basic questions: "What does the decision and its implementation do *to* people?" What does it do "*for* people?" and "How does it allow them to participate?"[1] In economic terms, the first two questions involve, respectively, measuring the costs of economic decisions and measuring their returns or benefits. These are standard questions which economists deal with every day, and which we focus on in this chapter. The third question is not easily dealt with in current economic thinking but seems to have something to do with the basic right of individuals and communities to substantively participate in their own economic destinies — that is, whether the process has some utility above and beyond the economic results per se.

Economic Issues

Productivity Enhancement in Surplus Situations

Should biotechnology innovations which enhance productivity be ignored, banned, or subsidized by the public sector in industries facing surpluses? The European Economic Community has banned growth hormones for a number of reasons, including the fact that they have large surpluses as a consequence of their highly protective agricultural policies. Some Wisconsin farmers have picketed at the University of Wisconsin, demanding no research on the bovine growth hormone. On the other hand, public university research on biotechnology is supported by public funds in many states in the United States.

Should the fact that an industry is facing surpluses deter support of biotechnology research or advances in that industry? Generally, large persistent surpluses are found only in industries with prices artificially maintained by government intervention in the market system. In some cases, those surpluses are maintained by design to protect against weather vagaries like those that created the world food crisis in 1973–1974. In other cases, they are the product of political power or mistakes.

Generally, one could argue that trying to hold back technological advance is like trying to hold back the tide. In addition, enhancing productivity and reducing costs generally allow an industry to become more competitive in serving its ultimate customer here and abroad. In situations where surpluses are present because of government intervention, reducing production costs should allow legislators to lower the level of price protection and maintain the same level of income protection, while reducing the consumer price, enhancing consumer demand, and improving the potential long-run viability of the industry. Usually surpluses are a temporary situation while biotechnology advances are long-term in nature, so the current situation should not be the primary determinant of the desirability of biotechnological research or products.

Structure of Agriculture

Regardless of existing surpluses, many question whether biotechnology applied to agriculture will automatically result in a reduction of small- to medium-size farms with concentration of production in the hands of fewer, larger-size farms. To the extent that biotechnology will speed the rate of farm productivity increase, it would seem likely that fewer producers would be required unless increases in demand occurred at an equal or faster rate. Who gets displaced will ultimately depend on who most quickly adopts these technologies. The expectation would be that the higher-cost, least-competitive producer would be the first victim. In those agricultural industries where economies of size are significant, the smaller, less technologically sophisticated operators would tend to be the last adopters and the first dropouts. A new round of structural adjustment seems inevitable.

If the exact character of agriculturally related biotechnology is capital-intensive, larger-size operations would be more likely to justify major capital expenditures than would small operations; this would encourage smaller farmers to get bigger or get out. If no big investments are required, size may not be as important as sophistication, but that still may favor larger farm operations to some extent since they can justify investing their time in gaining the required expertise to use sophisticated new technology. Yet, biotech companies trying to sell new innovations to farmers have a clear incentive to produce new products in the simplest form to expand their potential market; this incentive could lead to little size discrimination in new technology adoption by farmers. It is also possible that some advances will be labor-intensive, which could stimulate rural employment. Others may ultimately enhance consumer demand and market share, causing the producers of products with improved nutrition, taste, texture, or shelf life to increase in number at the expense of producers of other products. The only certain thing at this stage is that the current situation will change. As pointed out, the aggregate nature of that change is not necessarily clear at this time, but it is an important issue that might be analyzed as new biotechnologies emerge.

More interesting questions may arise if biotech companies find that new proprietary products must be grown and produced under control, either involving farm ownership or the more likely method of production contracting. In the latter case, the market for contracts would substitute for existing commodity-marketing systems.

Further changes in crop characteristics could cause both significant changes in climatic zones with comparative advantages in some crops and wholesale restructuring in production and related processing-industry location. While these are not likely in the near future, they certainly could be possible later in the evolution of biotechnology. Changes in livestock or crop product characteristics (e.g., leaner pork or vegetable oils with less saturated fatty acids) or in relative prices could also cause significant consumer product substitution and demand shifts as well as significant shifts in consumer market shares and related production patterns in some commodity market strategic groups. Significant product-characteristic changes or new, competitive products create significant shifts in various commodity production and processing industry volumes, location, and economic viability.

Primary Beneficiaries of Biotechnology

Many question who will be the primary beneficiaries of biotechnological advancements. The corollary question is, Who will be the primary losers as these advancements are realized? Clearly, the biotech companies that develop an innovative product with patent protection should be expected to share in the economic benefits that the new product offers its purchasers. In some cases, other companies selling complementary products may also benefit, especially if a plant is designed to use that product most effectively relative to its competitors (e.g., a plant resistant to a specific herbicide).

There is a large element of risk in long-term development, with few of the many projects begun by scientists ultimately succeeding. Large research and development costs over the previous decade, along with such factors as investments in production facilities, must be recaptured. The amount of payback will depend on economic benefit generated for users of the new products, the degree of competition currently or potentially in that market, and the share of the benefits necessary to pass on to the farmer in order to make the new product a viable commercial business.

Farmers are the likely target customers for most agricultural biotechnology products, but there will have to be a clear economic incentive for the farmer to buy the new product and modify the production process. Whether the new product is yield increasing, cost reducing, risk reducing, or quality enhancing, the farmer's perceived value of the performance changes will have to be greater than the associated costs to get any significant market volume. For example, animal health product companies typically find that beef or hog producers require a minimum of $2.50–$3.00 increase in revenue before they will pay $1.00 for the improved input, while broiler producers may require much less.

Farmers who adopt new biotech products first are likely to get the primary benefits of temporarily above-normal profits as their costs per unit decline or benefit from improved quality. When their neighbors also adopt the improvements, excess profits begin to erode as farmers expand production; the net result is falling prices for output and additional structural change due to another cost-price squeeze until near-perfect competitive levels are once again achieved. Even if a minority of farmers adopt the new technology, they may control enough production to shift the industry supply curve to affect market prices and profit levels significantly. Nonadopters would be more likely to have low profits or losses and may be forced to adopt significant technical advances or drop out. This is the so-called technological treadmill—taking on the newest technology just to maintain the profit status quo.

Consumers, not farmers, ultimately are the primary beneficiaries of technological improvements in agriculture. This benefit is realized in higher-quality food products and lower food prices. Industries within the agricultural sector which most improve their technology and their cost relative to perceived value also have a clear advantage in competing for the consumer's food dollar. The poultry industry is an excellent example of this. The tremendous increase in efficiency of poultry production and lower relative prices coupled with behavioral changes in consumer diets has resulted in a rapid increase in per capita consumption and market share in the last three decades.

Yet there are some aspects of biotechnological innovation in which the benefits or costs are not yet clear. For example, some crops will soon be genetically engineered to be resistant to pests, which may reduce the need for some pesticides with possibly detrimental environmental or ecological effects; this will result in significant economic benefits to the users. While the apparent effects of this development — or of the development of viral-disease–resistant plants — would appear to be beneficial, the effects of herbicide resistance which can now be genetically engineered into some plants (e.g., tomatoes) are less clear. Herbicide-resistant plants obviously could benefit farmers with weed problems and facilitate low-tillage farming methods. Will farmers be likely to spray more chemicals as a consequence of growing such plants? Or will yields be enhanced sufficiently that fewer acres are needed and fewer chemicals used? There are other biotechnological innovations being designed to break up and remove chemicals (like PCBs) from contamination sites. The individual and aggregate effects of some potential developments like these clearly will require further study, but the potential for biotechnological advances to replace some current chemicals used in food production and clear up some environmental problems seems potentially exciting.

If biotechnological advances expedite the rate of productivity improvement faster than demand for agricultural products improves, then the types of changes in the structure of agriculture and rural communities seen over the last few decades would be expected to continue, possibly at a more rapid pace (though not nearly as fast as change in the 1980s brought about by the farm financial crisis). For example, if the bovine growth hormone stimulates a 15 percent increase in milk production, that may actually mean a 10 percent increase per lactation and a reduction of the number of cows necessary to supply current milk needs (not nearly the 30–40 percent estimated in early studies) only if *every* farmer adopts the technology. This effect would obviously be less if many farmers did not adopt the technology or if prices were allowed to drop (but with incomes remaining protected) to enhance dairy-product demand. Some high cost dairy farmers will be squeezed out of the dairy enterprise if government income supports are reduced, supply quotas are imposed, or new dairy buy-out programs are initiated to restrict growing surpluses. While some, especially farm workers, might leave farming, others might stay in farming but shift their labor and investment into different enterprises. This is not likely to be as dramatic as the financial demise of some farmers in the Midwest in the 1980s, when farmers lost their farms. However, the extremes experienced in the 1980s can give some perspective on the types of change resulting from technological developments. These include (1) unemployment, at least transitionally, of dairy farmers and dairy-farm workers as less labor intensity is typically required in alternate farm enterprises; (2) the write-off of dairy-related specialized capital in agriculture; (3) the costs to local communities where dairy farming was displaced.

The next-best enterprise characteristics in areas where dairy farmers would be squeezed out or bought out are the primary determinants of the on-farm costs and the rural community costs. If fewer labor-intensive enterprises are second-best, greater underemployment or unemployment will result—typically most painful for older, less-educated persons who are less adaptable to training and new employment and who live in communities with few other employment alternatives. The social costs of the lower incomes or employment would be greatest for those rural communities perched precariously close to the edge of the critical mass of population and economic activity necessary to support major businesses, schools, recreation, and other amenities. The aggregate impacts of such technological change on the quality of life, the structure of rural communities, and other costs associated with changing forms of income and employment are clearly factors which ought to be considered and measured. How to minimize the costs of adjustment and capitalize on the opportunities presented by technological change is the challenge facing farmers and rural communities.

Change is a continual concern and does bear a key element in the United States and other countries. Change carries with it impacts which are born by society. Some impacts are in the form of benefits while others represent costs. Models are available for economic analysis of changes; however, many have focused on analysis of changes that have already occurred. It is much more difficult to evaluate a priori a change that is on the horizon. Data on elements such as expected adjustments can be very sketchy, making projections difficult. Nonetheless, it is imperative that a developed society have some notion of the consequences of forthcoming technologies. It is necessary to have an informed judgment of likely impacts.

Analytical Approaches

Previous Studies

Before considering possible procedures for economic evaluation of biotechnology, we briefly consider previous attempts to analyze technological advancements. We consider the research approach used and the types and availability of data used, and we assess how these procedures may be applied to the current problems of determining the social and economic impacts of biotechnology.

A key element in assessing the impacts of technology is estimating the rate of adoption and diffusion of the technology throughout the user industry. A new technology has zero impact if it is not adopted for use.

A landmark study which considered the mechanics of how technology is generated and propagated in U.S. agriculture was conducted by Zvi Griliches in 1957.[2] He evaluated the process of hybrid seed corn adoption in the United States using the data gathered after hybrid seed corn had been adopted. He plotted the adoption patterns and analyzed the cross-sectional differences in the adoption of this technology. The model developed by Griliches, which is now widely used, is

$$\ln (P/K - P) = a + bT$$

where,

P = adoption rate
a = intercept
b = coefficient for T
T = year trend (i.e., year = 1, 2, . . ., n)
K = upper limit of adoption rate (i.e., K = .95, .90, etc.)

Griliches reported that a new variety was adopted very slowly by producers at the beginning because of uncertainty associated with potential yields. However, the adoption took place at an increasing rate as others viewed the success of the early adopters. This continued until an upper limit was reached where some producers absolutely would not adopt the technology. The results form an S-shaped adoption trend over time (Fig. 12.1).

Griliches hypothesized that the rate of acceptance and the ultimate equilibrium level of acceptance were functions of the profitability of the technology. This hypothesis was confirmed by the data available. The "profitability" hypothesis is also relevant for biotechnology.

Following this initial development of techniques for analysis of adoption of new technologies, Griliches and others used the procedures to

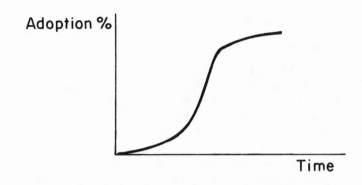

Fig. 12.1. Agricultural technology adoption trends over time.

further analyze the impacts of technology on society.[3] Griliches proceeded to evaluate the net social impacts of hybrid seed corn using the classical consumer-producer surplus techniques, which will be discussed in more detail in the "Ex Ante Research Approaches" section that follows. Societal impacts were based on data collected on research costs, displacement of resources, and the returns from planting hybrid seed corn. This same procedure has been applied to several other innovations over the years, including the development of a tomato harvester,[4] the development of hybrid strains of rice,[5] and the development of improved yields and increased fiber length and strength in cotton.[6]

Although consumer-producer surplus analysis has been widely used, in 1983 Ito, Grant, and Rister used dynamic econometric modeling techniques to assess the impacts of the development of a high-yielding rice variety.[7] They began by using the technique of adoption estimation developed by Griliches and then proceeded to use a dynamic, general equilibrium, deterministic simulation model developed by Grant, Beach, and Lin to analyze the rice supply-demand sectors and price relationships among producers, millers, and retailers.[8] To estimate the impacts of the new semidwarf varieties, the estimated adoption rates were integrated into the yield equations of the simulation model, and then simulations were made with and without the semidwarf varieties included. They estimated the impacts on acreage, yield, production, prices, and demand. Although the initial calculations were made using three years of previous data, the model was used to project the impacts through 1990. This type of estimation provides another interesting avenue for prediction of the economic impacts of relatively new but commercially untested biotechnology.

Although these studies all provide some useful insights into ways of estimating the economic impacts of biotechnology, it should be noted that each study cited was conducted on an ex post basis. That is, in every case the technology already existed and was being used at least on a test basis. Biotechnology does not afford us this luxury. Many of the exciting developments are still years from commercial availability, and, even when commercial availability is possible, clearance for use is uncertain since the techniques are subject to approval by the EPA, FDA, USDA, and/or state agencies. This suggests that some innovative analytical techniques will be needed for meaningful evaluation of the economic impacts of technology on an a priori basis.

A recent study which estimates the before-the-fact economic impacts of bovine growth hormone (bGH) on the dairy industry was completed by Robert Kalter.[9] In the analysis, Kalter, using a framework similar to that of Griliches, evaluated the adoption and diffusion pattern of bGH. However, Kalter used survey responses from dairy producers to estimate the producers' acceptance of the new technology. Once the adoption and diffusion analysis was completed, Kalter proceeded to estimate the economic impacts by using linear programming analysis, which will be discussed in more detail in the analysis section that follows. This is the first extensive study which has been completed with respect to the economic impacts of biotechnology, and it offers a useful format for future studies.

Ex Ante Research Approaches

As with the historical analysis of technological innovation, it is necessary to begin a meaningful analysis of the economic impacts of biotechnology by first addressing and estimating the rate of adoption of the various biotechnological innovations which may be introduced. Varying the extent of adoption will significantly change the magnitude of the economic and related social impacts.

As mentioned previously, one useful technique to estimate the adoption and diffusion of biotechnological innovations is Griliches's framework. However, Griliches's study was conducted long after the technology was introduced; thus it was possible to fit the model to the historical data available and then make estimates of the behavioral parameters of interest using the model. The innovations made possible by biotechnology typically have, in contrast, not occurred yet, and they are sufficiently important that before-the-fact estimates of their likely impact are necessary to contribute to the debate over their potential benefits and costs. The rate of adoption of the technologies will be dependent on the economic value of the benefits derived, the capital requirements and variable costs of using the technology, the management changes which may be required, and the perceived risk of using an unproven technology, as well as how the producer perceives consumer acceptance of the final product on which some type of biotechnology has been used. One possible way to address this problem is through subjective producer surveys, wherein the producer is presented with a likely scenario of the potential characteristics of the new technology and then is asked whether the technology would be acquired in various situations. Using this information, it may then be possible to provide preliminary estimates of the adoption rates and ultimate market penetration which can be anticipated for the given information.

Once a consensus estimate for the expected adoption of the innovation has been obtained, it will be possible to proceed with the process of estimation of the economic impacts of the innovation. Four methods that may provide useful information on which to base decisions include econometric analysis, welfare analysis, mathematical programming, and the Delphi technique. Each technique has its strengths and weaknesses. A brief description of each technique and the necessary alterations that must be made for its adaptation to an a priori analysis of the impacts of biotechnology follow.

Econometric Models

The first technique considered is econometric analysis. Econometrics is basically concerned with explaining statistical relationships among economic variables. These relationships are then used to explain production, allocation, and distribution decisions. But how might econometric techniques be applied to the issue of biotechnology?

Biotechnological innovations often are production-enhancing or cost-reducing processes using biological methods. This would imply that these techniques will increase the commodity supply. Econometric models that incorporate realistic biological and economic behavioral relationships of industry supply response would be an important component in estimating the impacts of technological change.

Nearly all major agricultural commodities have been modeled using some type of dynamic, general-equilibrium econometric model. These models often involve supply-demand relationships and price relationships among producers, processors, and retailers. When econometrics are used, it must be remembered that adoption of biotechnology may have a twofold effect on shifting historical trends. First, many innovations change the nature of production; that is, they change the biological constraints associated with production. Second, they may change the economic relationships such as cost of production and level and types of inputs for production. Thus, many of the available models of supply responses must be modified extensively in order to obtain reliable estimates of the impacts of the technology.

Although it is plausible to alter the specifications of the models to take account of the changes outlined, another problem arises in that it is not possible to test the reliability of the specifications of the model using standard statistical tests without historical data. However, an analyst has to settle for the best information available, and that is likely to be experimental data generated by universities and private companies as well as the expert opinions of the researchers developing and testing the biotechnology innovation. Using this data the analyst can build most-likely scenarios and can use sensitivity analysis to qualitatively determine the confidence level in the results.

Another consideration in using econometrics to model the impacts of biotechnology is the cross-product effects many of these technologies will incur. For example, the use of growth hormones in lean meat production not only affects the particular meat animal species in which it is used but may also have impacts on the demand for other substitute meat products. If dramatic changes in fat content of red meats occur, broiler and turkey demand clearly could change. Beyond that, since many growth hormones result in an increase in production efficiency, demand for the feed inputs may also be affected, resulting in changes in grain prices, government feed grain program costs, and similar factors. This example illustrates that it will be necessary to build into each existing econometric model the cross-product demand relationships and input-market linkages to generate a realistic dynamic model which considers the interrelationships across markets.

Although econometrics lends itself to analyzing microeconomic implications, it is rather difficult to extend these results to evaluate societal impacts of biotechnology. However, welfare analysis may aid in answering macro-type questions such as, "Do consumers benefit from the development of the technologies at the expense of the producer?" or, "Do these innovations increase the well-being of everyone?"

Welfare Analysis

An economic approach to measuring the relative gains and losses from alternate states of the economy is to determine changes in the consumer-producer surplus. As shown in Figure 12.2, consumer surplus is defined as that area (*b*) above the price line (*P**) and below the demand curve, while producer surplus is defined as the area (a) above the supply curve and below the price line.

Thus, the surplus areas depend on the relative positions of the supply and demand curves as well as the slopes of the supply and demand curves.

In order to meaningfully apply this approach it is first necessary to have estimates of the elasticities (slopes) of the supply and demand curves. Then, using data after the change has occurred, it is possible to reestimate the position of the curves and then evaluate the changes in producer and consumer surplus to see who has gained or lost under the new state.

Supply and demand curves are presently available for many of the commodities affected by biotechnological innovation, making the first part of the analysis relatively simple. The complicating problem is that historical data to estimate the effects on the supply and demand curves from a change in biotechnology are nonexistent. However, it is feasible to proceed with the estimation based on the technical experimental data from field testing done to date.

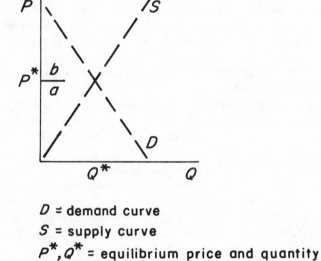

D = demand curve

S = supply curve

P^*, Q^* = equilibrium price and quantity

a = producer surplus

b = consumer surplus

Fig. 12.2. Consumer-producer surplus areas.

Another problem in welfare analysis arises on the consumer side. It is not immediately apparent how consumers will react to products which have been produced using genetic engineering or other methods. It is not known if they will react positively to a product available at a lower cost. Thus, it may be necessary to do extensive consumer surveying or market tests to determine what the effects of the use of biotechnology will have on product preference. Again, this will be difficult to do a priori. However, it is possible to develop alternative scenarios based on current information and present these to consumers to attempt to estimate how their product demand will shift. A better method would be to develop prototype products and labels and observe consumer choices in a slightly more realistic scenario.

Mathematical Optimization Programming Models

Mathematical programming is a tool frequently used to assess adjustments likely to result from technologies still in the developmental stages. With this technique, a base is developed for comparison. Information on the new technology is then incorporated to project likely impacts.

Mathematical programming techniques range from linear programming to integer programming to quadratic programming. Models are commonly developed around a profit-maximizing or cost-minimizing objective function.

Mathematical programming requires development of a reasonable set of production and/or cost parameters for the biotechnology being evaluated. As already suggested, this can be very difficult at an early stage of development. However, one advantage of mathematical programming is the ease of changing variables (expected yield levels, etc.). It conveniently allows for a sensitivity analysis of parameters to determine how sensitive results are to selected variables. Through this approach, a number of "what if" questions can be addressed. For example, presume a new technology is being developed that will have a cost-reducing impact on beef production. The level of cost reduction is not fully understood, but it is felt that it is somewhere in the range of $.50–$2.00 per cwt of beef. Once the model is developed, it does not require much time to run numerous scenarios. For example, scenarios for $.50, $1.00, $1.50, $2.00, $2.50, and $3.00 per cwt reduction in beef production can be run to see what the expected outcome is for different cost levels. This sensitivity analysis approach can be used for any variable in the model.

An additional benefit of sensitivity analysis is that it allows determination of which model parameters are sensitive to changes. For example, there may be disagreement on expected production impacts of a new technology. A programming model may indicate that the exact impact is not important because expected adjustments remain relatively stable over a wide range of production impact levels.

A key advantage of mathematical programming is that, unlike econometric estimation, it does not rely on historical data to establish expected trends or other projections. It represents a current model of the industry. However, results are only as good as the model and the embodied assumptions. It provides good evaluations of ultimate impact but often is not a good projector of the likely pattern of change over time. Recent examples of the use of mathematical programming include work at Cornell University,[10] University of Texas,[11] and Iowa State University.[12]

As with any approach, it is important not to view absolute numerical results from programming models as infallible. Their real contribution is in indicating the directions of change in the results. An observation of an increasing or decreasing trend is of much greater value than is the overall level of the change. Key model assumptions may have a significant impact on the indicated magnitude of the change, and these indications would be erroneous if the assumptions were flawed. While it is recognized that mathematical programming may not provide accurate information on absolute shifts, we must not lose sight of the fact that programming is quite efficient in projecting trends.

Also, impact evaluators must not fall into the trap of developing a complex model because everybody else develops complex models. Many times sophisticated models are used with unsophisticated data. Any model is only as good as the available data. When new technologies are evaluated on an a priori basis, much of the data has a degree of uncertainty around it. A simplified approach to development of expected trends using available theory may be as accurate and much less time consuming than a sophisticated model. It is not hard to think of studies that have utilized a complex modeling procedure to determine expected trends when a little foresight and thought, a piece of paper, and two hours of time would have forecasted the same trends.

Delphi Technique

The final approach to assessing expected impacts of biotechnological discoveries which is discussed in this chapter is that of expert opinion, or the Delphi technique. With the Delphi technique, information is usually obtained from individuals most knowledgeable about the technology. This information includes the probability of the technology being successfully developed and its impact on such factors as yield levels if adopted. This can be an effective method of establishing a range of values for uncertain data;

it may also be used within other techniques, such as mathematical programming. Through mail, telephone, or on-site interviews, experts can provide information on potential technological innovations, expected production impacts, and so forth. Some researchers have used a two- or three-stage Delphi technique. The first stage is an information-gathering phase; the following phases are used to gain information on the level of confidence responders have in their responses. With this technique, years of experience and knowledge from working with the product are captured, along with any biases that the experts may have.

Summary

In conclusion, we have identified a few of the issues or questions which often arise about biotechnology and traced some of the likely implications and unresolved issues regarding biotechnology. We suggest that short-term surpluses, especially those that are due to government intervention, should not be a major factor determining biotechnology advances or restricting research. Second, we expect that the structure of agriculture may continue to evolve toward larger-scale operations, but that biotechnology advances are not necessarily going to be significant contributors to speeding up that evolutionary change, though changes in regional or commodity competitive advantage could lead to significant structural changes in the future. The benefits of biotechnology are going to be shared, and the consumer, though often skeptical and alarmed about biotechnology, will be the primary beneficiary. The first-adopter farmer will be the temporary beneficiary, with other farmers continuing on the technological treadmill. The biotech companies will share the benefits for their research and development investments in a risky and uncertain new technology. Environmental benefits may also accrue, though the effects will have to be evaluated on an ad hoc basis.

Finally, we reviewed several approaches to evaluating the impact of new biotechnologies before they become available, and concluded that there are several approaches which can be used to estimate likely implications on new technologies before they are introduced. Very few of these approaches, however, are sufficiently comprehensive. During the next two decades, this will be an extremely important area for social scientists interested in the food and agricultural sector and rural communities.

Notes

1. National Conference of Catholic Bishops, *Economic Justice for All: Catholic Social Teaching and the U.S. Economy* (Washington, D.C., June 4, 1986).

2. Zvi Griliches, "Hybrid Corn: An Exploration in the Economics of Technological Change," *Econometrica* 25 (Oct. 1957):501–22.

3. Zvi Griliches, "Research Costs and Social Returns: Hybrid Corn and Related Innovations," *Journal of Political Economy* 66 (Oct. 1958):419–31.

4. Andrew Schmitz and David Seckler, "Mechanized Agriculture and Social Welfare: The Case of the Tomato Harvester," *American Journal of Agricultural Economics* 52 (Nov. 1970):569–77.

5. R. E. Evenson, F. M. Piedad, and H. Yujiro, "Social Returns to Rice Research in the Philippines: Domestic Benefits and Foreign Spillover," *Economic Development and Social Change* 26 (April 1978):591–607.

6. Harry W. Ayer and G. Edward Schuh, "Social Rates of Return and Other Aspects of Agricultural Research: The Case of Cotton Research in Sao Paulo, Brazil," *American Journal of Agricultural Economics* 54 (Nov. 1972):557–69.

7. Shoichi Ito, Warren R. Grant, and M. Edward Rister, "Impacts of Technology Adoption on the U.S. Rice Economy—The Case of High Yielding Semi-dwarfs," paper presented at Agricultural Economics Association Meetings, Reno, Nevada, July 25, 1986.

8. Warren R. Grant, John Beach, and William Lin, "Factors Affecting Supply, Demand, and Prices of U.S. Rice," NED, ERS, U.S. Department of Agriculture, October 1984.

9. Robert J. Kalter et al., "Biotechnology and the Dairy Industry: Production Costs, Commercial Potential, and the Economic Impact of the Bovine Growth Hormone," *A.E. Research* 85, no. 20 (Cornell University, Ithaca, New York, December 1985).

10. Kalter et al., "Biotechnology and the Dairy Industry"; Robert J. Kalter and Robert A. Milligan, "Factors Affecting Dairy Farm Management and Profitability," National Invitational Workshop on Bovine Somatotropin, St. Louis, Missouri, September 21–23, 1987.

11. Robert D. Yonkers, Ronald D. Knutson, and James W. Richardson, "Dairy Farm Income as Affected by BST: Regional Characteristics," National Invitational Workshop on Bovine Somatotropin, St. Louis, Missouri, September 21–23, 1987.

12. Michael Boehlje, Gary Cole, and Burton English, "Economic Impact of Bovine Somatotropin on the U.S. Dairy Industry," National Invitational Workshop on Bovine Somatotropin, St. Louis, Missouri, September 21–23, 1987; James B. Kliebenstein and Seung Y. Shin, "Impact of Bovine Somatotropin (BST) on Dairy Producers," Chapter 14 of this volume.

13 Who Will Gain from Biotechnology?

JACK DOYLE

The question of who will gain from biotechnology is an especially important one, since this is a technology that will fundamentally alter every industry now based on biology and chemistry. Agriculture, food processing, medicine and pharmacology, silviculture, aquaculture, energy, and natural resources industries – all stand to be directly affected, if not dramatically transformed.

Biotechnology, in the context of these industries, will have a profound impact on the production of certain "life necessity" products, as well as other critical materials such as fibers, ores, energy, and forest products. Therefore, it bears stating with some emphasis that wielding genetic technologies in these industries will also mean considerable political influence.

The question about who will gain from biotechnology also implies that there will be winners and losers with the use of this technology. But it may be presumptuous at this stage of the technology's development to definitively point the finger and say who will win and who will lose. The science and the technology are still young. Biotechnology does hold promise, and society as a whole could gain significantly from its careful application. Yet the most intriguing questions have to do more with how the knowledge behind this technology will be used. But more on that later.

There are, however, some patterns emerging in the recent history of this technology that enable us to see where it is headed. Society could decide that these patterns are not in its best interest – that is to say, there is still time to make some choices.

First, a historical perspective on biotechnology is needed to show how we got to where we are today. After this technology has been laid out in the economic and legal contexts of the last fifteen years or so, the winners and

177

losers, "the haves and the have-nots," begin to emerge. A review of some of the key events as they have occurred in three areas—breakthroughs in the science, corporate and investor activities in the commercial sphere, and corollary developments in law and government policy—is appropriate.

The 1970s: The Recombinant DNA Era

First, we will look at the developments in the science itself, beginning arbitrarily in 1970 with the first electron microscope photograph of DNA. This is significant in a layman's sense because it marks the point where we are literally inside the cell, where we can actually *see* DNA.

In 1971, Stanford biochemist Paul Berg was the first to combine DNA from two viruses. In one sense, "recombinant DNA" was born with this development, but the techniques Berg used with viruses were cumbersome and did not really unlock the technology's full potential. This was left to the discoveries and synergy of two other scientists, Boyer and Cohen.

In November of 1972, Herbert Boyer, a thirty-six-year-old bacteriologist working at the University of California at San Francisco, and Stanley Cohen, a forty-six-year-old biochemist at Stanford University, had a chance late-night meeting in a Waikiki, Hawaii, delicatessen. The two were in Hawaii attending a scientific meeting and came to the restaurant for a late-night sandwich. In conversation, they soon discovered they were doing complementary work.

Boyer was one of the first scientists to purify restriction enzymes, which could cut DNA at precise locations. Cohen was working with DNA in bacteria. He was particularly interested in moving new genes into bacteria for experimental purposes and was working with free-floating, circular bits of DNA called plasmids. Plasmids existed outside of the bacterium's single chromosome strand and were small, comprised of three to four genes each. Cohen thought all along that plasmids would be wonderful vehicles for moving desired genes into bacteria because he knew that occasionally bacteria exchanged the genes on their plasmids. Yet, he did not know what to do with the plasmid. Now, with Boyer's "chemical scissors," he had a good idea of where to start.

By 1973, Cohen and Boyer and their associates had successfully used plasmids to clone genes in bacteria and move genes between two different species of bacteria.[1] The Cohen/Boyer plasmid method of rDNA meant that any gene from anywhere could be spliced into a bacterial plasmid, then incorporated into a bacterium whose multiplication would make endless quantities of the gene product. From that point on, genetic engineering was on its way.

At that time, however, there was not much of a biotechnology industry. Indeed, only a very few people were beginning to see the commercial import of what Cohen and Boyer had done.

There was commercial activity in the 1970s, however, even before the Cohen/Boyer breakthrough. As early as 1971, for example, Cetus, one of the first start-up biotechnology companies, was founded in Berkeley, California. Cetus was then in the business of improving strain selection of microbes used in standard fermentation systems. Other privately created start-up firms followed: Genentech in 1976 (in which Boyer was a principal), Genex in 1977, Biogen in 1978, and Molecular Genetics in 1979, to name a few. In almost every case, these firms were founded by one or two university scientists experienced in molecular biology or recombinant DNA work.

This was also a period when a few major corporations began to put large amounts of money into universities for molecular biology R & D, as Monsanto did at Harvard in 1975 with a twelve-year, $25 million contract for research. Other corporations also began to venture some capital in a few start-up companies, as Chevron did in 1978 with a $1 million equity investment in Cetus.

But in the 1970s, it was the scientific activity that was most exciting, with major laboratories and start-up companies competing feverishly with one another to achieve new breakthroughs and newspaper headlines.

In August 1976, for example, after some nine years of painstaking work, a group of scientists at MIT working under the direction of the Indian Nobel laureate Har Gobind Khorana, successfully constructed a complete gene—in this case a small, relatively well-known gene for a nucleic acid known as transfer RNA. Other advances were also made.

In 1978, Harvard researchers using genetic techniques produced rat insulin; Genentech produced human insulin; Stanford University researchers transplanted a gene for hemoglobin from a rabbit into monkey cells; and scientists at Cornell University implanted a gene in a yeast cell that produced the amino acid leucine.

In 1979, Genentech produced a human growth hormone and in 1980 two kinds of interferon. Biogen also produced human interferon.

The First Regulatory Debate

By 1974, however, the advance of the emerging DNA technologies was far enough along to raise the specters of accidents, escaping organisms, and runaway science. At that time, scientists themselves were the first to raise questions about recombinant work and, particularly, work with cancer cells. Paul Berg decided to abandon plans to introduce genes from a tumor virus into *E. coli* bacteria after his colleagues suggested that the resulting organisms might spread cancer to humans.

In July of that year, Berg published a letter in *Science* calling for a self-imposed moratorium on certain recombinant experiments. This led to the creation in October 1974 of the National Institutes of Health (NIH) first "DNA Molecule Advisory Committee" which began to draw up safety guidelines. In February 1975, 150 scientists from sixteen countries gathered at the Asilomar Conference in Pacific Grove, California, and called for self-regulation. A subsequent report from the Asilomar gathering became the basis for the NIH guidelines which were formally published in June 1976, emphasizing adherence to a regimen of containment procedures.

It is important to remember that biotechnology is a *laboratory-based,* academic practice for the most part, and that public concerns are primarily focused on the accidental escape of microbes from laboratory facilities.

During 1976 and early 1977, a public controversy erupted over the siting and operation of some recombination DNA laboratories in local communities. Local ordinances were adopted in Cambridge, Massachusetts, and regulatory bills were introduced in the New York and California legislatures. In the U.S. Congress, some sixteen bills for regulating genetic research were introduced between 1977 and 1978. Although a few of these measures were actually passed by both the House and Senate, no final regulatory legislation ever emerged from Congress — an outcome largely the result of lobbying by scientific groups. By 1978, NIH began to relax its guidelines, and, for most practical purposes, the first controversy over biotechnology began to fade.[2] During this period, scientific advances continued to move biotechnology forward.

As the 1970s came to a close, three American scientists shared the 1980 Nobel Prize for chemistry for their work with recombinant DNA techniques and for deciphering the genetic code.

1980: A Key Year for Commercial Biotechnology

In 1980, two very important developments occurred that propelled biotechnology into the national limelight and into the realm of big business: a U.S. Supreme Court decision and the Wall Street debut of a new biotech company. The first development was the Supreme Court decision, *Diamond* v. *Chakrabarty,* a case which had its origin in the early 1970s.

In 1972, a General Electric research scientist named Ananda Chakrabarty created an oil-digesting *Pseudomonas* through the cell-fusion of plasmids. G.E. and Chakrabarty promptly filed a U.S. patent application for a genetically altered bacterium—a product of four fused strains—that would eat oil spills. This patent application immediately raised questions about whether a living microorganism could be patented under the existing statutes. The patent was denied, and the decision was appealed. Finally, in June 1980, the U.S. Supreme Court ruled, in a landmark 5-to-4 decision, that Mr. Chakrabarty's bug was "patentable subject matter," that it was "a product of human intervention."

In October of 1980, Wall Street investors responded in a frenzy to the first public stock offering of a biotechnology company—Genentech, then an unknown company. At the time, Genentech was a small company doing some interesting science, but it had no commercial products. But that did not matter. The stock soared from $35/share to $89/share within minutes of the opening offer, and the company raised millions. At the end of the day, the company's market value was placed at $529 million, and partners Robert Swanson, a young businessman, and Herbert Boyer, of the Boyer/Cohen plasmid method of rDNA, had shares worth about $82 million apiece.

The combined effect of the *Chakrabarty* decision and Genentech's dramatic showing on Wall Street was to bring corporate America charging into biotechnology. Genes could be patented, money could be made, and investors were interested. Within months of these events, several giant pharmaceutical houses, including Abbott Laboratories, Hoffman-La Roche, Bristol-Meyers, and Upjohn, all began recombinant DNA programs for interferon.

During the 1980s, the pace of scientific advance quickened, and genetically engineered products began to make their way to market.

In 1981, for example, reports of new advances included a genetically engineered vaccine for hoof-and-mouth disease; a new growth hormone for dairy cows and beef cattle; and the first transgenic plant, the "sunbean," which is the product of moving one gene from a French bean plant to a sunflower cell.

By March 1981, *Time* was featuring Herbert Boyer of Genentech on its cover, and the lead story was "The Boom in Biotech." In 1981 alone, Atlantic Richfield established the multimillion-dollar Plant Cell Research Institute; Du Pont stunned its competitors by announcing it would spend $120 million on "life sciences" R & D and biotechnology; and the Hoescht Corporation made a $70 million research deal with Massachusetts General Hospital. At the state level, North Carolina and Michigan established biotechnology centers.

By this time, there were about eighty new biotechnology companies operating in the United States, and a new trade association called the Industrial Biotechnology Association was established in Washington, D.C.

In 1982, the first herbicide-resistance genes were cloned, and Humilin, a genetically engineered version of insulin, was approved for commercial sale by the U.S. Food and Drug Administration.

In 1983, "supermouse" appeared on the covers of *Science* and *Nature;* bacterial genes were moved into plants; and the first proposed field test of a genetically altered microbe was hauled into court for environmental reasons.[3]

By this time, companies as diverse as Kellogg and Martin Marietta had invested in new agricultural biotechnology companies; Kemira Company, Finland's largest chemical concern, contracted with Calgene to engineer herbicide resistance into rapeseed crops; and Heinz and Campbell were pursuing the genetics of the tomato.

Throughout the 1980s, major corporations continued buying into small biotech companies, establishing research contracts with them, striking deals with universities, or setting up sizeable in-house projects.

For example, corporate officials at Kodak, realizing that they had a chemical "library" of some 500,000 molecules that might offer possibilities beyond photographic film, set up a Life Sciences division in 1984 and moved into biotechnology. Since then, the company has spent more than $200 million on in-house research and contracts with at least six biotech companies for research on cardiovascular therapeutics, food processing enzymes, and plant and animal agriculture.

In 1984, Du Pont and Monsanto also unveiled new multimillion-dollar research centers, demonstrating their continued commitment to biotechnology in a number of business areas.

But something else began to occur: major corporations began acquiring biotechnology companies wholesale.

In 1984, W. R. Grace struck a $60 million deal with Cetus to establish AgraCetus, and Lubrizol acquired the Agrigenetics Corporation. In the following year, Bristol Meyers paid $395 million for Genetic Systems, Inc., and Eli Lilly bought Hybritech for $350 million.

Early in 1985, PPG of Pittsburgh, a Fortune 100 chemical company, made a fifteen-year, $120 million agreement with the Scripps Research Center in La Jolla, California, for basic biotech research in plant science for herbicides and other agricultural products. Later that year, in November, Texas Tech University announced that a team of scientists who were culturing plant cells in the laboratory discovered that some of those cells had produced cotton fibers.

In 1986, a genetically engineered version of interferon was approved by the FDA; a team of scientists at the University of California at San Diego succeeded in moving a firefly's light-emitting gene into tobacco plants; and a thirty-five-year-old plant geneticist produced orange juice from test-tube cultures of orange cells.

Biotechnology companies at this time were still making very successful public offerings on Wall Street. In July 1986, for example, Calgene, a California biotechnology company making its first outing on Wall Street, sold out in one day, with 2.25 million shares selling at $14 per share, a price that was three times more than the company's book value.

The seed industry should also be mentioned here. In 1986, Imperial Chemical Industries (ICI), England's largest company, acquired the Iowa-based Garst Seed Company, one of America's leading hybrid corn seed firms. "ICI believes the seed industry is going to be revolutionized in the next 10–15 years by the application of genetic engineering and plant microbiology in the same way the microchip has revolutionized the computer industry and the electronics industry," said ICI's Richard Wood when his company acquired Garst. Monsanto, Ciba-Geigy, Pfizer, Sandoz, and others also own seed companies, while giants like Mitsubishi and Standard Oil are now involved in plant genetics research. Hundreds of seed companies have been acquired worldwide in the last decade or so.

In December 1986, a team of researchers from the University of Nottingham in England succeeded in regenerating rice plants from single-cell protoplasts. Before this development, attempts to grow rice plants from single, nonseed cells had repeatedly failed. This ability to regenerate whole plants from rice protoplasts was hailed as a development that removed a significant barrier to genetic engineering in the cereals.

And 1988 was an eventful, if not important, year for biotechnology. There was the first legal field test of a genetically altered organism—that being the infamous ice minus field test in California; there was a decision in April from the U.S. Patent Office that genetically engineered animals were patentable; and finally there was an illegal release of a genetically altered microorganism by a university researcher in Montana.

What This History Tells Us

What does all of this history tell us? What does it add up to, and how does it instruct us about winners and losers, about who will gain from biotechnology, and about political power?

When you look back at these seventeen years of science, commercial investment, and public policy, certain trends begin to emerge. In fact, there are four major trends at work here, each interacting with and reinforcing the others: biotechnology, patenting, mergers, and R & D collaboration (government/industry/university).

First, there is the technology itself—proceeding more rapidly than even the scientists ever thought possible.

Second, there is patenting, the legal basis to have a property right and to maintain a limited monopoly in the market. As it stands today in the United States, this property right is available for seeds, plants, microbes, animals, and also, it appears, individual genes.

Third, there are mergers. Since 1981, there have been mergers of huge proportion, as in the pairings of Du Pont/Connoco, Nestle/Carnation, Gulf/Chevron, and Hoechst/Celanese, to name a few. Many of these new superconglomerates are heavily involved in biotechnology research.

And finally, there is a growing trend toward collaboration in research and development between government, industry, and the university.

In total, these four interactive trends—biotechnology, patenting, corporate mergers, and R & D collaboration—present some new and worrisome problems: (1) they present a powerful new economic and political combination with the potential to shape public policy in new and fundamental ways; (2) they blur the traditional roles and responsibilities of some of our key institutions, such as the role of government as regulator and the university as society's neutral arbiter and advisor; and (3) they raise questions about our traditional checks and balances on economic power and the use of scientific knowledge.

For example, consider the recent corporate mergers in the context of DNA.

DNA and the "Life Sciences" Conglomerate

There used to be clearly drawn lines of distinction in the United States between chemical companies, pharmaceutical companies, energy companies, and food companies. These industries more or less operated separately. Today that is changing rapidly for two reasons: (1) mergers across these lines, and (2) DNA.

We have always had mergers and acquisitions, and just because a few major corporations are buying each other up across traditional lines of business activity does not mean that anything new or profound is occurring. What is happening *is* new and profound, and it is especially profound because of the reach and power of DNA.

DNA will magnify the mergers of recent years many times over and will boost the economic prowess of the resulting conglomerates in fundamental ways. Emerging out of this mixture of mergers and biotechnology are huge multinational "life sciences" conglomerates, unprecedented entities that will use genes just as earlier corporate powers used land, minerals, or oil.

The special quality about DNA that empowers these larger corporations so broadly, and like no other raw material before it, is its universality. DNA is a common language to chemistry, pharmacology, energy, food, and agriculture. It is a malleable and multifaceted raw material that can be used broadly and very efficiently, particularly by a large corporation involved in many fields of biological and chemical research. What does not work in medicine may be useful in agriculture or in some fermentation process. Little DNA research, in this regard, can be "wasted."

In many ways, DNA is the ideal corporate resource: it can be patented and wielded as a property right; it can be manipulated and "instructed" in the laboratory; it can replace or reduce the reliance on "old," cumbersome raw materials like farmland or bulk feedstocks; it can eliminate or reduce labor costs; it can help to circumvent worrisome or erratic variables such as weather; it can produce copious quantities of scarce and hard-to-make products for pennies; and finally, because of its value-added possibilities and high potential margin, it can be very profitable. In short, DNA can both stretch and consolidate capital while diminishing risk and exposure in labor and resources.

What we are seeing here is the empowerment of the private sector — and especially of multinational corporations — in a fundamentally new way, with little regulation or restraint because governments everywhere are involved in high tech expansionism. Whether for international political reasons in pursuit of technological supremacy or economic competitiveness, or for simple tax-base and revenue-generating considerations at the state and local levels, or for the oft-heard rhetorical promise of creating jobs, government is on the biotech bandwagon. This brings us to the next issue.

The Problem of R & D Collaboration

Biotechnology research is increasingly being conducted jointly and cooperatively through tripartite arrangements between industry, university, and government. These ventures are being officially encouraged by government and often incorporated into public policy. Congress has specifically adopted new measures or changed old laws that allow for more industry/ university or industry/industry collaboration and has provided tax incentives for private investment in research.

At the state level, a number of legislatures and/or governors' offices, often working with universities, have created new, specially funded "biotechnology centers" or "centers-of-excellence." At least thirty-one states have now created some form of biotechnology program.[4]

At the center of many of the new state and federally encouraged biotechnology research programs is the university. Since the onset of the biotechnology boom that began in the mid-1970s, universities have been losing their best biologists and geneticists to a surge of new entrepreneurial business ventures and corporate research laboratories. In addition, corporate funding of university biotechnology research has skyrocketed, often coming in huge, multiyear commitments with strings attached.[5] All of this has meant a very fundamental change in the way biology research is being conducted on campus as well as a further erosion in the traditional role of the university.

Not long ago, the university and all publicly supported institutions were believed to carry a broad social responsibility inherent in their charters. Public dollars meant public responsibility and, presumably public benefit. In one sense, the university was looked upon as society's conscience on a broad range of questions and was expected to weigh and carefully consider the best uses and applications of knowledge. Historically, the public has looked to the university and public-sector scientists for guidance and direction — and sometimes, when faced with new information, even to articulate new values. Generally, we have tried to create universities and public research institutions as places of neutral inquiry and analysis, where society could turn for unbiased advice, guidance, and answers.

The university and public-sector research institutions have also played a crucial role in conducting research that has been "orphaned" by the private sector or that has not been pursued because the perceived market share or economic pay-back was too small. In this sense, university research can help to stimulate competition and innovation as well as provide stimulation for products or strategies that may be of social benefit for environmental or public health reasons. Yet, all of this is now changing, and the alternatives and social conscience roles are being gradually lost in the biological sciences.

What is emerging on campuses now and, in fact, in a larger national context is a new ethos, framed primarily by government and business, that says it is all right for the public and private sectors to form a mutual embrace in research and development. The problem with this is the blurring of roles, the elimination of checks on technology, and the erosion of a neutral science and technology assessment process essential to public well-being.

This loss of neutrality, and the opportunity for the alternative research to flourish, is critical for biotechnology because there is still time to make some fundamental choices about what kinds of research are in the best public interest.

It is absolutely imperative, therefore, especially with the onset of biotechnology, to reinstate in the strongest of terms the broad social contract inherent in university and all publicly funded research. The aim here is not to stifle industry/university exchange and learning but to maintain the integrity of the scientific process, to avoid commercial intrusion into that process, and to insure that we have neutral scientific institutions and unaffiliated scientists to evaluate *all* of the costs and benefits of new technologies.

This is especially important given the choices that lie ahead.

"Biotechnology" or "Biotechknowledgy"?

There are two kinds of biotechnology out there—the one kind emerging now, which might be called *business-as-usual* biotechnology for the lack of a better term, and the other kind, which will be called biotech–*knowl-ed-gy.*

The first kind of biotechnology is gene centered and engineer oriented—more or less on the same trajectory as the chemical industry was forty years ago. It is a product-based technology, driven by quick-fix/ "silver bullet" thinking, by what can be treated or applied. It is also preoccupied with securing a legal property right or patent on the biological thing to be sold—whether that be a new wonder drug, an engineered microbe, a new crop variety, or a genetically engineered animal. (Patents have less to do with real innovation than they have to do with exclusive marketing rights and securing market share.)

Biotechknowledgy, on the other hand, is a knowledge-based approach to the biology of agriculture, medicine, and natural resources management. It is not driven by the need to fashion a product or to secure a patent. The approach here is more preventative or systems in orientation, relying on ecological principles in agriculture and natural resources management, for example. This kind of "techknowledgy" will use the new information and breakthroughs that come from the workbenches of molecular and cellular biology to bring about a better understanding of natural systems, whether those be in the human body, in our forests, or on the agricultural landscape.

In farming, for example, biotechknowledgy would place a premium on information to the farmer about crop rotations, new disease- or insect-resistant varieties, and/or scouting and monitoring pest populations to reduce exposure and/or prevent certain kinds of pest problems. A biotechnology approach, on the other hand, might emphasize a genetically engineered microbe to fight a particular bad pest or advocate the use of a particular genetically engineered herbicide-resistant crop variety (and the compatible chemical herbicide) for solving a weed problem.

In biotechknowledgy, the farmer is using and applying the knowledge, perhaps with the introductory assistance of an extension agent. The farmer's labor and skill are very much involved in the management process. In biotechnology, however, the farmer is using and purchasing a product in which the knowledge has been internalized, so his labor and skill become less important. This is what was meant earlier in the discussion about how DNA, employed as a malleable raw material to expand the "efficiency" of capital, can be used to thoughtlessly reduce or eliminate labor when actually expanding the labor opportunity may be the more powerful economic engine that comes with the discovery and use of new biological knowledge.

The Opportunity for Research Innovation in Iowa

This discussion of two kinds of biotechnology is used only to illustrate the point that we have some choices to make with this technology, and if they are not made soon, we may miss the opportunity to put this new knowledge base to uses that could be truly beneficial for agriculture, medicine, and the environment.

Here in the state of Iowa, and at Iowa State University, there exists an opportunity to make an important stand for *biotechknowledgy* in agriculture. A slender foothold has been achieved in the Leopold Center for Sustainable Agriculture, a commitment that makes eminent sense given the economic and environmental state of conventional agriculture. Yet, at the same time, the state legislature has made a multimillion-dollar commitment for agricultural biotechnology research which seems to be heading off in the business-as-usual direction, contradicting in a way what the Leopold Center was established for.

So, this is where a stand should be made, calling to task any biotechnology research and funding that contravene the objectives of the sustainable agriculture program. Further, this should be the juncture at which biotechnology dollars are used to achieve sustainable agriculture objectives.

This stand on behalf of sustainable agriculture is much broader than just researching and using sustainable agricultural practices on the farm. It is also a public health imperative, a strategy that makes good economic sense in terms of return on investment for the farmer, and is at the very center of what the land-grant university and extension service should be about.

A September 1986 *Des Moines Register* poll shows that 58 percent of those surveyed in Iowa believe farm chemicals are the biggest threat to water quality and that 78 percent favor limits on farm chemicals. Even farmers are beginning to take a hard line. In the *Des Moines Register* poll, for example, more than half of the Iowans responding from farm households favored placing limits on farm chemical use.

In a similar public opinion poll of four hundred Iowans taken in late September and early October of 1986 for the Iowa Department of Natural Resources (conducted by Lord, Sullivan and Yoder, Inc. and paid for by the U.S. EPA), some 65 percent of farmers and retired farmers favored tighter restrictions on farm pesticides, while 52 percent favored tighter restrictions on fertilizers.

There is a groundwater law in Iowa that has some strong wording about no further pesticide contamination of groundwater supplies. All of these public concerns coupled with the public policy initiatives that are already underway create a strong political base to direct agricultural research — including biotechnology research — in ways that focus on eliminating, and adopting alternatives to, chemical pesticides.

Beyond this, there is a case to be made for using the "new biology" to help agriculture do a whole range of positive economic things — from the introduction of new crop varieties and livestock breeds, to decoupling farming from its present reliance on capital- and energy-intensive inputs.

And further, this is a grand opportunity to revive the historic mission and purpose of the land-grant university, the experiment stations, and the extension service. Our land-grant university system is a huge national system, consisting of some 72 universities, 146 agricultural experiment stations, and thousands of local extension agents. In recent years this system has been targeted for consolidation and defunding, but it could play a leadership role in revitalizing agriculture and cleaning up the environment with a little creative policy making and a commitment to new research initiatives.

In 1986, in a speech at Cornell University celebrating the Hatch Act Centennial, I suggested that the land-grant complex use the opportunity of biotechnology to make disease- and insect-resistance work in crops and livestock a major national research priority. Since then, I have gone even further, suggesting that such a priority be justified in the interest of phasing out pesticides. It is long past the time when we should have set a national goal for eliminating pesticides from agriculture. Now there are both a growing public concern and the knowledge base to fashion acceptable alternatives. All we need is an institutional vehicle with a public policy base.

Yet, researchers at a number of universities—including Auburn, UC Davis, the University of Chicago, Cornell, Harvard, LSU, Michigan State, and Rutgers—as well as at USDA research centers are using biotechnology to make crops and forest trees resistant to chemical herbicides. In addition, at least thirty chemical, pharmaceutical, and biotechnology companies are also working to make crops herbicide resistant. Others are working on ways to match chemical plant growth regulators to particular genes in crops so they can be "signaled" or turned on or off at the right time to do one or more specific metabolic things. Similar work is going on with livestock. This kind of research investment will only prolong the pesticide and supplement approaches in agriculture.

Within the land-grant complex, there is a need to reinstate the role of providing agricultural technology of broad public benefit. There is also a need to demonstrate consumer and environmental benefits in agricultural research initiatives. What better way to do this than through a major public sector research effort aimed at reducing public health risks and the farmer's cost of production through the development of nonchemical pest-management strategies? The idea here is not to exclude the private sector but for the public sector to "prime the pump"—to take leadership in, and absorb the risks of, a major reorientation in agricultural production research that could have potentially wide public benefit.

Public sector agricultural research institutions could be directed to prioritize research that focuses more on a low-input/high-rate-of-*farm*-return system. This system would stress farmer profitability through low-cost production; biological processes and cultural management in the field; new crops diversification; and multigene, host-based disease and insect resistance for crops and livestock. Such a system, if introduced in a concerted and creative way, could result in less surplus, fewer environmental side effects and public health risks, improved farm-sector stability, and greater rural employment opportunity.

Such a program would elicit widespread public support and accomplish several things simultaneously. First, it would reduce public, farmer, and farmworker exposure to pesticides. Second, it would reduce the cost of production for farmers and thereby improve farm income and profitability.[6] Third, it should improve consumer faith in the agricultural system and possibly reduce prices once all associated pesticide costs were reduced throughout the system. And fourth, it could provide a powerful basis for rejuvenating the land-grant universities and agricultural experiment stations.

With such a program, however, the definition of biotechnology and what is pursued in the name of nonchemical and/or biological alternatives will be absolutely crucial. What is needed, as outlined above, is a biotechnology that embraces "commonsense biology" and "commonsense genetics," yet eschews the genetic engineering of organisms simply for the sake of making new products. We need to approach biotechnology as a *knowledge* opportunity, not simply as a manufacturing opportunity. If we use this new biology only to divide up the natural world into its smallest possible commercial parts rather than to improve our ability to work *with* the biological realm, we may only create further economic and environmental problems for the future.

There is a lot of good biology already "on the shelf," so to speak. And there is a lot of good, innovative work going on throughout the country by researchers at our land-grant universities and agricultural experiment stations.[7] We have extensive information about soil tilth, plant breeding, crop rotations, intercropping, insect adaptability, agricultural diversification, and other fields—all of which has application today. We should not become so enamored of biotechnology that we go out of our way for a high-tech solution when a commonsense alternative is right in front of us.

The Challenge Ahead

There is now a tendency in Washington to tilt research dollars away from the "old" USDA land-grant/experiment station complex in favor of "new" recipients, including universities with a so-called critical mass of interdisciplinary science and/or specifically dedicated biotech centers. Much of this money and research appears to be moving in the direction of *biotechnological product development* in agriculture, rather than *biological process understanding.*

Public-sector agricultural research programs need to be revitalized and recredited in the public's eye. Yet, using biotechnology to do that may backfire if the new technologies are used to pursue the same old goals or to deliver to agribusiness products that spawn the same or a worse set of side effects. What is needed is a bold reorientation of existing programs to capture benefits for farmers, consumers, the environment, and public health and safety in a way that has not yet been attempted on any meaningful political scale.

The State of Iowa and Iowa State University are in a good position to show the rest of the nation what can be done with creative biology in agriculture. The question is, will they take up that challenge?

Notes

1. Later, the Cohen and Boyer teams would demonstrate that even animal DNA (ribosomal DNA) from a South American toad could be spliced into a bacterial plasmid that would form recombinant DNA molecules in the bacterium *E. coli.*
2. This period has its own detailed history, which is too involved to include here. There are several excellent books on the subject, one of which is Sheldon Krimsky's *Genetic Alchemy: The Social History of the Recombinant DNA Controversy* (M.I.T. Press, 1982). (See especially p. 445.)
3. The year 1983 was a very significant time for the biotechnology industry, as this was when the first outdoor experiments of genetically engineered organisms were proposed. This was also the point at which the judicial, regulatory, and Congressional series of debates began over the safety of introducing genetically altered organisms into the environment—the "deliberate release" debate, as it was being called.

Although this aspect of biotechnology has its own specific history, suffice it to say here that one of the most important issues to emerge on this front so far was the contention by the Reagan administration and industry that existing laws were sufficient for regulating biotechnology in the open environment. Some ecologists, environmental organizations, and members of Congress disagreed and argued that new laws, or at least Congressional clarification of existing laws, were needed.

A second corollary issue is that of being able to predict the environmental fate and ecological effects of genetically engineered organisms released into the environment. A number of ecologists have expressed concern that the existing ecological data base is incomplete and inadequate and that we lack the proper experience needed to make such predictions about what may or may not happen with genetically altered organisms.

So today, the debate on the deliberate release issue, four years after it began, shows no sign of resolution, and, in fact, has been escalated in the last two years by illegal and out-of-country releases. Congress and state legislatures, meanwhile, seem increasingly interested in the regulatory question. For a historical overview of the deliberate release issue and an environmental viewpoint on the regulatory question, see "Biotechnology Regulation: An Environmentalist's Perspective," an address by Jack Doyle at the American Chemical Society Conference on Biotechnology, Snowbird, Utah, 3 July 1987, available from the Environmental Policy Institute, Washington, D.C., 202-544-2600.

4. Since 1985, large sums of money have been coming to Iowa State University for biotechnology research. In December 1985, for example, Pioneer Hi-Bred International, Inc., made a $500,000 matching grant to ISU to fund the Pioneer Endowed Chair in Molecular Biology of Maize. Since then, the university has received approval to use academic revenue bonds to build a $30.5 million state-of-the-art biotechnology research building. ISU is also the recipient of a four-year, $18 million agricultural biotechnology research program funded by state lottery and bonding funds, which is designed to create new opportunities for agricultural product development and manufacturing in Iowa. And Iowa, of course, is not alone.

Ohio State University, with appropriations from the state legislature, believes it can become "the Silicon Valley of agricultural biotechnology." And according to literature from North Carolina's Biotechnology Center, "all programs of the Center encourage statewide collaboration between universities and industry, and are designed to further research significant to the development of useful products."

5. Today, corporate research grants to universities are increasing in number and are coming in larger amounts. In 1956, business provided colleges and universities with about 13 percent of total gifts; now it provides nearly one quarter. Since 1976, total giving to colleges has tripled, but business contributions have risen fourfold. In the academic year 1984–85, corporations gave universities more money than did the entire national alumni body. In 1985–86 corporations gave a total of $1.7 billion to American universities. Of this share, research arrangements between corporations and American universities constitute more than one-third, now running at some $600 million annually.

The following are just a few of the hundreds of new biotechnology-related research grants that universities have received from private business:

- 1981. West Germany's Hoechst A.G., one of the world's largest chemical corporations, made a ten-year, $70 million grant to Harvard-affiliate Massachusetts General Hospital for the creation and operation of a department of molecular biology.
- 1981. Du Pont made a five-year, $6 million contract with the Harvard Medical School for genetic research.
- 1982. Celanese made a three-year, $1.1 million contract with Yale University for basic enzyme research.
- 1983. A $7.5 million package of gifts and grants was made to Cornell University by Union Carbide, Corning, and Eastman Kodak to help establish a new biotechnology institute.
- 1983. Standard Oil of Ohio made a five-year, $2 million grant to the University of Illinois to establish the Center in Crop Molecular Genetics and Genetic Engineering (at that time, the SOHIO grant was the largest private grant ever made to the university's College of Agriculture).
- 1984. Rhom & Haas made a $1 million gift toward the University of Pennsylvania's Plant Science Institute.
- 1985. PPG, Inc., made a fifteen-year, $120 million contract with the Scripps Research Center of La Jolla, California, for basic biotechnology research on herbicides and other agricultural products.
- 1986. Monsanto renewed its biotech research contract with Washington University, agreeing to grant $26 million in additional funds thru 1990, making a total of more than $50 million since 1983.

6. Farmers currently spend an estimated $18 billion annually for purchased feed, $7.4 billion for fertilizer, $4 billion for pesticides, and $4 billion for seed. That adds up to national cost-of-production bill of at least $33 billion without including other related input costs. Any reduction in these costs would certainly improve farm income and, presumably, U.S. agricultural competitiveness. Moreover, the public health and environmental costs of using many of these farm inputs are also high and could be decreased accordingly with a reduction in use.

7. For example, Donald Barnes, a plant breeder at USDA's research center at the University of Minnesota, has developed a new variety of alfalfa called *Nitro* that produces good livestock fodder and puts high amounts of nitrogen back into the soil. When plowed back into the soil, *Nitro* puts ninety-four pounds of nitrogen into the ground (compared with fifty-nine pounds for the top standard variety). *Nitro* represents the kind of genetic improvement in agriculture that can save growers money on the input side of their operation and thus increase their profitability.

14

Impact of Bovine Somatotropin (BST) on Dairy Producers

JAMES B. KLIEBENSTEIN and SEUNG Y. SHIN

The use of bovine somatotropin (BST) in dairy herds has been shown to increase productivity levels. Bovine somatotropin is a biotechnology that, when injected into dairy cows, results in increased milk production levels of anywhere from 10 to 40 percent.[1] Presently the injections are given daily and have been shown to have a rather quick and dramatic increase in production levels.

Bovine somatotropin is a technology that is entering an industry that is experiencing excess resource utilization, overproduction, and an excess commitment of land, labor, and capital. Additionally, the technology is entering an industry where the government is making strong moves and commitments to reduce production levels. With these characteristics, questions of dairy farm business survival, industry and firm level adjustments, and management strategies that increase survival chances become important.

Survey results have indicated that the rate of adoption of BST is likely to be very rapid, with 70 to 80 percent of the milk producers using it within the first three years of availability.[2] Given this rapid adoption rate and the type of industry, adjustments also are likely to be very dramatic. These adjustments have implications for dairy farmers and the entire dairy processing industry.

It is probable that adoption of BST will cause a critical examination of current U.S. dairy policy. Past dairy policies have had a somewhat stabilizing influence on milk price and dairy farm income.[3] These policies were in existence in the absence of the potential for dramatic increases in milk production. Bovine somatotropin may be the catalyst that will either force a movement of the dairy industry to a free-market policy or a movement to the other extreme, a highly controlled dairy industry with production quotas. Both views are currently being discussed within the dairy industry. Should BST be adopted, the current attitude in Washington would not support the burdensome surpluses of milk that would likely result with current milk price support levels. Program costs would be too great.

Adjustments resulting from BST will most likely not be uniform. Some segments of the dairy industry will be impacted to a greater extent than others. Regional impacts will not be a zero-sum game. Some regions will gain, while other regions will be net losers.

The purpose of this study is to investigate the impacts of BST on management needs, farm profitability, and dairy production regions.

Individual Farm Firm Impact of BST

In a recent study, Kalter and Milligan outlined some factors that are necessary to determine the overall impact of BST on dairy farm firms.[4] Those factors are (1) success in adoption, (2) dairy management skill, (3) quality of available resources, (4) business management skills, (5) business financial health, and (6) herd size.

Successful adoption, especially for the early adopters, will be very important in the overall use of BST by the dairy industry. For successful adoption over time, it is necessary that the income level through adoption exceed that which would be in effect without adoption. In effect, the increased efficiency in milk production through use of BST would need to be greater than the expected reduction in milk price that is likely to come as a result of increased production levels. Moreover, income levels of the early adopters would need to exceed the income levels of those individuals observing the use of the technology.

Effective adoption of many new technologies requires above-average management. BST is not unique. Dairy farm operators will be faced with a number of questions for which they have little data upon which to base an analysis. They will have to make subjective judgments on how some variables will affect their individual farms and farm profitability levels. Before adoption of BST, it would be advisable to make sure that producer management practices and strategies are above average.

Information presented in Table 14.1 shows the type of production increase which is necessary to justify use of BST in a dairy herd.[5] A productivity gain of at least 5 percent is needed to justify BST use. For rates below 5 percent, producers are better off not adopting BST. Additionally, the information in Table 14.1 shows that productivity increases through BST usage would need to be about 25 percent to place dairy producers on the same income productivity level reached without the availability of BST. However, in this comparison, the critical assumption is that the dairy program in the future would be maintained as it was in 1985, the base period.

Table 14.1. BST response and farm profitability at two initial herd production levels

| | Current Milk Sold per Cow Category[a] | |
	15,000 to 15,999	18,000 and over
Mean milk sold	15,500	18,500
Estimated net farm income		
1985[a]	$22,406	$47,577
Without BST usage[a,b]	10,167	27,530
5% productivity response to BST[b,c]	9,226	27,237
15% productivity response to BST[b,c]	16,744	39,552
25% productivity response to BST[b,c]	24,262	51,866

Source: Kalter and Milligan, Cornell University, "Factors Affecting Dairy Farm Management and Profitability." National Invitational Workshop on Bovine Somatotropin. September 1987. Reprinted by permission.

[a]Categories and baseline profitability are from Smith, Knoblauch and Putnam, "Dairy Farm Management Business Summary," Cornell University, 1985, p. 24.

[b]Milk price response is $.84/cwt based on a 10 percent aggregate increase in milk production.

[c]Profitability response is based on an $11.00 milk price net of marketing costs, a $5.00/cwt marginal cost of milk production, and a BST cost of $.25/cow/day per 200 days per year.

Another important variable to successful implementation of BST is the quality of resources available. With the use of BST, dairy cows will be placed under additional stress. With this additional stress may come such elements as breeding problems and increased level of disease. BST mandates changes in ration formulation and feeding level. Cows receiving BST treatment actually have increased nutrient requirements because milk production levels have increased. Rations need to be higher quality.[6] The main gains in feed efficiency through BST usage result because nutrients used for maintenance of the animal comprise a smaller percentage of the total consumption. The net effect of a higher quality ration is that the cost per pound of ration increases. The Cornell study pointed out that when this is taken into account, increase in production from BST will need to be about 10 percent for breakeven production. Moreover, even with a high response rate of 25 percent, the farm business will not be able to maintain its previous level of profitability.

Business management skills and the financial position of the business will need to be in good order for successful use of BST. First of all, the introduction of BST is likely to introduce a higher level of instability into the dairy industry. To effectively manage this instability will take top-level management. The business will also need a sound financial position to survive the instability likely to occur during the adjustment process.

Another factor felt to be important in the use of BST is herd size. While BST has been argued to be size neutral, it has many associated factors that are more common in large farms than on small or medium-size farms. One of these factors is the level of management capability. Management capabilities tend to be higher on the above-average-size farms than they are on moderate to small-size farms. Thus, while BST is a technology that may not require a large outlay of capital investment, it is a type of technology that may be more beneficial to large operators than to small operators. Additionally, if there is a cost to acquiring BST information, this cost, on a per-cow unit basis, will be lower for the large operators than the small operators.

Farm Level Impacts

A recent study looked at farm level impacts of BST under three different scenarios: (1) a milk price support, (2) a free-market price, and (3) a milk-production quota plan.[7] Herd size was eighty cows. Information on farm level returns is presented in Table 14.2. The first scenario looks at return levels with milk price supported at the level of $10.50 per cwt. The second half of Table 14.2 presents the impact of a free-market dairy policy. It should be no surprise that with a constant milk price, as the response

Table 14.2. Comparison of government price support and free market results

	ROVC[a]	
	12,500 lbs.	15,000 lbs.
Milk price supported[b]		
No response	$62,701	$71,376
10% response	68,969	79,025
25% response	78,638	90,642
Free market policy		
No response[c]	$62,701	$71,376
10% response[d]	60,032	68,300
25% response[e]	55,990	63,465

[a]Return over variable costs.
[b]Price support is $10.50 per cwt.
[c]Milk price = $10.50 per cwt.
[d]Milk price = $9.50 per cwt.
[e]Milk price = $8.27 per cwt.

level from a technology increases, the return to the farm operator increases. This is due to the fact that price of the output remains constant while the cost per unit of production declines, thus leaving a larger share of money in the form of profit. However, in reality, this is not likely to be the case as the current trend is toward less government support and reduced levels of government funding. With the large and growing level of national debt, pressures for reduced levels of federal outlay are likely to be even greater in the future.

Under a free-market policy, returns to the farm operator decline as the level of response to BST increases. A study by Boehlje and Cole projected that a 10 percent BST response would reduce milk prices by 9.5 percent, while a 25 percent BST response would reduce milk prices by 21.2 percent.[8] These prices were used in the Table 14.2 analysis. Under a free-market condition without BST, the overall return level for the farm operation was $71,376 for an eighty-cow herd that averages 15,000 pounds per cow per year. In comparison, a 10 percent response to BST results in a return level of $68,300, and a 25 percent response results in a return level of $63,465. These results are somewhat consistent with those shown in Table 14.1 in the Cornell study.

Impacts resulting from a voluntary milk production quota plan are shown in Table 14.3. This plan is a voluntary plan in that producers can decide either to participate in the government voluntary milk production control or not to participate. A milk price support level of $10.50 per cwt would be in effect for participating farmers while nonparticipating farmers

Table 14.3. Return over variable costs for quota and nonquota producers under selected quota participation levels (15,000 lbs production level)

% of Producers in Milk Quota	ROVC[a]	
	Quota Farm[b]	Nonquota Farm[c]
	10% Production Increase	
0	. . .	68,300 (9.50 $/cwt)
25	75,085	70,982 (9.75 $/cwt)
50	75,085	73,663 (10.00 $/cwt)
75	75,085	76,344 (10.25 $/cwt)
100	75,085	. . .
		25% Production Increase
0	. . .	63,465 (8.27 $/cwt)
25	79,634	69,680 (8.78 $/cwt)
50	79,634	76,018 (9.30 $/cwt)
75	79,634	83,087 (9.88 $/cwt)
100	79,634	. . .

[a]Return over variable costs, expressed in dollars per herd per year.
[b]Milk price is $10.50 per cwt.
[c]Milk price is shown in the parentheses.

would receive the free-market price. The table reflects returns to partici-pants and nonparticipants with a fifteen-thousand-pound per cow produc-tion level. Breakeven level of participation is about two-thirds or 67 percent of the producers. Whenever participation levels are below this amount, the participating (quota) farm receives an income level which exceeds that of a nonparticipating farm. When participating levels exceed 67 percent, the nonparticipating farm receives a higher income level than the participating farm receives. This result holds for both the 10 percent and 25 percent production response levels.

Regional Impacts

Recent studies by Yonkers, Knutson, and Richardson and by Boehlje and Cole examined regional distributional shifts expected as a result of adoption of BST.[9] Yonkers outlined four factors which will likely affect shifts in the dairy industry: cost of production per unit of milk, milk pro-duction per cow, size of typical dairy farm in the region, and management practices currently in effect. These are similar to the factors that Kalter and Milligan developed.[10] These factors are key components for individual farm adoption of BST and resulting production levels and thus are key components in regional shifts that would likely occur.

The Yonkers study used a simulation approach to determine the proba-bility of farm survival through adoption of BST. Table 14.4 presents infor-

Table 14.4. Probability of farm survival, 1987–96, adoption of BST in 1990, re-sponse to BST as indicated, no change in milk price

Region/State/Farm Size	No BST Used	Level of Response to BST		
		10%	15%	20%
Upper Midwest				
Minn., 52 cows	54%	72%	86%	96%
Minn., 125 cows	100	100	100	100
Northeast				
Pa., 52 cows	50	70	88	98
N.Y., 200 cows	100	100	100	100
Southwest				
Ariz., 350 cows	84	98	100	100
Calif., 1450 cows	100	100	100	100
Southeast				
Ga., 350 cows	100	100	100	100
Fla., 1450 cows	100	100	100	100

Source: Yonkers, R.D. et al., Texas A&M University, "Dairy Farm Income as Affected by BST: Regional Characteristics." National Invitational Workshop on Bovine Somatotropin, St. Louis, Missouri, September 21–23, 1987. Reprinted by permission.

mation on probability of dairy farm survival when milk prices are assumed to remain constant. Table 14.5 presents that information assuming milk prices decline by $1.00 per cwt when BST is adopted. Information presented in Table 14.4 shows that the Upper Midwest and Northeast regions of the United States will be adversely impacted as compared to the Southwest and Southeast regions. The Southeast appears to be in the strongest competitive position, followed by the Southwest, the Upper Midwest, and finally the Northeast. It should be no surprise that the probability of farm survival declines as the price of milk decreases, as shown in a comparison of results in Tables 14.4 and 14.5. However, for the Southeast, the probability of survival remains at 100 percent with all the scenarios studied.

Table 14.5. Probability of farm survival, 1987–96, adoption of BST in 1990, response to BST as indicated, milk price decreases $1.00 in 1991

Region/State/Farm Size	No BST Used	Level of Response to BST		
		10%	15%	20%
Upper Midwest				
Minn., 52 cows	6%	8%	20%	46%
Minn., 125 cows	84	88	96	100
Northeast				
Pa., 52 cows	0	4	24	56
N.Y., 200 cows	100	100	100	100
Southwest				
Ariz., 350 cows	28	82	88	98
Calif., 1450 cows	100	100	100	100
Southeast				
Ga., 350 cows	100	100	100	100
Fla., 1450 cows	100	100	100	100

Source: Yonkers, R.D. et al., Texas A&M University, "Dairy Farm Income as Affected by BST: Regional Characteristics." National Invitational Workshop on Bovine Somatotropin, St. Louis, Missouri, September 21–23, 1987. Reprinted by permission.

The information in Tables 14.4 and 14.5 also presents the likely effects of BST adoption on farms of various sizes. For example, in Minnesota and Pennsylvania the medium- to small-size farm will likely be more adversely impacted than will the larger farm. Differing impacts according to farm size are not as evident for the Southwest and the Southeast regions when milk price remains constant. However, with a milk price decline, the smaller Southwest operators will feel the impacts of BST to a larger extent than will the smaller operators in the Southeast. This is especially so if the response rate to BST is 15 percent or less. Another interesting comparison is that the large Midwest or Northeast dairy operations are in effect smaller in terms of cow numbers than are the small Southwest or Southeast dairy operations. Additionally, the large Midwest and Northeast dairy operation has a higher probability of survival than does the small Southwest dairy farm operator when milk price declines $1.00 per cwt. This is true with a herd size of only about half the level of the Southwest herds. It seems clear that availability of BST will accelerate pressures to increase the average size of the dairy farm in all areas of the country.

Implications and Conclusions

The trend to larger and fewer dairy farms will likely continue into the future as has happened in the past. BST will continue and possibly accelerate the trend that has been occurring for many years. In general, this technology is similar to many other past technologies where production decisions become more complex and interactive.

A key component of successful use of BST appears to be the overall level of dairy farm management. A key variable in many of the economic studies that have been completed to date has been the management component embodied in the analysis. It appears that effectively managed farm operations with higher production levels will reap the major benefits from the adoption of BST. Farm operations with below-average management and below-average production levels will likely lose. It seems imperative that management strategies be developed that can be effectively utilized with BST adoption. This management development focus is especially important for regions of the country that are currently projected to be negatively affected by BST. A method by which these regions of the country may remain competitive in the dairy industry is to increase the management qualities and capabilities that are on the dairy operations.

The same conclusion also applies to small versus larger dairy farm operators. If it is a judgment of political forces and society in particular areas of the country that dairy farm size should be held relatively constant or that dairy farm size should be at the level that would support a family farm type of operation, it seems rather clear that a major focus is needed on developing the management capabilities and potential of small-farm operators so that they can effectively compete with the much larger operators located in the Southwest and the Southeast regions of the United States. The bottom line is that funding needs to be placed into programs that will develop the potential for this type of dairy industry to survive in the face of technologies that are advantageous to the larger operators. Along these lines, BST is not alone; there are many technologies that can be applied to this same argument. All farms have the ability to adopt BST. The key is what is necessary for successful use. A main distinguishing characteristic to successful use is whether or not the management capability is in place on the farm to effectively use BST and remain competitive in the dairy industry. The extension service can play a critical role in developing the management potential that is needed to survive in the modern-day dairy industry. The role of extension is especially important in those areas that are dominated by moderate- to small-size dairy operations.

Notes

1. R. J. Kalter, R. Milligan, W. Lesser, W. Macgrath, and D. Bauman, *Biotechnology and the Dairy Industry: Production Costs and Commercial Potential of the Bovine Growth Hormone* (Department of Agricultural Economic, Cornell University, A.E. Res. 84–22, December 1984).

2. R. J. Kalter and R. A. Milligan, "Emerging Agricultural Technologies: Economic and Policy Implications for Animal Production," paper presented at National Academy of Sciences Conference on Technology and Public Policy, Washington, D.C., December 11–13, 1986.

3. U.S. Department of Agriculture, Economic Research Service, "Dairy: Background for 1985 Farm Legislation," Agriculture Information Bulletin No. 474 (September 1984).

4. R. J. Kalter and R. A. Milligan, "Factors Affecting Dairy Farm Management and Profitability," National Invitational Workshop on Bovine Somatotropin, (Cornell University, St. Louis, Missouri, September 21–23, 1987).

5. Kalter and Milligan, "Factors Affecting Dairy Farm Management and Profitability."

6. D. E. Bauman, P. J. Eppard, M. J. DeGeeter, and G. M. Lanza, "Response of High-Producing Dairy Cows to Long-Term Treatment with Pituitary Somatropin and Recombinant Somatropin," *Journal of Dairy Science* 68 (1985):1352–62.

7. S. Y. Shin, "The Effects of Bovine Growth Hormone on the Profitability of Missouri Dairy Farms" (masters thesis, Department of Agricultural Economics, University of Missouri, Columbia, 1986); M. Bennett, *1984 Dairy Enterprise Business Earnings and Costs* (Fm 85–4, Missouri Cooperative Extension Service, September 1985).

8. M. Boehlje and G. Cole, "Economic Implications of Agricultural Biotechnology," presented at the Iowa Academy of Science 97th Annual Meeting, Central College, Pella, Iowa, April 1985.

9. R. D. Yonkers, R. D. Knutson, and J. W. Richardson, "Dairy Farm Income as Affected by BST: Regional Characteristics," National Invitational Workshop on Bovine Somatotropin, (Cornell University, St. Louis, Missouri, September 21–23, 1987); Boehlje and Cole, "Economic Implications of Agricultural Biotechnology."

10. Kalter and Milligan, "Emerging Agricultural Technologies."

15

The Socioeconomic Impact of Biotechnology on Agriculture in the Third World

HOPE SHAND

Introduction

New developments in agricultural biotechnology are now being promoted as the newest hope for feeding the world's hungry. While biotechnology offers opportunities to increase production, reduce costs, and improve the quality of food crops in the Third World, it is also likely to accentuate inequalities in the farm population, increase the vulnerability and dependence of farmers, aggravate the problems of genetic erosion and uniformity, and further concentrate the power of transnational agribusiness.

The introduction of agricultural biotechnologies in the Third World brings to mind inevitable comparisons to the "green revolution" of the 1950s and 1960s. While the green revolution affected only two major cereal crops (wheat and rice) in limited areas of the Third World, biotechnology has the potential to affect all crops, as well as livestock — in virtually any corner of the globe. Unlike the green revolution, which was introduced to the Third World by public plant-breeding institutions, today's "gene revolution" is firmly in the hands of the private sector.

With few exceptions, the future direction and benefits of biotechnology research are now controlled by a handful of international companies, with little or no discussion of the socioeconomic implications. New developments in biotechnology thus threaten to repeat the mistakes of the green revolution on an even more massive scale.

Last year in a speech before the Industrial Biotechnology Association,

Roger Salquist, president and CEO of Calgene, made some important observations about the impact of commercial biotechnology on agriculture:

> The major thing that's going to happen in terms of biotechnology in agricul-
> ture, I believe, the single most startling thing is a strategic restructuring of the
> industry to vertical integration. . . . Historically the processors of products
> from agriculture have purchased them on the commodity markets. *What's
> going to happen with biotechnology is that you're creating proprietary prod-
> ucts out of commodities.* [emphasis added][1]

Mr. Salquist's remarks are especially relevant to agricultural biotechnology in the Third World, where the biotechnology industry is already
focusing on the development of proprietary products from traditional agricultural commodities. This trend has serious implications for Third World
agriculture.

Case studies prepared by the Rural Advancement Fund International
(RAFI) indicate that new biotechnologies have the potential to eliminate or
displace traditional botanical exports on a massive scale resulting in the loss
of foreign exchange earnings, displacement of agricultural workers, and
economic instability in many Third World nations. In the following pages,
several different commodities are examined, and three major trends are
identified:

1. *Transfer of Production*—resulting in the elimination or displacement of traditional Third World export crops.

2. *Overproduction*—resulting in sharp decreases in commodity prices
and loss of foreign exchange earnings.

3. *Product Substitution*—the substitution of one natural product for
another, eliminating traditional markets for certain agricultural commodities.

Transfer of Production—The Case of Vanilla

The case of vanilla illustrates the potential of biotechnology to displace
or eliminate traditional botanical exports and to transfer agricultural production from the Third World to laboratories and factories in the industrialized world.[2]

Two U.S.-based companies are now attempting to produce a natural
vanilla product in the laboratory through a process known as *phytoproduc-
tion*—a type of plant tissue culture. Escagen, a small biotechnology company based in San Carlos, California, is already producing natural vanilla
in the laboratory, and hopes to have a product on the market by mid-1989.

Natural vanilla is an expensive flavoring which comes from the bean of the vanilla orchid. It can be grown commercially only in a few Third World countries. Today, 98 percent of the world's vanilla crop is produced by four island nations: Madagascar, Réunion, the Comoros, and Indonesia. Madagascar alone, where up to 70,000 small farmers are engaged in production of this labor-intensive crop, accounts for three-quarters of the world's vanilla production. The economies of these countries depend on the export of vanilla beans, valued at approximately $67 million annually.

The U.S.-based companies that are now culturing vanilla cells to produce vanilla flavor in the laboratory are not manufacturing an "artificial" product. Their product would be a natural, plant-derived flavoring. If commercially successful, this new technology would have the potential to displace vanilla bean exports on a massive scale. The production of natural vanilla would likely shift from Third World island nations to laboratories and factories in the industrialized world, eliminating the need for traditional cultivation of the vanilla orchid and many thousands of jobs related to vanilla bean cultivation and harvest.

Viewed in terms of world agricultural trade, vanilla export earnings are relatively small and insignificant. But vanilla is just the tip of the iceberg — it represents only one of thousands of plant-derived substances (flavors, fragrances, nutrients, pharmaceuticals, dyes, etc.) which may be future targets of biotechnology research. The worldwide market for all plant-derived products is approximately $10.5 billion (U.S. dollars).[3]

Overproduction—The Case of Cocoa

Cocoa is another target of current biotechnology research.[4] Unlike vanilla, cocoa represents a major agricultural crop — the second most important agricultural commodity produced from tropical regions in the international trade market. Worldwide, annual exports of cocoa beans are valued at $2.6 billion. Africa accounts for 57 percent of world production, Central and South America account for 34 percent, and East Asia accounts for 9 percent.

Various techniques of biotechnology are being applied to cocoa in the United States, Europe, and Japan. Here in the United States the largest research effort focusing on biotechnology and cocoa is underway at Pennsylvania State University, where over fifteen chocolate manufacturers are supporting a multimillion-dollar research program on the molecular biology of *Theobroma cacao,* the cocoa plant. At Penn State, researchers are using both tissue culture and genetic engineering to create higher-yielding and higher-quality cocoa beans, as well as plants which have greater disease and insect resistance.

The goal of Penn State's research is "to stabilize the export crop for manufacturing countries."⁵ Using genetic engineering, for example, scientists will someday be able to form new cocoa plants tailored to meet the specific needs of industry. One long-term project is to engineer a cocoa variety containing a gene for thaumatin, a super-sweet protein derived from an African shrub. The end result would be a sugarless, sweet-tasting chocolate product—eliminating the need to add sugar in the manufacture of chocolate.

In the shorter term, Penn State researchers hope to develop higher-yielding cocoa varieties. Using biotechnologies, scientists predict that it will be possible to obtain future yields of up to three thousand pounds of beans per acre—*an increase of 750 percent above today's average yields.*

It is likely that the benefits of advanced technologies and high-yielding cocoa varieties will be skewed toward large-scale cocoa growers. As a result, cocoa production will shift from small-scale producers to large-scale cocoa plantations. Small-scale producers in Africa, where the majority of the world's cocoa is now produced, will be at a particular disadvantage. According to Dr. Russell E. Larson, science advisor of the American cocoa research association:

> Probably 50 percent or more of the cocoa in the world is produced on small holdings. For economic reasons, it is not feasible for these growers to apply some of the advanced technologies. . . . Brazil and Malaysia have a higher proportion of large size plantations and are able to apply advanced technologies quickly. It is probable that African growers will be hard-pressed to achieve the high production levels of Brazil and Malaysia in the near future.

The application of new biotechnologies to cocoa will thus facilitate a fundamental shift in the world production of cocoa from small-scale producers to large-scale plantations. Future cocoa production will likely be concentrated in Brazil and Malaysia, where advanced technologies and large-scale plantations are now in place. Ultimately, dramatic yield increases will result in overproduction of cocoa and a sharp decline in cocoa prices—a trend which will affect all cocoa producers, large and small, and the economies of all major cocoa-producing countries.

Overproduction—The Case of Oil Palm

Oil palm offers another example of biotechnology and overproduction. In the case of oil palm, yield increases are the result of a new method for cloning high-yielding palms developed by Unilever, the world's largest vegetable oil buyer and largest food enterprise. Unilever is introducing cloned oil palms throughout the tropical world – from Colombia to Brazil, West Africa, Indonesia, Malaysia, and the Philippines.

In Malaysia, Unilever's plantations are already well established, and cloned oil palms may increase yields by 30 percent.[6] Malaysian palm oil exports in 1985–86 exceeded 4.8 million metric tons and now constitute about one-fourth of vegetable and marine oils traded in international markets. Almost half of the world's increase in edible oil trade during the past five years is due to increased exports of Malaysian palm oil.[7]

Despite the impressive yields, however, the oil palm boom has not benefitted Malaysian producers. Because of enormous surpluses, they are now producing below the cost of production. Virtually all Malaysian oil palm manufacturers produced at a loss in 1986.[8]

The Malaysian oil palm glut is also affecting producers of other edible oils. Small-scale palm oil producers in Africa, for example, may lose their markets because Malaysia is exporting palm oil to Africa. In the Philippines, 700,000 small-scale coconut farmers have already suffered sharp declines in exports of coconut oil because of the world glut of low-priced oil palm.[9]

The impact of high-yielding oil palms is not limited to Third World farmers. Malaysian exports of palm oil now exceed total world exports of soybean oil – a situation which has resulted in large stocks of U.S. soybeans and a loss of markets for U.S. soybean farmers.[10] Early in 1988, the U.S. Soybean Association launched a full-scale offensive against palm oil – labeling it as "tropical fat" and claiming that it is unhealthy for American consumers.[11] The full effects of overproduction are not yet known, since thousands of acres of Unilever's high-yielding oil palm clones are now being planted throughout the Third World.

Product Substitution – The Example of Oil Conversion

Biotechnology makes it possible to substitute one agricultural product for another in modern food processing, thus destabilizing or eliminating the demand for certain raw commodities. Current research on the genetic modification of oil seed plants as a means of converting cheap oils (such as palm or soybean oil) into high-quality cocoa butter is one example.

According to Bioprocessing Technology, April 1987, "New technologies have potential to overturn oils and fats markets by reducing reliance on higher priced imports such as cacao butter. Discontented with the need to import, companies will produce similar oils from domestic sources, in the process even creating oils not found in nature" (p. 1).

Several companies in the United States and Japan are pursuing this goal. One California-based company, Genencor, has filed patents on a process which could be used to convert cheap palm oil into expensive cocoa butter. Fuji Oil Co., Ltd. (Osaka, Japan) has also patented a process to develop cocoa butter substitutes from olive, safflower, or palm oil.

The use of biotechnology to develop cocoa butter substitutes from lower-quality oils illustrates the impact that biotechnology may have in altering or disrupting traditional markets for agricultural products produced in the Third World. If a process to synthesize cocoa butter using biotechnology is commercially successful, the worldwide glut of cheap palm oil and other low-priced edible oils would undoubtedly replace a large share of the cocoa butter market.

Product Substitution—Thaumatin and Other Natural Sweeteners

Biotechnology offers the potential to displace sugar as an industrial sweetener through the development of new, natural sweeteners from plants. One of the most promising natural sweeteners, the protein thaumatin, is extracted from the fruit of a West African plant, *Thaumatococcus daniellii.*[12] Thaumatin is generally recognized as the sweetest substance known to man—about 100,000 times sweeter than sugar.

Several major corporations and small biotechnology firms in the United States and Europe are now attempting to use recombinant DNA technology to produce thaumatin protein in the laboratory. In 1985–86, the intensely sweet thaumatin protein was successfully cloned by scientists at Unilever (The Netherlands) and Ingene (Santa Monica, California).

If the thaumatin protein can be economically produced using genetic engineering, thaumatin could capture a substantial share of the sweetener market, particularly for low-calorie sweeteners in the United States, Europe, and Japan. (In the United States alone, the sweetener market is now worth $8 billion, of which $900 million is low-calorie sweeteners.)

In recent years, other types of substitute sweeteners have already eroded traditional sugar markets. The introduction of high fructose corn syrup (HFCS), a sweetener manufactured from corn using immobilized enzymes, is the most dramatic example. U.S. consumption of HFCS grew from 1.35 million tons in 1978 to 4.3 million tons in 1984, while U.S. sugar imports dropped from 6.1 million tons in 1977 to 1.5 million tons in 1985–86. According to Dutch researchers, the livelihood of an estimated eight to ten million people in the Third World is threatened by the loss of traditional sugar markets and the drop in world sugar prices.

If commercially successful, the thaumatin sweetener will not single-handedly displace traditional markets for sugar. However, thaumatin is only one of several plants which produce naturally-occurring, sweet-tasting compounds. These plants and other sweetener sources will undoubtedly be the focus of further biotechnology research. The development of a thaumatin product via biotechnology is just the beginning of a transition to alternative sweeteners which will displace Third World sugar markets in the coming years.

Addressing the Impact of Biotechnology

Biotechnology will have a fundamental impact on agricultural production in all areas of the world, but it is likely that producers and consumers of the Third World will be among the first and most profoundly affected. These findings are amplified in a recent statement by the under secretary general of the United Nations' Economic Commission for Africa:

> Biotechnology, like all technological breakthroughs before it, will lead to considerable structural change in production, international trade and cooperation. Above all, it poses the greatest challenge to the African economies, with their monocultural production system and their excessive dependence on export earnings derived from one or two commodities. All the tropical crops of primary interest to Africa are at risk. Given the current collapse in commodity markets and prices, biotechnology will simply be the last straw. It will ring their death tolls.[13]

Ultimately, biotechnology cannot be labeled as *good* or *bad*. Emerging biotechnologies offer the potential for many wonderful and beneficial products — insect- and drought-tolerant food crops or a new vaccine for malaria, for instance. The possibilities are virtually unlimited. But the real issue is who will control the products of biotechnology and who will benefit from them. As mentioned earlier, agricultural biotechnology is clearly dominated by the private sector. And when research priorities are determined by profit-making potential, the needs of Third World farmers are likely to receive a low priority.

There is no doubt that powerful new biotechnologies could be used to address the real and basic needs of society, both in the United States and abroad. Steps can be taken to address the possible negative impacts of new, emerging technologies on agriculture in the Third World and to ensure that biotechnology is not used to widen the gap between north and south and rich and poor. The following are a few suggestions for moving in that direction:

1. *Access to Information.* Based on RAFI's studies on the socioeconomic impact of biotechnology and the response to that information when disseminated in Third World countries, it is clear that those nations and individuals who will be most immediately and dramatically affected by emerging biotechnologies have little or no prior information about products or potential products which will fundamentally alter their lives and livelihoods.

It is absolutely vital that Third World nations have access to information about the social and economic consequences of introducing biotechnologies into developing countries. In order to develop strategies to cope with these consequences, Third World planners and policy makers must be able to monitor the activities of biotechnology companies, new scientific developments and potential products.

2. *National Biotechnology Programs.* Third World nations must also have an opportunity to develop their own priorities and goals for establishing national biotechnology programs. Biotechnologies could thus be used to meet the basic needs of the local population and help to diminish or avoid dependence on imported or inappropriate technologies.

The establishment of national biotechnology programs in the Third World requires financial assistance and access to scientific training. Due to limited funds and political pressures, the United Nations may offer little support for such programs. Recent events are not encouraging. The United Nations Industrial Development Organization (UNIDO) previously supported the International Center for Genetic Engineering and Biotechnology (ICGEB), a program designed to train Third World scientists in new biotechnologies in both India and Italy. Earlier this year, UNIDO announced that it would withdraw support for ICGEB, reportedly because of pressure from the United States and Japan, the world's leaders in commercial biotechnology.

3. *Regulating Biotechnology in the Third World.* Third World countries must also formulate national laws to regulate the use and testing of biotechnology products within their borders. Since regulations are still ill-defined or controversial in the industrialized countries, many biotechnology researchers are becoming more aggressive about taking experimental testing overseas to locations where regulations do not exist.[14]

In November 1986, it was disclosed that the Philadelphia-based Wistar Institute in cooperation with the Pan American Health Organization had conducted experiments with a genetically engineered rabies vaccine in Argentina—without the knowledge of the Argentine government or many of the workers who were handling the vaccine.[15] It is impossible to know how many other biotechnology products are now being developed or tested in the Third World without official sanction from national governments.

4. *International Code of Conduct.* Many of the principles and guidelines for the development, use, and testing of biotechnology in the Third World could be embodied in an international code of conduct on biotechnology adopted by the appropriate United Nations agencies (i.e., Food and Agriculture Organization [FAO], UN Conference on Trade and Development [UNCTAD], World Health Organization [WHO], and UN Industrial Development Organization [UNIDO]. A similar code of conduct on pesticides, designed to establish a minimum international standard for measuring pesticide-related practices of both governments and the pesticide industry, was approved by the UN Food and Agriculture Organization in 1985.

Because of the importance of the United Nations as a global forum for Third World countries, in particular, discussion and debate of an international code of conduct on biotechnology could give important visibility to the issue of biotechnology and lead to greater awareness of the socioeconomic consequences of biotechnology in the Third World. An international code of conduct on biotechnology would enable researchers, producers, and consumers, particularly in developing nations, to develop and procure safe and useful techniques and products of biotechnology in full recognition of the contribution and needs of all parties involved.

5. *Conservation of Genetic Resources in the Third World.* Agriculture originated in the Third World. It is in the Third World that the greatest genetic diversity of agricultural crops is found. For decades, plant breeders have used the genetic resources found in traditional crop varieties and their botanical relatives to help agriculture evolve and adapt to ever-changing needs and conditions. These Third World resources are essential for maintaining pest and disease resistance in modern crops. It was traditional varieties found in the Third World that made it possible for plant breeders to develop new, high-yielding varieties of wheat and rice. Ironically, the introduction of modern varieties throughout the Third World accelerated the pace of "genetic erosion." Even now, traditional crop varieties are being replaced and driven into extinction by new varieties.

Genetic resources found in the Third World also provide the essential building blocks and raw materials for biotechnology. Biotechnology will replace some crops. And it will produce remarkably different and attractive new crop varieties to compete with traditional types. In the absence of effective collection and genetic conservation programs, will the new crop varieties of biotechnology usher in a new and even more destructive era of genetic erosion?

With the advent of biotechnology, the value of genetic diversity seems even more immediate and the need to conserve the natural diversity of the Third World even more pressing. Without conservation safeguards, biotechnology is doomed to repeat the mistakes of the green revolution by ignoring the need to conserve the raw materials upon which its future and the future of agriculture depend.

Notes

1. Roger Salquist, "The Future of Biotechnology in Agriculture," Presentation at the Industrial Biotechnology Association Annual Meeting, Washington, D.C., October 24, 1986.

2. All information on vanilla comes from Hope Shand, "Vanilla and Biotechnology," *RAFI Communique* (January 1987).

3. "Commercializing Plant Tissue Culture Processes: Economics, Problems and Prospects," *Biotechnology Progress* 1, no. 1 (March 1985): 1.

4. Unless otherwise noted, all information on cocoa comes from Hope Shand, "Biotechnology and Cacao—A Report on Work in Progress," *RAFI Communique* (May 1987).

5. Tracy Walmer, "Cracking the Cocoa Bean," *Penn State Agriculture,* (fall 1986): 14.

6. Kees van den Doel and Gerd Junne. "Product Substitution through Biotechnology: Impact on the Third World," *Trends in Biotechnology,* (April 1986): 89.

7. *Foreign Agriculture* (March 1987): 13.

8. Ibid.

9. Van den Doel and Junne. "Product Substitution through Biotechnology."

10. *Foreign Agriculture* (March 1987): 13.

11. "Trading Blows Over the Fat of the Land," *South Magazine* (July 1987): 111.

12. All information on thaumatin comes from Hope Shand, "Biotechnology and Natural Sweeteners," *RAFI Communique* (February 1987).

13. Letter from Adebayo Adedeji, UN Under Secretary General, Economic Commission for Africa, 25 September 1987, to Mr. Sven Hamrell, Dag Hammarskjöld Foundation.

14. "Overseas Testing of Genetically Engineered Microbes Called into Question," *Genetic Engineering News* (April 1987): 3.

15. "Argentines Angry over U.S. Rabies Vaccine Test," *New York Times,* (November 11, 1986): 1.

PART

V

Social
Considerations

16

Toward an Assessment of the Socioeconomic Impacts of Farmers' Adoption of Porcine Growth Hormone

ERIC O. HOIBERG and GORDON L. BULTENA

Introduction

The history of American agriculture richly demonstrates how technological innovation has produced fundamental, and sometimes disruptive, changes in the social and economic fabric of society. The mechanical revolution in farming earlier in this generation and the petro-chemical revolution following World War II brought increased production efficiencies and steady surpluses of farm commodities. But these revolutions spurred a massive restructuring of the agricultural industry.

American agriculture is presently on the threshold of a third major technological revolution, one that portends further alterations in food production and agricultural structure. Based upon advancements in molecular biology, this biotech revolution is expected to bring impressive benefits to farmers and consumers. But these benefits will be partly offset by adverse effects of the new technologies on farm families and rural communities.

In this chapter, we examine how the impending introduction of a new biotechnology, porcine somatotropin, could affect pork production and general farm structure. To date, little information is available about the potential impacts of this soon-to-be-released technology. Drawing upon previous assessments of technology-induced change and studies of a comparable biotechnology in the dairy industry (bovine somatotropin), we identify some important changes in production and farm structure that may accompany farmers' use of porcine somatotropin.

Since porcine somatotropin won't be commercially available until after 1990, it is necessary in projecting likely impacts to draw upon information

215

about the technology itself, the nature and speed of its likely diffusion in farm populations, and general changes incurred from earlier agricultural advances. Unlike ex-post assessment of the new technologies (that is, assessment after changes have occurred and can be studied), the ex-ante approach used here requires the prediction of future events. While lacking the certainty of ex-post analysis, a forward-looking approach to impact assessment offers an important advantage in informing public debate and policy-making about innovations prior to their release and imposition of potentially disruptive effects upon society.

Technological Innovation in Pork Production

There are many examples of technologically driven changes in livestock production. New technologies have brought improved feeding programs and feed additives, better health care and diagnostics, new approaches to animal housing, and improved breeding techniques. Technological innovation has been generally attractive to farmers because of the link to improved production efficiencies and, consequently, to more profitable livestock operations.

Recent breakthroughs in biotechnology are now spawning a new generation of innovations for livestock production. One of the most important has been the ability to "manufacture" growth hormones. Manufactured hormones, which are expected to become commercially available for dairy cattle and swine within the next few years, promise to revolutionize meat and milk production and to alter the existing structure of livestock enterprises.

Growth hormones (scientifically called somatotropin) are proteins produced naturally in animals by the pituitary gland. Hormones function to increase animal metabolism and growth rates. Scientists have been interested in growth hormones since the 1930s when it was shown that injections of somatotropin led to accelerated growth rates in laboratory animals.[1] A major barrier to the commercial use of hormones, however, was that only small amounts of somatotropin could be extracted from the pituitary glands of slaughtered animals. The inability to obtain a large supply of somatotropin made it expensive and commercially unattractive.

Recent advances in biotechnology have now made the production of growth hormones possible in laboratories. Through recombinant DNA technology, in which an amino acid sequence is spliced into bacteria, scientists are able to inexpensively produce large quantities of somatotropin. However, a perplexing problem to be solved with this technology is how somatotropin is to be given to animals. The benefits of somatotropin cannot be obtained through a feeding regime since it is a protein and consequently is inactivated by digestive enzymes. Instead, it must be delivered either through frequent injections or implants. Presently, the state-of-the-art technology for both cows and swine is biweekly injections.

Trends in Hog Production

The potential impacts of porcine somatotropin on American agriculture must be viewed against a backdrop of established trends in the pork industry. At the turn of this century, nearly 4.5 million, 93 percent, of all U.S. farms, produced hogs for commercial or home consumption. By 1978, this number had fallen to about 500,000 farms, or 20 percent of all farming operations.[2] Figures for 1986 show the number of hog operations at just under 350,000.

The sharp downturn in the number of hog operations has been accompanied by steady increases in the herd size of the surviving units. Between 1969 and 1978, there was a 22 percent decline in farms reporting annual sales of fewer than 100 hogs, and a 139 percent increase in those with annual hog sales of 1000 or more.[3] By 1986 only 7 percent of the nation's medium and large hog operation (those with 500+ hogs) accounted for 56 percent of total hog inventories.

Growth in the number of large-scale hog producers has been most pronounced in the Southwest and the Southeast. The Midwest has lagged, due largely to the existence of more diversified farming operations, established production patterns, and constraints imposed by existing facilities.[4] Nationally, the typical hog production unit has moved steadily away from a small, diversified operation to a system characterized by substantial size and specialization. Underlying this change are technological innovations in health and nutrition, improved production facilities, favorable government policies, and increased managerial sophistication.[5]

Other recent changes in hog production include a shift from farrow-to-finish operations to more split-phase operations, the use of more specialized buildings for different phases of production and a corresponding decline in open pasture operations, and a more receptive attitude by farmers toward marketing hogs on a carcass grade and weight basis.[6]

The steadily increasing concentration of production in the hog industry, while not as dramatic as in other livestock production sectors, has nonetheless brought substantial homogeneity among hog producers. The emerging dominance of large, highly specialized and commercialized hog operators stands in stark contrast to the structural diversity of the past. The increased homogeneity of this industry will likely accelerate the rapidity with which future biotech innovations are adopted by hog producers.

Use of the Adoption Model to Predict Impacts

The "adoption/diffusion model" provides a useful tool for gauging the likely socioeconomic impacts of porcine somatotropin.[7] This model, used in an ex-ante manner, permits estimation of the speed with which a new technology is likely to be diffused in the farm population and provides insight into the likely adoption patterns of different categories of farm operators.

The speed with which an innovation is diffused is affected by many variables. Research has demonstrated that there is a battery of factors that commonly affects the speed with which farmers adopt new technologies, including personal and farm characteristics such as age, education, and farm size.[8] Usually early adopters of new technologies are younger and better educated than their peers, operate larger units, and are more concerned about keeping current with new production techniques in order to remain competitive.

Characteristics of a technology itself—including its initial financial cost, projected profitability, relative advantages over established production practices, observability of benefits, and social acceptability in the local community—can be important to speed of its adoption. Porcine somatotropin has several characteristics which suggest it may diffuse rapidly following its commercial availability. First, costs of the required biweekly injections should be within the financial capabilities of most hog producers. Also, use of the technology will seemingly provide an attractive financial return to adopters (currently estimated at about 3:1).

Two other facets of new technologies that commonly affect the speed of their diffusion are trialability and observability. Porcine somatotropin seems well suited to adoption on a trial basis. By administering this hormone to only a few animals and making comparisons with a control group, a potential adopter can readily gauge the profitability of the technology. The initial capital investment required to implement this technology is low enough that the trial can be discontinued if unprofitable. Thus, the financial risks involved in the trial and eventual adoption of porcine somatotropin are not great.

If hogs continue to be sold at their present weight, the observability of benefits will be most apparent to those marketing on a grade and weight basis. Given the potential for significant improvements from porcine soma-totropin in carcass composition, marketing strategy could be an important factor influencing adoption, especially if a premium price is paid for leaner carcasses.

Recent studies of a similar technological advancement in the dairy industry (bovine somatotropin) provide clues to the likely adoption of por-cine somatotropin. It has been shown that many dairy farmers will quickly adopt growth hormones once they become commercially available.[9] But, as with past technologies, some farmers (the so-called late adopters and lag-gards) will plan to delay their adoptions until bovine somatotropin has been well "proven" on neighbors' farms. In fact, some operators, perhaps as much as a third of all dairy farmers, are adamantly opposed to adopting this new technology.[10]

To date, study hasn't been made of the propensity of hog producers to adopt porcine somatotropin. However, given its similarities with bovine somatotropin and the promise of attractive financial benefits, rapid adop-tion seems likely, especially among larger operations. In fact, porcine soma-totropin offers advantages over bovine somatotropin in that it will not only bring higher output and production efficiencies but also a better quality food product (i.e., less carcass fat).

One factor that may negatively influence the attractiveness of porcine somatotropin to farmers is its delivery system—that is, the requirement of periodic injections. The most acceptable system—delivery through feed ad-ditives—isn't possible since porcine somatotropin can't be broken down through the normal digestive process. Periodic injections of this hormone will be required, probably on a biweekly basis during the final two months of the hog feeding cycle.

Another possible impediment to farmers' adoption of porcine somato-tropin is that its use may be more compatible with a hog confinement system than with an open pasture system. In fact, persons with pasture systems may have to convert to confinement operations as part of their adoption decision. The requirement of such collateral adoptions under-scores the need to look beyond an innovation itself (e.g., its purported scale neutrality regarding financial commitment) to the related changes and in-vestments that may be required for its successful implementation and operation.

Established management practices also may have to be upgraded with farmers' adoption of porcine somatotropin. Among other things, hormone injections will likely require a more sophisticated system of record keeping for many farmers.[11] Also, a more complicated feeding program will be required to take full advantage of the benefits of this technology.

Along with identifying traits of an innovation that are important to its diffusion, the adoption model sets out stages through which farmers progress in making decisions about the use of a new technology. The first stage, *awareness,* is that period in which an individual first becomes conscious of the availability of a new technology. For porcine somatotropin, information necessary for achieving awareness is, as yet, very sparse.

In reviewing some general and commodity-specific farm journals covering the past two years, we found expanded coverage of potential agricultural applications of biotechnology. Much of the content of these articles, however, dealt with plant genetic engineering and bovine somatotropin, which is not surprising given the recent public controversy and debate surrounding these topics. We found very little attention being paid in the farm press to application of biotechnology in the pork industry. Where coverage occurred, it generally dealt with genetic engineering and advances in reproduction. Also, since porcine somatotropin isn't yet being marketed, there has been little information forthcoming from commercial sources that might raise farmers' awareness levels.

Once aware of an innovation, farmers typically amass factual data about it to facilitate their decision making (*information* stage). The fact that porcine somatotropin is still being perfected complicates the securement of reliable information. A yet-unanswered question is the role that research and extension components of the land-grant university system will play in acquainting the public with this technology. Will this system fade in importance, as some have suggested, to play an increasingly ancillary role to the private sector in information dissemination? It seems likely that as agribusiness firms come to play a more central role in the development of biotechnology innovations, they will also emerge as the primary sources of consumer information about these products.

After obtaining information, farmers typically enter what has been called the *evaluation* stage. In this stage, they make mental evaluations about the appropriateness or fit of an innovation to their personal operations. As regards porcine somatotropin, these evaluations must encompass both new managerial demands imposed by the technology and its compatibility with existing facilities and farming procedures. Other factors that may be pertinent to an adoption decision are likely consumer reactions to the use of hormones and to the changed food product (which affect future price structures), the possibility of greater government regulation of production (which affects farmer autonomy), and how diffusion of the technology will affect collective, versus individual, interests of hog producers.

Finally, some farmers will proceed to a *trial* stage of decision making in which they experiment with an innovation on part of their operations. In this stage, comparative observations can be made as to the on-site applicability and likely profitability of the innovation, information which is crucial to making the final adoption decision.

Anticipated Socioeconomic Impacts of Porcine Somatotropin on Pork Production

The adoption model discussed in the previous section provides a general framework for gauging the likely impacts of farming innovations. Drawing upon this model, we set out below some anticipated impacts of porcine somatotropin, with attention paid to impacts on production processes and farm structure.

Recent studies of the likely effects of growth hormones on the dairy industry offer valuable insight into how hormone use may eventually affect the pork industry. But while the dairy findings are useful, it is necessary to recognize that the dairy and pork industries display important differences which may affect adoption patterns. For one thing, dairying usually occupies a more central position in farm operations than do hogs, which tend more often to be part of a diversified production system. Also, unlike hogs, the production and pricing of milk is tightly controlled by federal programs. But perhaps the most important difference is that use of bovine somatotropin doesn't change the nature of the food product itself but only brings about production gains. A distinct advantage of porcine somatotropin is that it facilitates production of a leaner, better quality carcass, thus enhancing the attractiveness of pork to consumers and permitting potential market expansion.

Changes in Productivity

Technological innovation has permitted American farmers to boost their production capabilities to the point that, despite a sharp decline in farm numbers, there are sizeable surpluses today in many food commodities. The recent expansion of production capabilities is well demonstrated in the dairy industry. Over a thirty-year period, 1955 to 1985, milk production per cow increased about 240 pounds per year as a result of improved feeding programs, better health care, and scientific breeding. Continued advancements in these areas are expected to sustain this annual increase into the future, resulting in a total annual production of about 17,000 pounds of milk per cow by the turn of this century. However, with the use of bovine somatotropin, these production figures could be substantially higher, perhaps as high as 24,000 pounds per year.[12] Given a static per capita demand for milk, any increased production capability will necessitate a marked reduction (perhaps as much as 30 percent) in dairy cattle by the year 2000 and an even greater decline in the number of dairy farms.[13]

Farmers' use of porcine somatotropin should bring about similar, but perhaps less dramatic, gains in pork production. Production-related benefits from this technology include (1) marked increases in average daily weight gain of animals; (2) increased feed conversion efficiency, thus less feed required per animal; (3) a decline in backfat and increase in lean meat, resulting in higher quality carcasses; and (4) no danger of residual chemicals in the meat or concomitant withdrawal period prior to marketing.

Projected production-related disadvantages of porcine somatotropin are severalfold. First, its adoption will require frequent (biweekly) injections during the last two months of the feeding period. (It has been predicted that there may be little net change in labor requirements because of the shorter time that animals are on feed.) Second, animal health problems may develop, such as undesirable abscesses from injections. Third, because of frequent injections, it may not be feasible for use in open housing (pasture) situations. Fourth, related feeding and animal health problems will likely necessitate the use of more sophisticated farm management techniques.

Structural Changes

The adoption model suggests that it is generally the larger, better capitalized farm operators who are the first to implement new technologies.[14] As a result of their early adoption, these persons are generally able to increase output, drop per-unit production costs, and secure higher profit margins in the marketplace. But as the pool of adopters expands over time, continued increases in aggregate production generally bring lower commodity prices and diminished profit margins. Those adopting later are commonly denied the financial benefits (windfall profits) obtained by persons who adopted quickly. But as their competitive edge erodes with continued diffusion of the production-boosting technology, early adopters are spurred to a new round of innovation to maintain their profit margins and the impetus for technological change (technological treadmill) is repeated.

The adoption model indicates that technological change, while beneficial to the economic well-being of some farmers (especially early adopters), can be detrimental to the interest of others. The continued survival of farmers who are slow to implement more efficient production technologies is now typically contingent upon their willingness to accept low returns on their labor and/or the availability of off-farm income.

The biotechnology revolution will probably accelerate the process by which new technologies are rendering farmers obsolete. As concluded in a recent national study, "the combination of future yield increases from new technologies and current economies of size in these commodities means that there will be substantial incentives for farms to grow in size".[15] This growth is typically achieved by purchase of neighboring units. With continued technological progress, it is projected that the number of farms will decline to about 1.2 million by the year 2000, down nearly 50 percent from present levels.

Consistent with historic trends, it is the largest farms that will be the major beneficiaries of the new biotechnologies. Whereas over 80 percent of the largest livestock operations are expected to incorporate one or more important new animal technologies by the turn of the century, 50 percent of medium-size farms, and only 10 percent of the smallest farms, will make such adoptions.[16]

The magnitude of potential impacts on farm structure from the biotechnology revolution is well demonstrated in recent scenarios of developments following the introduction of bovine somatotropin. Despite its promotion as a "scale neutral" technology, the diffusion of bovine growth hormone is expected to bring an annual decline of over one thousand dairy farms in New York state alone; this is twice the present attrition rate.[17] The potential for even larger adjustments is suggested in the recent assessment that fewer than five thousand well-managed dairies of fifteen hundred cows each could provide all of the milk needed in the United States.[18]

As is true of the impact of bovine somatotropin on dairying, the availability of porcine somatotropin should contribute to a continued, if not accelerated, displacement of pork producers. These losses, however, may not be as dramatic as for dairy farmers, partly because hog production usually is part of a more diversified farm operation. But like dairying, porcine somatotropin will disproportionately benefit the largest operators; small, inefficient producers are unlikely to find this technology an automatic path to more profitable operations.[19]

Upgrading of Management Skills

Farmers' adoption of porcine somatotropin will require them to achieve greater efficiencies in their management techniques. Among other things, this means an upgrading of their information processing and decision making skills, a need for systematic monitoring of operations, better record keeping, and adoption of new marketing strategies. In fact, economic benefits of the new biotechnologies could be lost if they are not skillfully applied. Unfortunately, many pork producers, in the press to remain competitive, will probably feel compelled to adopt growth hormones without a full comprehension of the collateral skills and farm changes that are required to make this innovation financially rewarding.

Shifts in Production Locales

Technological change can bring major shifts in the locales of commodity production. The development of center pivot irrigation, for example, permitted an extension of corn production into the Great Plains. New biotechnology innovations hold promise of prompting further shifts in where some food items are grown. The restructuring of genetic codes and development of genetically produced frost retardants, for example, may make it possible for some crops to be adapted to harsher environments. The science that opens up opportunities to grow oranges in Iowa may also permit Utah to grow corn, suggesting that terms such as the *corn belt* and *cotton belt* may someday be inaccurate descriptions of regions where specific commodities are grown.[20]

Regional shifts in commodities will sometimes occur because of the disproportionate benefits that technological change confers on large, efficient producers. For example, impending biotechnology developments in the dairy industry are expected to accelerate a shift in milk production to regions with large dairy herds.[21] Similarly, the availability of porcine somatotropin could shift the impetus of pork production away from some midwestern states which, although currently accounting for a large percentage of total marketings, are not leading states in the relative size of production units.

Agribusiness Impacts

American agriculture is characterized by an ever-increasing concentration of corporate control over many inputs (such as machinery, seeds, fertilizer, and pesticides) essential for food production. These inputs, as well as the marketing of farm products, are commonly controlled by a handful of national and multinational agribusiness corporations.[22]

The large financial investments required in biotechnology research and product development will likely make many small agribusiness firms uncompetitive and bring an ever-growing domination in agriculture of large, multinational firms. Farmers, in securing needed production inputs, will be required to deal in a monopolistic sector for the products essential to their economic survival. As petty capitalists, American farmers are finding themselves increasingly sandwiched between monopolistic farm input and marketing sectors, with an associated erosion of their managerial autonomy and earning capabilities.

Conclusion

Technology has been characterized as being "dual-edged,"[23] that is, as offering rich promise of serving societal needs while at the same time causing significant social disruption. In the past, use of the adoption/diffusion model has led social scientists to concentrate on the positive side of technological innovation by identifying and overcoming obstacles to the rapid diffusion of technology. The model largely ignored, until recently, potentially negative effects of changes caused by technological advancement.

It is argued in this paper that the adoption/diffusion model, cast in an ex-ante perspective, contributes to an enhanced understanding of farmers' likely reactions and adjustments to impending biotechnologies. Using the model to identify profiles of likely adopters, compatibility of innovations with extant agricultural structures and procedures, and farmers' accessibility to alternative information sources has potential for informing policy-making before the release and consequent impacts of biotechnological innovations. The influence of technology on society need not be a one-way street. Rather, as Summers has pointed out, "social structures have a degree of independence which permits them to exert reciprocal influences on technology and to influence its creation, adoption, diffusion and its effects on the future course of social history."[24]

Notes

1. R. D. Boyd, "Somatotropin and Productive Efficiency in Swine," *Animal Health and Nutrition* (February 1987):23–26.
2. Marvin Hayenga, V. James Rhodes, Jon A. Brandt, and Ronald Deiter, *The U.S. Pork Sector: Changing Structure and Organization* (Ames, Iowa: Iowa State University Press, 1985), p. 21.
3. Ibid., p. 22.
4. Roy N. Van Arsdall and Henry C. Gilliam, "Pork," in *Another Revolution in U.S. Farming?* ed. Lyle P. Scherz et al. (Agricultural Economic Report no. 441. Washington, D.C.: Department of Agriculture, 1979).
5. Ibid.
6. Ibid.
7. Everett M. Rogers, *Diffusion of Innovations* (New York: The Free Press, 1983).
8. Ibid.
9. U.S. Congress, Office of Technology Assessment, "Technology, Public Policy, and the Changing Structure of American Agriculture." (OTA-F-285. Washington, D.C.: U.S. Government Printing Office, 1986; W. Lesser, W. Magrath and Robert Kalter, "Projecting Adoption Rates: Application of an Ex Ante Procedure to Biotechnology Projects," Report No. 85–23, Department of Agricultural Economics (Ithaca, New York: Cornell University, 1985); Peter Nowak and Roy Barnes, "Potential Social Impacts of Bovine Somatotropin for Wisconsin

Agriculture: A Survey of 270 Wisconsin Dairy Producers," paper presented at the meeting of Rural Sociological Society, Madison, Wisconsin, 1987.

10. Nowak and Barnes, "Potential Social Impacts of Bovine Somatotropin for Wisconsin Agriculture."

11. Boyd, "Somatotropin and Productive Efficiency in Swine."

12. William Hansel, "Big Milk Productivity Increases in Sight with More to Follow," *Choices* (2d Quarter 1986):26–27.

13. U.S. Congress, Office of Technology Assessment, "Technology, Public Policy, and the Changing Structure of American Agriculture."

14. Rogers, *Diffusion of Innovations.*

15. U.S. Congress, Office of Technology Assessment, "Technology, Public Policy, and the Changing Structure of American Agriculture," p. 15.

16. Ibid., p. 131.

17. W. Magrath and Loren Tauer, "The Economic Impact of bGH on the New York State Dairy Sector," Agricultural Economics Staff Report 85–22 (Ithaca, New York: Cornell University, 1985).

18. U.S. Congress, Office of Technology Assessment, "Technology, Public Policy, and the Changing Structure of American Agriculture," p. 53.

19. Martin Isaac Meltzer, "Repartitioning Agents in Livestock: Economic Impact of Porcine Growth Hormone," Masters thesis, Cornell University, Ithaca, New York, 1987.

20. Paul Lasley and Gordon Bultena, "Farmers Opinions About Third-Wave Technologies," *American Journal of Alternative Agriculture* 1, no. 3 (1986):122–26.

21. Lewellyn S. Mix, "Potential Impact of the Growth Hormone and Other Technology on the United States Dairy Industry by the Year 2000," *Journal of Dairy Science* 70 (1987):487–97.

22. Jack Doyle, *Altered Harvest: Agriculture, Genetics, and the Fate of the World's Food Supply* (New York: Viking, 1985); Jack Kloppenburg, "The Social Impacts of Biogenetic Technology in Agriculture: Past and Future," in *The Social Consequences of New Agricultural Technologies,* ed. Gigi Berardi and Charles Geisler, (Boulder, Colorado: Westview Press, 1984); Martin Kenny, "Agriculture and Biotechnology," *Biotechnology: The University-Industrial Complex* (New Haven, Connecticut: Yale University Press, 1986).

23. Sheldon Krimsky, "Biotechnology and Unnatural Selection: The Social Control of Genes," in *Technology and Social Change in Rural Areas: A Festschrift for Eugene Wilkening,* ed. Gene F. Summers (Boulder, Colorado: Westview Press, 1983).

24. Gene F. Summers, "Introduction," in *Technology and Social Change in Rural Areas: A Festschrift for Eugene Wilkening,* ed. Gene F. Summers (Boulder, Colorado: Westview Press, 1983), p. 2.

17 Biotechnology, Agriculture, and Rural America: Socioeconomic and Ethical Issues

FREDERICK H. BUTTEL

Agronomists and other agricultural scientists, along with engineers and health scientists, have been the true revolutionaries of the 20th century. But they are reluctant revolutionaries!

They have wanted to revolutionize technology but have preferred to neglect the revolutionary impact of technology on society. They have often believed that it would be possible to revolutionize technology without changing social institutions. . . . The rising value that society places on the health of workers and consumers, and on environmental amenities such as clean water, clean air and clean streets, will continue to lead to a demand for effective social control over the development and use of agricultural technology.

—V. W. RUTTAN

Introduction

These words of Vernon W. Ruttan, the well-known economist of agricultural research, perceptively grasp why coming to grips with the "bioethics" of agricultural research has been so difficult.[1] It is significant that a bioethics symposium sponsored by and held at a land-grant university would have been inconceivable prior to the current age of biotechnology and remains uncommon to this day. Perhaps the only paper by a representative of the land-grant establishment to address such matters in any manner prior to the biotechnology era, Boysie E. Day's American Society of Agronomy (ASA) presidential address, "The Morality of Agronomy," indicated clearly the politics that pervaded the land-grant system a decade ago.[2] Day argued that the only morally defensible direction for agronomy and agricultural research in general must be to pursue single-mindedly continued productivity increase in agriculture with its concomitant benefits to farmers and consumers—that is, to be the "revolutionaries" of whom Ruttan has spoken; conversely, it would be immoral for the agronomy profession to respond to the new agendas for agricultural research being promulgated by the likes of Jim Hightower.[3]

The sociopolitical milieu of land-grant research is now far different from when Day wrote his "man-the-barricades" version of the ethics and morality of agricultural research in his ASA presidential address a decade ago. There have been several major changes in this milieu that have made discussions of bioethics not only possible but also necessary.

The first set of factors in the reshaping of the sociopolitical milieu of public agricultural research has to do with the emergence of international technological competition in the late 1970s and early 1980s.[4] To understand the significance of international technological competition, we have to go back more than a decade and a half ago. The state agricultural experiment stations (SAES's) had been criticized for nearly a decade, beginning with the Pound Report, for their lack of attention to basic research.[5] The Pound Report, however, fell largely on deaf ears in the SAES community until a successor report, the so-called Winrock Report, jointly sponsored by the Office of Science and Technology Policy of the White House and the Rockefeller Foundation, raised the possibility that federal formula funds for research might be reduced if the SAES's failed to move more aggressively into basic biological research.[6] There was another key difference between the Pound and Winrock Reports which was reflective of the new era of international technological competition and which led to a rapid response by the SAES system to the latter report. As opposed to the emphasis of the Pound Report on the need to conduct basic biological research in order to do good science, the authors of the Winrock Report saw the purpose of basic research in far more instrumental terms: it was to enhance the competitive position of U.S. firms and the U.S. economy in global technological competition. Attracted by the carrot of increased industrial and state government funding and by the prestige of basic research, and fearful of the stick of reduced formula funding and continued criticism by federal policymakers and influential corporations, the SAES's have very rapidly embraced basic biological research (or "biotechnology" as it is more commonly known) since 1982. There is scarcely an SAES now that does not purport to have a biotechnology program of some sort (though twenty or so of these programs have very small staff commitments and are primarily paper programs).

Second, the increased land-grant commitment to biotechnology has not always gone smoothly. The increased emphasis on biotechnology has often generated considerable internal opposition within colleges of agriculture, especially on the part of scientists whose work was of the relatively applied sort that was typically sacrificed in the reallocation of research resources into biotechnology. This internal opposition, however, is arguably less than it was three or four years ago. Nonetheless, the land-grant system has also suffered from adverse external publicity relating to its development research on bovine growth hormone (bGH), one of the most promising new biotechnologies for animal agriculture.[7] There have also been concerns expressed about the increased industrial influence on land-grant research and about the possible environmental consequences of genetically modified life forms.[8] Finally, in some states farmer groups have been hesitant about the increased emphasis on basic research and the decreased SAES commitment to doing applied research geared to the technical needs of these state-level groups.

Third, the U.S. and world farm economies have moved into a long period of stagnation and crisis, and the agricultural research community has at times found itself on the defensive about its new technologies having contributed to the crisis. With many farmers struggling with high debt-asset ratios and the prospect of long-term low commodity prices, farmer groups, often of relatively large commercial operators, have been influential in urging the agricultural experiment stations to explore low-input alternatives to capital-intensive production technologies. More shortsightedly, some national commodity groups have mobilized to emasculate federal funding of international agricultural research when it might lead to competition with U.S. farmers' export sales. Nonetheless, as conventional farmer groups have pushed successfully for some new priorities in public agricultural research, organic farmer and related public nonfarm interest groups have intensified their efforts to influence SAES priorities. A number of land-grant universities—including the University of Minnesota, University of Wisconsin, University of Nebraska, University of Maine, University of California, Iowa State University, and Cornell University—have responded by announcing major new efforts to develop low-input technology. Further, one might argue that another reason for land-grant institutions having changed their minds about low-input research over the past few years has been to shore up their base of support among farmer groups that had begun to erode during the biotechnology thrust of the early 1980s.

These and other factors—stagnation of public funding of agricultural research, very rapid changes in intellectual property law relating to biological innovations, the rapid restructuring of the agricultural input industries, and so on — have led to a distinctly new milieu of SAES research.[9] Virtually all of these changes can be connected with the rise of biotechnology in SAES institutions — as antecedent, consequence, or both. With this background in mind, I will explore two sets of issues relating to biotechnology and rural America. The first will be the matter of whether there is anything new or distinctive in the bioethics of agricultural biotechnology as opposed to the techniques and social relations of research that prevailed prior to the 1980s. The second will be to set forth some highly tentative guesstimates of the impacts that biotechnology might have on agriculture and rural America.

Is Biotechnology Distinctive?

[T]here can be no questions about society's right to hold the science community responsible for the consequences of the technical and institutional changes set in motion by research.
— V. W. RUTTAN
"Agricultural Scientists as Reluctant Revolutionaries"

The Distinctiveness of Biotechnology: An Assessment of Prevailing Arguments

In the brief period since the late 1970s, *biotechnology* has been transformed from an obscure expression coined by start-up firm entrepreneurs in their quest to attract investors to a term that is increasingly well known in virtually all quarters of agriculture.[10] There is now a notion broadly shared by both defenders and critics of biotechnology that the era of agricultural biotechnology will be a distinctive one, mainly because of two *biological* aspects of the emerging technologies. The first notion concerns the dramatic potentials for productivity increase that these new technologies will make possible. Due to widely circulated claims that bGH will increase milk production per cow by 25 percent or more in a short period of time, many see bGH as a harbinger of the potentially massive productivity impacts that biotechnology will have in national and world agriculture.[11] The second notion is that the life forms made possible by biotechnology are unique — that is, they will be qualitatively different from those that can be developed through "conventional" research methods.[12]

These claims about the uniqueness of biotechnology on essentially biological grounds can be criticized in several respects. First, as will be discussed at more length below, it is unlikely that the productivity increases that will be made possible by biotechnology over the next two to four decades will be revolutionary, at least in terms of accelerating national and world productivity and output trend lines beyond those achieved during the post–World War II period.

Second, I will attempt to make the case below that to the degree biotechnology is unique–and hence must have a distinctive bioethics–this uniqueness is more a function of its social relations than of its biology.

Third, as my introductory comments have suggested, the social and bioethical issues raised by biotechnology are, in substantial measure, ones generic to agricultural research, biotechnological or not. Many of the issues raised by the biotechnology boom–debates over the social justice and environmental responsibilities of land-grant institutions, over whether these institutions should place greater stress on basic or fundamental research, and over how these institutions should be responsive to farmer needs–are by no means new. These issues are generic to land-grant research and extension programs. Each, in fact, was already a major issue (though in more muted or restricted forms) *a decade prior* to the explosion of interest in biotechnology.

Is biotechnology unique? Must there be a distinctive bioethics of agricultural biotechnology? A reasoned response would be "no" *and* "yes." The answer must be no in the sense that there is a basic set of social and ethical issues raised by the development and deployment of *any* technology (agricultural or nonagricultural, "traditional" or biotechnological): Who benefits? Who loses? What are the environmental consequences? What have been the alternatives foregone? To whose needs does the technology development process respond, and why? How does the technology affect control over the production and distribution system? What are the social goals and ethical criteria that should guide research problem choices?

The answer must also be yes, but not primarily because of the biology of recombinant DNA and associated techniques and methods. Rather, the distinctiveness of biotechnology, in my view, lies primarily in its associations with major changes in the *social relations* of research, technology development, and innovation.

It is my observation that most persons concerned about the impacts and bioethics of biotechnology inevitably tend to begin with data or guesses about how these technologies will affect farmers.[13] I would be the last to argue that the impacts of biotechnologies on farmers are unimportant. In particular, biotechnology portends major impacts on Third World farmers, who will probably be the agriculturalists most affected by biotechnology over the next twenty to thirty years.[14] But it is my view that premising bioethical inquiry, in the first instance, on how farmers and farm structure will be affected serves to deflect attention from some fundamental aspects of biotechnology. The impacts of biotechnology thus far have been almost entirely confined to two realms: (1) impacts on the structure of public research institutions and their relationships with private research and development (R&D), and (2) impacts on the structure of the agricultural inputs industry and private R&D, which will likely continue to shape public research institutions. Farm structure has been virtually unaffected by biotechnology as of this point due to the fact that there have been essentially no biotechnologies that have been commercialized. Finally, as suggested above, it is by no means clear that the impacts of biotechnologies on farmers and farm structure will be so qualitatively different from those of biochemical technologies after World War II as to justify seeing biotechnology as being unique in its impacts on farmers.

How, then, is biotechnology unique in terms of the social relations of the research and innovation process? In a word, the overarching change revolves around the *privatization* of public research.[15] I will now suggest four major respects in which biotechnology has led to or represents a privatization trend and, as well, four other concomitants of privatization for public research institutions.

The Privatization Trend

The first aspect of the privatization trend that deserves mention is the emergence of international technological competition referred to earlier. More specifically, competition over high-technology R&D has become a central policy focus of the governments of most of the advanced industrial countries. Biotechnology is recognized as being one of the two cornerstones—along with computers, semiconductors, telecommunications, and other information technologies—of the future .high-technology world economy.[16] As a result, molecular and cellular biology—which was the epitome of an ivory-tower basic research discipline as recently as fifteen years ago—has become transformed into a commercially relevant, if not commercially oriented, field in essentially all the advanced industrial countries. Accordingly, there has been set in motion a global competitive process which has dictated that only at its future competitive peril can any particular nation-state allow its basic biology research programs to remain insulated from commercial objectives. International competitiveness is thus an objective force affecting the structure of global R&D. But it should also be recognized that competitiveness is now a full-blown *ideological* form as well; increasingly we witness the specter of the "imperative of meeting international competition" being used to justify all manner of changes in funding, organization, and goals of agricultural and other biological research.

A second important, but often unrecognized, aspect of biotechnology has been its premature commercialization.[17] By this I mean that biotechnology was commercialized by private firms at a point at which significant product sales were ten to twenty years down the road and at which its commercial research frontiers have typically been similar to its basic research frontiers.[18] By comparison, most other high-technology industries are based more on applied research (and applied science disciplines such as in engineering) than on Nobel-caliber basic research. Market horizons of ten-plus years are very uncommon in the other high-technology sectors. Among the implications of this premature commercialization have been the short "distance" between basic and applied genetic engineering research and the corollary need of private firms to achieve access to basic research results in universities.

A third aspect of privatization has been the development of closer industry-university relationships in agricultural research. These relationships include increased levels of industrial sponsorship of research, the increased prevalence of equity ownership of private firms by university scientists and long-term consulting relationships between "biotechnology" faculty and private firms, increased attention to university patenting of research results for remunerative transfer to private industry (i.e., receipt of royalty income), and the creation of industrially funded biotechnology institutes.[19] It is important to stress that there has been a long history of industrial sponsorship of SAES research well prior to the biotechnology era. Industrial sponsorship of research in the past, however, was of a greatly different character than it tends to be at present. There is evidence that industrial grants and contracts to SAES institutions in the early 1970s tended to be more numerous, for very small amounts (typically $3,000 to $5,000 per year), and for more highly applied—virtually product-development—research than is generally the case at present. Also, it should be stressed that one should not exaggerate the extent to which land-grant institutions find themselves being the recipients of large industrial grants and contracts. Most SAES's have found that their biotechnology programs have attracted less industrial funding than they originally anticipated. The bulk of land-grant biotechnology program support currently comes from state governments and other public sources rather than from industry.

A final aspect of the privatization of biotechnology research concerns the increased role of intellectual property restrictions— particularly industrial patents and trade secrecy—in both public and private research. The most important changes—*Diamond* v. *Chakrabarty, ex parte Hibberd,* and a federal law permitting universities to patent federally funded research results—have occurred during the current decade and were connected to the commercialization of biotechnology and to international technological competition.[20] The principal implications for SAES research of the elaboration of intellectual property restrictions are (1) the incentive that exists for public researchers and research institutions to direct basic or fundamental inquiry toward patentable or otherwise protectable results, (2) the changing character of contractual relations between universities and industry—particularly the longer delays in publication and the increased prevalence of labs having to maintain trade secrecy—as private firms seek to withhold results from competitors, and (3) the increased climate of secrecy that will result when research findings have a potentially high market value.

Responses of Public Research Institutions to Privatization

The privatization of basic biological research has been both an opportunity and a problem for SAES's and other public research institutions. The key opportunity for the SAES's has been to capitalize on federal and especially state government funding for establishing new biotechnology programs. In the main, this has been driven by technological competition in its subnational forms—that is, by the actions of roughly twenty U.S. state governments to capture high-technology investments, much as national governments are trying to do. Land-grant universities have benefited most from the state-level expression of technological competition; most of the land-grant universities with major agricultural biotechnology programs are in states where state governments have been keen to attract high-technology investment by supporting university research. This has been a significant windfall for SAES's, especially at this time of stagnation of public funding of agricultural research.

While land-grant universities have clearly benefited from international technological competition and from the hope of state governments that basic research results can be readily privatized by new high-technology companies in their states, there have been some problems as well. One is that SAES agricultural biotechnology programs have by and large been established with no particular socioagronomic goals. For many such universities, the goal in having a biotechnology program has been confined to wanting to have one![21] With only a few exceptions, SAES biotechnology programs have been established before asking the crucial prior questions: Biotechnology for what? Biotechnology to benefit whom? Biotechnology for achieving what social and agronomic goals? This lack of deliberate goals, as suggested below, has become part of a posture that may threaten the public support base of land-grant universities in the future.

A third aspect of the public response to privatization concerns the fact that biotechnologies have become controversial well before commercial introduction. Virtually unprecedented, for example, is the fact that large segments of the public — including but not limited to farmers and agribusinesses — have become aware (though often in hazy terms) of the pluses and minuses of bGH two or more years prior to the time it will be commercially available. Also, unlike previous forms of agricultural technology, agricultural biotechnology already has formidable critics many years before its technologies will be transferred to farmers' fields and feedlots. Most significant for land-grant institutions is the fact that in some land-grant universities bGH has generated a considerable amount of opposition to and scrutiny of SAES research by *farmers,* who in the bulk of the post–World War II period were fairly accepting of new agricultural technologies.

The final implication of the privatization of biotechnology for SAES institutions concerns the quote with which I began this chapter. Ruttan has stressed what actually has long been known about agricultural research and technological change — that agricultural research is not unambiguously of benefit to all farmers.[22] Agricultural research, by increasing productivity and output, tends to reduce commodity prices and set in motion a technological treadmill that results in some fraction of farmers being forced out of business or attracted to higher-paying off-farm employment.[23] This tendency is largely accounted for by the fact that agricultural commodities tend to have low price and income elasticities of demand, which causes product prices to fluctuate disproportionately in response to changes in supply. Some fraction of farmers does benefit from productivity-increasing agricultural research — namely, early adopters of technology who receive "innovators' rents" until a point at which a certain proportion of farmers has adopted the new technology, caused declining product prices, and precipitated a technological treadmill. These early innovators, however, are a minority of the agricultural population while those who receive little or no benefit are generally a majority.

The result of the nature of the distribution of the benefits and costs of technological change among farmers is that farmers historically have been ambivalent about research — neither actively supporting nor opposing research appropriations at the federal level.[24] The historic resolution of farmer ambivalence about research at the federal level has been substantially mitigated at the state level, however. At the state level, farmers can be induced to support research which enables them to compete more effectively with farmers in other states. Thus, the raison d'etre of the land-grant system has come to be one of doing *applied, locally adapted research,* the benefits from which are at least partially capturable by state-level farmer groups. Accordingly, the funding backbone of the SAES system, even in this era of biotechnology, remains that of state appropriations. For most SAES's state funds are at least three-fold those of federal funds. In the larger SAES's, which are generally those that have been most successful in attracting state government funding of new biotechnology initiatives, state funds are ten- to thirty-fold those of federal monies.

The significance of this long digression on the political support base of public agricultural research lies in the fact that biotechnology research, at least in its early years, will consist primarily of developing *generic* technologies that are applicable well beyond the boundaries of a state.[25] Thus, the new biotechnology thrust runs counter to the raison d'etre of state-government–supported agricultural research. The fact that some of the new biotechnologies, especially bGH, have become controversial is, in part, related to the generic character of this technology: bGH is just as applicable to dairying in California, Florida, New Zealand, and the Netherlands as it is to dairying in New York and Wisconsin, the two states in which the most bGH research has been done by SAES faculty and where bGH has been most controversial. It is also because of the wide geographic applicability of bGH that bGH has been projected to have such substantial dislocating impacts on dairy product markets and the structure of dairy production. There probably would have been far less controversy over bGH in Wisconsin and New York if bGH would have been a locally adapted technology from which farmers in these states were in a privileged position to benefit.

Perhaps, then, a crucial characteristic of biotechnology – and hence of its bioethics – will be the fact that the social relations of agricultural research are leading to this technology being developed primarily in the form of broadly applicable, generic technologies (at least in the early stages before technology becomes routinized). An increased emphasis of the SAES system on generic technologies, rather than on locally adapted research that is more consistent with farmer organization support of land-grant appropriations, could have far-reaching implications for the system.[26]

Biotechnology, Privatization, and the "Triangular Squeeze"

Biotechnology and the trend toward privatization of research which it has engendered have served to greatly complicate the politics of SAES research and to render its bioethics far more complex. A decade ago land-grant institutions faced a fairly straightforward political situation – to continue to keep state-level clientele groups (mainly farmer commodity organizations) happy in order to sustain political support for state appropriations. The challenges of basic-science–oriented critics, such as the authors of the Pound Report, and of social justice and environmental critics, such as Hightower,[27] were of some importance then, but were sufficiently minor so that business as usual could continue.[28]

The impact of biotechnology on SAES politics, however, has been to vastly complicate this situation by placing land-grant institutions in a "triangular squeeze." The SAES system now finds itself cornered within a triangle of forces over which it has little leverage. These triple forces are as follows. (1) National-level institutions and organizations, primarily agribusiness and the executive branch of the federal government, want the land-grant system to pursue basic research, largely for generic application across the country. The results of this research into new opportunities for increased agricultural productivity are to be transferred to industry for agribusiness' own benefit and to assist in the U.S. effort to compete in high-technology with other countries. (2) Public interest groups and other social justice and environmental critics of SAES research, whose concerns have been given unexpected credence by the bGH situation and related controversies, want the land-grant system to restrain the development of (actual or potentially) socially and environmentally destructive technology and to put the needs of people above the imperative of efficiency. These groups prefer that efficiency and productivity in the traditional sense be downplayed as criteria for research problem choice and that new criteria—ecological sustainability, the needs of small and family farmers, human health and nutrition—be pursued with equal or greater vigor.[29] (3) State-level groups and institutions, historically the major constituencies of the land-grant system, want the system to continue to give highest priority to the needs of traditional state client groups. In particular, state farmer groups want the SAES's to emphasize locally adapted technologies and to be cautious about diverting land-grant resources to pursuing national agendas. Each corner of the triangle, if it perceives land-grant indifference or hostility, has the potential to severely harm the system.

The Future Impacts of Agricultural Biotechnologies on U.S. Farm Structure: A Tentative Assessment

They [agronomists and other agricultural scientists] have been pleased to accept credit for reducing the cost of crop and animal production while avoiding the responsibility for lower commodity prices.

—V. W. RUTTAN
"Agricultural Scientists as Reluctant Revolutionaries"

Long years ago the National Grange took exception to the slogan so widely populated in that period about "making two blades of grass grow where only one grew before," and vigorously warned against the danger of overproduction of food.

—CHARLES M. GARDNER
The Grange: Friend of the Farmer (1949)

In world history there have been four "agricultural revolutions"—that is, relatively widespread and profound socioeconomic and technical transformations of agriculture that resulted in major productivity increases.[30] In each, there was a major shift in agricultural production techniques (rather than mere evolutionary changes in then-current production systems). We are currently at the end of the most recent revolution, in which by historical standards there has been a brief but very dramatic increase in global productivity. Will the biotechnology era be a fifth agricultural revolution, following so closely on the heels of the one characterized primarily by plant breeding and petrochemical inputs?

On one hand, this question is a scientific and technical one—a question that I have no particular expertise for answering. Yet despite my admitted lack of expertise on the biological potentials of biotechnology to enable major, revolutionary productivity increases, evidence of a social scientific nature suggests that we are unlikely to have before us the fifth world agricultural revolution.[31] I make this observation for several reasons. One is that agricultural biotechnology R&D appears primarily to involve a set of new products that follow—in evolutionary, not revolutionary, fashion—from the petrochemical/green revolution. That is, the goals that biotechnology is being used to pursue primarily involve either patching up the problems that have been caused by previous agricultural technologies (e.g., salinization, pest resistance, the expense of nitrogen fertilizer, and biocides) or obviating diminishing returns and productivity plateaus that have become manifest in current green-revolution technologies (e.g., herbicide tolerance to make weed control more efficient).[32] A second reason for doubting the revolutionary character of biotechnology is that the track record of the immediately preceding agricultural revolution—a compound annual growth rate of world agricultural output from 1950–1985 of 2.43 percent, representing roughly a 150 percent increase in output during this time[33]—will be difficult to sustain, let alone exceed, over the next thirty-five years.

Much of the received knowledge about the socioeconomic and productivity impacts of biotechnologies has come from a fairly circumscribed area—that of the bGH controversy. The narrowness of this base of evidence is of considerable significance. It is likely to be the case that bGH and the animal growth hormones in general will prove to be biotechnologies exceptional in the immediacy and extent of their productivity impacts.[34] In particular, the backbone of the world agricultural system—cereal grain and other basic food crops such as the roots, tubers, pulses, grain legumes, and so on—is likely to be affected by biotechnology more slowly and to a far smaller degree than either livestock or high-value horticultural commodities, for both technical and socioeconomic reasons.[35] For the cereal grains in the advanced industrial countries, the productivity impact of biotechnologies will probably be little more, and could perhaps be less, than sustaining the post–World War II pattern of 2+ percent annual compound rates of increase in yields and output.

The debate over the prospective impacts of bGH on farm structure has crystallized along lines that seemingly could not be more polarized.[36] On one hand, the critics of bGH see that this new technology will hasten the demise of the family farm—a process that, in their view, has already gone too far. On the other hand, proponents of bGH suggest that family farmers will have no choice but to cast their lot with productivity-increasing technological change, even if many of these farmers will not survive the process. In other words, the survival of the family farm as an institution is seen to depend on increasing its efficiency and competitiveness through adopting new technologies.

There are two ways of approaching this debate over the impacts of bGH on the structure of the dairy production sector (and, by inference, the impacts of biotechnologies as a whole), bearing in mind, of course, that bGH may prove to be an exceptional biotechnology. One tack has been to assess critically, based on historical evidence about technological change in dairying, whether the dramatic productivity increases anticipated by both defenders and critics of bGH are likely to be manifest.[37] There is now an accumulation of evidence, both social scientific and technical, that these prospective productivity, aggregate output, and structural impacts of bGH have indeed been exaggerated.

Second, even if we make generously "optimistic" assumptions about the productivity and aggregate output impacts of bGH, as both the critics and defenders of the technology have tended to do, there are reasons for seeing that the ultimate socioeconomic impacts of the technology may not be nearly so dramatic as often supposed.[38] In particular, I will focus on the implications of bGH for the family farm, which tends to be the bottom line of most discussions of the structural impacts of new technologies such as bGH. Indeed, the two rival positions in the debate, while seemingly in diametrical opposition to each other, reflect astonishingly well the contradictory reality of family farming in U.S. and Western economic history: The very precariousness of family farming tends to ensure its survival. Direct production in agriculture in the United States and elsewhere generally involves a predominance of family-proprietor units. This is in large part due to the fact that for most agricultural commodities direct production is insufficiently profitable to be of interest to large corporations.[39] Also, in contemporary American agriculture many medium-size and especially small (or "subcommercial") family farmers are able to stay in business because of the ability of one or more family members to earn off-farm income. But, in the main, the absence of predictably high returns from farming combined with the modest level of scale economies that can be realized from horizontal and vertical integration have enabled farming to remain relegated to family-type or family-proprietor operations. There has thus been a dual reality of structural change in U.S. agriculture since World War II. There are now about 70 percent fewer farms than there were in the early 1940s—a rather breathtaking decline. Average farm size has more than tripled. The 1982 census of agriculture revealed that about 1.2 percent of farms, those with gross annual farm sales of $500,000 or more, accounted for nearly one-third of gross farm sales. According to OTA, by the year 2000, 50,000 large farms, about 4 percent of the total, will account for 75 percent of U.S. gross farm sales.[40] But this rapid pace of concentration has not led to a widespread corporate agriculture. Even the majority of the elite 1.2 percent of farms in the 1982 Census—a mere 27,800 farms out of a total of 2.2 million—are family proprietorships. To be sure, they do not resemble our image of the traditional American family farm. In particular, most of these very large farms rely primarily on hired labor. Nonetheless, more than 60 percent of the total labor used on U.S. farms is still provided by farm family members.

Thus, each of the rival images of the impact of bGH on the structure of dairy production and on the family farm will prove to contain a kernel of truth. Bovine growth hormone will, in all likelihood, lead to substantial dislocation among and loss of family dairy farms, even if this process occurs more slowly than indicated in the widely cited ex-ante assessments of bGH.[41] At the same time, for the family and larger-than-family dairy farmers who survive bGH-engendered increases in aggregate output and product price declines, adoption of bGH will no doubt help in meeting competition from producers of other beverages and foods. It seems unlikely, however, that a sector like dairying plagued by overproduction and low prices will be an attractive investment outlet for large-scale nonfarm investors. The survivors of bGH are likely to be just that: survivors, rather than industrial-farmer beneficiaries of a sectoral shakeout.

In my view, biotechnologies, including but not limited to bGH, will in all likelihood deepen the now longstanding pattern of structural change in American agriculture. This will be the case for three reasons. First, the new biotechnologies will, on the whole, be relatively capital intensive—that is, they will be made available to farmers as relatively expensive proprietary inputs, such that these inputs will be most readily afforded by larger operators.[42] Second, biotechnologies will tend to be management intensive. Farmers (generally those with larger farms) with specialized operations, superior management ability, and access to information and data processing will be able to utilize the new biotechnologies most effectively.[43] Third, the new biotechnologies will tend to increase output—though probably no more and perhaps less rapidly than during the four decades after World

War II — and will, like most earlier technologies, tend to result in downward pressure on commodity prices.[44] Biotechnologies, therefore, will probably give us more of the same: increased concentration of agricultural production in the hands of very large family proprietorships (and some corporate-industrial farms); a disappearing middle of medium-size, full-time family farms; and relative stability in the numbers of very small, subcommercial, part-time farms. Large corporations will probably continue to be disinterested in direct production of most agricultural commodities. Put most bluntly, the way to make money in agriculture will continue to be selling inputs to and merchandising the outputs of farmers, rather than in farming itself.

Biotechnologies will probably neither destroy nor save the family farm. The truth will likely lie somewhere in between — in particular, in that the new biotechnologies will change the structure of family farming. Determining just how this will be the case will require comprehensive, rigorous socioeconomic impact assessment, both ex ante and ex post.

This final section of the chapter, in which I have argued against notions that the productivity and socioeconomic impacts of biotechnology will be of an exceptional or revolutionary character, should not be misinterpreted. I do not wish to convey the impression that these impacts will be trivial; I have suggested that they are likely to exhibit continuity with those of the green revolution (as defined broadly above to encompass the most recent agricultural revolution to affect the industrial world). The dislocations as well as the output increases of the most recent agricultural revolution were far from trivial! I do not wish to leave the impression that, because the impacts of biotechnology may be comparable to those of the green revolution, they deserve the same (virtually no) socioeconomic, environmental, and bioethical scrutiny that the previous impacts garnered both before they occurred and during the time they were most intense (circa 1940–1970). Finally, I am not making an implicit argument that the overall impacts of biotechnology are already so set in stone that it makes little sense to affect their course by influencing public (or perhaps private) agricultural research policy.[45] I will return to this issue shortly.

Concluding Comments: Implications for a Bioethics of Agricultural Biotechnology

I would like to conclude by venturing into some unfamiliar but vitally important territory: the implications of the foregoing for a bioethics of agricultural biotechnology. Though I am not trained or experienced in theoretical or applied ethics, I feel it is nonetheless possible to set forth some observations that can become grist for the mills of persons with the appropriate expertise.

First, I am of the feeling that the starting point of a bioethics of biotechnology must be where biotechnology is having its most immediate impacts: public agricultural research institutions.

Second, this will not be an entirely unique bioethics. The major bioethical issues in biotechnology—who benefits and loses, its implications for control of the production and distribution systems, what groups should be the major clientele of the land-grant system, and so on—will in substantial measure be the generic ones that pertain to all technologies.

Third, there are nonetheless some distinctive bioethical issues relating to biotechnology. The distinctive bioethics of biotechnology should begin with the first- and second-order implications of the privatization of publicly funded, basic biological knowledge and of associated phenomena such as the ideology and reality of international competitiveness.

Fourth, a sophisticated bioethics must steer between two major barriers to understanding. On one hand, many critics of the public agricultural research system have portrayed research administrators and scientists as witting or unwitting accomplices to evils perpetrated on family farmers, the environment, consumers, and so on. On the other hand, defenders of the historic role of the public research system see SAES personnel as being consummate, dedicated public servants—virtual heroes in pursuit of the public good. There is a need to develop a more fine-grained view of the structure of SAES's to transcend these unilluminating stereotypes.

One notion that may be helpful in deepening this more fine-grained analysis is that of the triangular squeeze briefly sketched out above. Public research administrators, according to this analysis, are alternatively powerful figures—people whose decisions have a major impact on the world—and powerless functionaries—persons who must juggle a set of pressures and forces over which they have little control. Also, many analysts of the public agricultural research system can benefit from better understanding the dynamics of contemporary research universities, scientific organizations, and scientific careers.[46] For example, U.S. research universities, perhaps more so than ever before, are implicated in national—and sometimes international—competition for prestige, recognition, and resources. Universities are thus far less autonomous than we often assume them to be. Further, within land-grant universities colleges of agriculture are, virtually without exception, minority university components and no longer play the major role in dictating overall university policy. College of agriculture administrators, for example, frequently must fight protracted battles with central administrations over whether faculty who do applied research or extension work deserve tenure in modern research universities. Finally, there is often a tendency to overlook the role of scientific organizations as loci of power and influence. A scientist's career is, for example, heavily influenced by the structure of scientific organizations and by the types of research that are valued by senior peers. It is no exaggeration to say that a scientist's prestige, tenure, promotion, and salary are typically determined more by his or her scientific organization—especially by the judgments of senior peer reviewers of the organization's journals—than by his or her land-grant employer. Yet scarcely a bioethicist has sought to examine scientific organizations as sources and arbiters of professional values. In sum, I feel that social science knowledge on the structure and dynamics of R&D institutions, as well as on the impact of technological change on farmers and the larger society, must be a major building block in a mature bioethics of biotechnology and agricultural research.

This said, why, in addition to scholarly curiosity, should we strive to develop a bioethical perspective? In other words, we must also ask: Bioethics for what? One possible answer, which is often raised in red-herring fashion by those fearful of bioethics initiatives, is that philosophers, sociologists, anthropologists, and political scientists merely want to control research policy-making. In my view, this is a misinterpretation of the intentions of bioethicists. Nonetheless, for the record, bioethicists imposing their values and goals, whatever they might be, on researchers and research administrators is no better than the status quo in which researchers have pursued their own interpretations of land-grant university goals innocent of knowledge on the likely impacts of these research priorities on farmers and other social groups.

A second possible purpose of a bioethical perspective is to develop a set of guidelines for scientists about what is or is not ethical conduct, much as biomedical and legal ethicists now do. This purpose, however, while it may be appropriate in some instances, should be seen as a small part of a bioethics of biotechnology and agricultural research. There are, to be sure, growing possibilities of conflict of interest problems on the part of scientists as the privatization of biotechnology proceeds. But I would argue that the key issues in biotechnology and agricultural research are *institutional* rather than individual issues; these issues have little to do with temptations by scientists to engage in unethical conduct, and hence these matters cannot be dealt with through developing codes of "ethical" conduct. There is, in fact, probably no group of scientists in the United States more committed to public service and to the public good than public agricultural researchers. Much the same can be said of the intentions of land-grant universities. Yet as the quote by Ruttan at the outset of the chapter suggests, there has been a systematic tendency for agricultural scientists and institutions to ignore all but the positive impacts of their work (and, I would add, a reluctance to grapple with alternative conceptions of public service and the public good). It is these value issues — not some abstract standard of ethical conduct — which should be the focus of a mature bioethics of agricultural research.

I would argue that the purpose of a bioethics of biotechnology and agricultural research, other than to produce sound scholarship in its own right, is to bring a broader range of information and a broader set of values and goals to bear on research decision making and thereby to improve the capacity of research institutions to serve human needs. The goal of bioethical inquiry should ultimately be mutual understanding, mutual respect, and sustained dialogue between bioethicists and agricultural-biological researchers and research administrators. Both groups, however, must be realistic about the promise and pitfalls of a prospective new era of bioethically informed research decision making. Bioethicists must recognize that establishing the needed dialogue will be a long-term process and that the structural realities of modern agricultural research in a milieu of international technological competition will often militate against their goals being readily implemented. Agricultural scientists and research administrators must understand that bioethicists, rather than representing a unified conspiracy to reduce their decision-making prerogatives, are a diverse lot. The research establishment must also recognize that bioethical dialogue cannot help but involve (mutual) criticism.

In this chapter I have argued that international technological competition, the privatization of public agricultural-biological research, and the triangular squeeze on land-grant institutions represent the master forces affecting land-grant institutions. These forces have led to and portend further change in the division of labor between public and private agricultural research.[47] I would suggest that the role of public research institutions in this division of labor should be the overriding issue for debate and dialogue among bioethicists and agricultural researchers. What should be the distinctly public role of public agricultural research? Should it primarily be to do research that is not privately profitable for industry to undertake and to transfer the results to private sector firms? Should it primarily be to develop technical systems that will not be in the interest of industry to pursue (for example, the use of biotechnology and more traditional methodologies to develop production systems that reduce the use of purchased chemical inputs)? Should it be to respond to farmers' immediate technical needs or to what might be their needs one or two decades in the future? What should be the mix? These division-of-labor issues, in my view, will be the major ones to be dealt with by bioethicists and agricultural researchers as the age of biotechnology comes to fruition.

Left unexamined in this chapter have been frameworks for the assessment of value issues in agricultural research, a matter that is well beyond my span of expertise.[48] I hope, however, that this chapter, as a social-science contribution, can shed some light on these value issues and make possible a more relevant bioethics of biotechnology and agricultural research.

Notes

1. V. W. Ruttan, "Agricultural Scientists as Reluctant Revolutionaries," *Choices: The Magazine of Food, Farm, and Resource Issues* (Third Quarter 1987):3.

2. B. E. Day, "The Morality of Agronomy," in *Agronomy in Today's Society*, ed. J. W. Pendleton (Madison, Wis.: American Society of Agronomy), 19–27.

3. J. Hightower, *Hard Tomatoes, Hard Times* (Cambridge, Mass.: Schenckman, 1973).

4. This is not to suggest that technological factors were unimportant in the world economy prior to the late 1970s; in particular, technological superiority and inferiority have long been major factors in shaping the fate of countries in the world economic system. As the world economy moved into a prolonged phase of stagnation after 1973–74, the initial response of industrial-country (particularly U.S.) firms was to bolster their sagging profit levels by seeking out cheaper sources of labor, both in low-wage regions of their countries and, in particular, in the Third World. This phase ultimately yielded to an emphasis on technological innovation as a competitive strategy once access to cheap labor became relatively generalized, thereby reducing its advantages as a firm-level competitive strategy. Developed country states accordingly began to shift their economic policies so as to encourage linkages between "basic" or "fundamental" research and private research and development (R&D) as part of national strategies to adapt to the decline of the traditional "sunset" industries.

5. National Research Council, *Report of the Committee on Research Advisory to the U.S. Department of Agriculture* (Washington, D.C.: National Research Council, National Academy of Sciences, 1972).

6. Rockefeller Foundation, *Science for Agriculture* (New York: Rockefeller Foundation, 1982).

7. There is a certain irony in the fact that the land-grant system has suffered from adverse publicity from bGH; the key technical breakthrough that has made bGH a commercial possibility was recombinant DNA work initially done by a biotechnology start-up firm in California. Animal scientists had known for decades that bGH produced by the pituitary gland was integral to milk production, and there was a considerable basic research literature on the topic prior to the biotechnology era. Land-grant research on bGH since 1980 has largely been of a product-testing nature—hardly basic biological research of the sort advocated by basic science critics of the land-grant system such as the authors of the Winrock Report. It seems to have been the case, however, that major land-grant universities involved in the bGH controversy have been so intent to demonstrate to state legislatures that biotechnology can "deliver the goods" that they have placed themselves in the difficult position of defending a technology they played a minor role in developing and one that promises to result in substantial dislocation in the dairy industry. (F. H. Buttel, "Agricultural Research and Farm Structural Change: Bovine Growth Hormone and Beyond," *Agriculture and Human Values* 4 [Fall 1986]:88–98.)

8. J. Doyle, *Altered Harvest* (New York: Viking, 1985).

9. See Hadwiger (1982), Busch and Lacy (1983), Kenney (1986), Kloppenburg (1988), Buttel and Busch (1988), and Hansen et al. (1986) for a further discussion of major forces that have shaped the structure of the land-grant system over the past fifteen years. D. F. Hadwiger, *The Politics of Agricultural Research* (Lincoln: University of Nebraska Press, 1982); L. Busch and W. B. Lacy, *Science, Agriculture, and the Politics of Research* (Boulder, Colo.: Westview Press, 1983); M. Kenney, *Biotechnology: The University-Industrial Complex* (New Haven: Yale University Press, 1986); J. Kloppenburg, Jr., *The Political Economy of Plant Breeding and Biotechnology* (New York: Cambridge University Press, 1989); F. H. Buttel and L. Busch, "The Public Agricultural Research System at the Crossroads," *Agricultural History,* forthcoming; M. Hansen, L. Busch, J. Burkhardt, W. B. Lacy, and L. R. Lacy, "Plant Breeding and Biotechnology," *BioScience* 36 (January 1986):29–39.

10. *Biotechnology* can be defined as the commercial or commercially relevant application of a set of subcellular biological research techniques such as recombinant DNA, immobilized enzymes, cell and tissue culture, protoplast fusion, and so on.

11. As will be discussed briefly below, however, there are reasons to believe that the prospective productivity and socioeconomic impacts of bGH have been exaggerated. Nonetheless, the productivity impacts of bGH and the other animal growth hormones will be substantial and will probably prove to be larger and more immediate than biotechnology advances in other commodity sectors such as the cereal grains, pulses, grain legumes, and roots and tubers. See R. Barker, "Impact of Prospective New Technologies on Crop Productivity: Implications for Domestic and World Agriculture," paper presented at the Conference on Technology and Agricultural Policy, National Academy of Sciences, Washington, D.C., December 1986; F. H. Buttel, "Agricultural Research and Farm Structural Change: Bovine Growth Hormone and Beyond," *Agriculture and Human Values* 4 (Fall 1986):88–98; F. H. Buttel and C. C. Geisler, "The Social Impacts of Bovine Somatotropin: Emerging Issues," paper presented at the National Invitational Bovine Somatotropin Workshop, St. Louis, Mo., September 1987.

12. This argument, to be sure, is made most clearly by critics of biotechnology, such as Jeremy Rifkin, who see the technology as unwarranted, dangerous tampering with the delicate balance of nature. This notion, however, is equally important to corporate investors in biotechnology who must claim in patent applications that the organisms they have "engineered" are new and distinctive in order to receive industrial patent protection. Corporate biotechnologists, however, must make an entirely different—inherently contradictory—claim before regulatory agencies: that biotechnology-derived organisms are no different than those developed through "conventional" techniques, and, accordingly, do not pose distinctive threats to environmental quality or human and animal health.

13. A personal experience along this line might be instructive. Roughly a year ago I prepared an overview paper on the impacts of biotechnology for a USDA/Economic Research Service periodical. I began the substantive part of the paper with a section on the impacts of biotechnology on public research and extension institutions. The editor, though an open-minded and helpful person, felt that material on research and extension institutions would not be of much interest to the readership; the readership, it was said, would not be especially interested in how biotechnology would affect farm structure and the rural economy. The editor thus suggested that the material on research and extension institutions should be dropped entirely for lack of interest. I was able to convince the person otherwise, but this is a useful example of how bioethical discussions often tend to ignore the areas in which biotechnology thus far has had its major impacts (and will likely continue to have its major impacts into the future).

14. M. Kenney and F. H. Buttel, "Biotechnology: Prospects and Dilemmas for Third World Development," *Development and Change* 16 (1985):61–91.

15. M. Kenney, *Biotechnology: The University-Industrial Complex* (New Haven: Yale University Press, 1986).

16. A. J. M. Roobeek, "The Crisis in Fordism and the Rise of a New Technological Paradigm," *Futures* 19 (April 1987):129–54.

17. M. Kenney (*Biotechnology: The University-Industrial Complex* [New Haven, Yale University Press, 1986]) has been among the few to note the importance of the very early commercialization of biotechnology.

18. This premature commercialization was rooted in the emerging crisis of the traditional industries during the 1970s and in a recognition that technical innovation, rather than mere access to cheap labor, would be necessary for firms to survive the competition engendered during a long economic downturn. The initial investments were made by small venture capital or start-up firms beginning in the late 1970s and early 1980s. Large multinational chemical, pharmaceutical, and petroleum companies soon followed suit—many apparently having been motivated not only by the profit potentials of biotechnology but also out of a defensive reaction to the investments made by small biotechnology companies. (See M. Kenney, *Biotechnology: The University-Industrial Complex.* [New Haven: Yale University Press, 1986].)

19. F. H. Buttel, M. Kenney, J. Kloppenburg, Jr., and D. Smith, "Industry-University Relationships and the Land-grant System," *Agricultural Administration* 23 (1986):147–81; M. Kenney, *Biotechnology: The University-Industrial Complex* (New Haven: Yale University Press, 1986); F. H. Buttel, M. Kenney, J. Kloppenburg, Jr., J. T. Cowan, and D. Smith, "Industry/Land-grant University Relationships in Transition," in *The Agricultural Scientific Enterprise,* ed. L. Busch and W. B. Lacy. (Boulder, Colo.: Westview Press, 1986), 296–312. As my colleagues and I have pointed out elsewhere, most observers of industry/land-grant university relationships have tended to exaggerate the importance of industrial sponsorship of research and biotechnology institutes and have downplayed the importance of long-term consulting relationships with large biotechnology firms and equity ownership relationships with biotechnology start-ups. These two observations are closely interrelated. Because large chemical-pharmaceutical firms can gain access to publicly funded, university-based knowledge through relatively inexpensive consultancy relationships and through investments in start-up companies (which have prominent university researchers as consultants, members of scientific advisory boards, and equity owners), there has been little incentive to invest large sums in sponsoring university research. Accordingly, most land-grant biotechnology programs have found that they have been able to attract far less industrial support than they anticipated when the programs were established.

20. J. Kloppenburg, Jr., *The Political Economy of Plant Breeding and Biotechnology* (New York: Cambridge University Press, 1989).

21. That is, having a biotechnology program is considered to be prestigious, to be important in retaining prestigious scientists, and to enable universities to generate grant and contract revenues that will help to increase institutional prestige.

22. V. W. Ruttan, "Agricultural Scientists as Reluctant Revolutionaries," *Choices: The Magazine of Food, Farm, and Resource Issues* (Third Quarter 1987):3.

23. W. W. Cochrane, *The Development of American Agriculture* (Minneapolis: University of Minnesota Press, 1979).

24. As I have noted in other papers with colleagues (J. Kloppenburg, Jr., and F. H. Buttel, "Two Blades of Grass: The Contradictions of Agricultural Research as State Intervention," *Research in Political Sociology* 3 [1987]:111–35; F. H. Buttel and L. Busch, "The Public Agricultural Research System at the Crossroads," *Agricultural History* 62[1988]:303–24), the base of support for public agricultural research at the federal level has been meager, mainly because its actual or presumed beneficiaries—farmers, consumers, and rural residents—for

various reasons have little or no incentive to support a national program of research. The only consistent supporters of public agricultural research at the federal level have been agribusiness firms and researchers and research administrators themselves.

25. Also, SAES biotechnology programs have generally been established, in part, by phasing out applied, locally adapted research such as traditional plant breeding (M. Hansen, L. Busch, J. Burkhardt, W. B. Lacy, and L. R. Lacy, "Plant Breeding and Biotechnology," *BioScience* 36 [January 1986]:29–39; J. Kloppenburg, Jr., *The Political Economy of Plant Breeding and Biotechnology* [New York: Cambridge University Press, 1989]).

26. The shift from locally adapted to generic technology would accordingly reflect, de facto if not de jure, a change in the principal clientele of the SAES system: from farmers to agricultural input firms. Since the bulk of these firms are out-of-state if not foreign enterprises, there could be a reinforcement of the perception that citizens of the state are not the primary beneficiaries of SAES research.

27. J. Hightower, *Hard Tomatoes, Hard Times* (Cambridge, Mass.: Schenckman, 1973).

28. D. F. Hadwiger, *The Politics of Agricultural Research* (Lincoln: University of Nebraska Press, 1982).

29. The traditional conceptualization of productivity has primarily revolved around yields and aggregate output, while the traditional notion of efficiency has placed particular emphasis on labor productivity. It can be argued, however, that the notions of productivity and efficiency can be made more meaningful if it is recognized that efficiency can be increased by reducing input usage just as well as by increasing yields and output and that efficiencies in the use of capital and natural resource inputs may be as important or more important than labor efficiency.

30. These include (1) neolithic settlement (3500–700 B.C.), (2) the medieval agricultural revolution (A.D. 600–1200), (3) the eighteenth century agricultural revolution in northwest Europe, and (4) the "green revolutions" of the advanced industrial countries after World War II and in the Third World after 1960. For a useful discussion of previous agricultural revolutions in the context of the debate over the green revolution, see M. Lipton, with R. Longhurst, *Modern Varieties, International Agricultural Research, and the Poor* (Washington, D.C.: World Bank, 1985).

31. This is not to say that biotechnology's impacts will be small, particularly in the pharmaceutical and chemical sectors. But to the degree that biotechnology will be revolutionary, it will probably not be for agricultural reasons *tout court*. For example, it is possible that the most revolutionary changes in agriculture that will result from biotechnology will be very rapid product substitutions and the shift of food and fiber production from the farm to the factory (e.g., substituting "vegetable casein" for milk casein in producing cheese and substituting single-cell proteins for soybean-derived protein) (Kenney, *Biotechnology: The University-Industrial Complex* [New Haven: Yale University Press, 1986]). Biotechnology may well very substantially reduce the role of land as a factor of agricultural production and deepen the role of the "agro-industrial chain" (D. Goodman, B. Sorj, and J. Wilkinson, *From Farming to Biotechnology* [Oxford: Basil Blackwell, 1987]).

32. Here the notion of *green revolution* is used broadly to depict the recent agricultural revolution based on petrochemical inputs and plant breeding that occurred in the industrial and developing worlds since World War II.

33. U.S. Department of Agriculture (USDA), *World Indices of Agricultural and Food Production, 1976–85* (Statistical Bulletin No. 744, Washington, D.C.: Economic Research Service, USDA, 1986).

34. R. Barker, "Impact of Prospective New Technologies on Crop Productivity: Implications for Domestic and World Agriculture," paper presented at the Conference on Technology

and Agricultural Policy, National Academy of Sciences, Washington, D.C., December 1986.

35. For example, improved transgenic cereal grain varieties are far more technically difficult to develop than are the animal growth hormone technologies (which basically involve an extension of a massive amount of human growth hormone research, and which are based on genetic manipulation of single-celled organisms). Socioeconomically, in a world characterized by already-substantial overproduction of the cereal grains, there are likely to be market barriers to further technical innovation in these crops. Also, many of the basic noncereal grain crops such as the roots and tubers are grown primarily for subsistence in developing countries so that technical changes involving purchased inputs are not likely to be widely adopted.

36. F. H. Buttel and C. C. Geisler, "The Social Impacts of Bovine Somatotropin: Emerging Issues," paper presented at the National Invitational Bovine Somatotropin Workshop, St. Louis, Mo., September 1987.

37. F. H. Buttel, "Agricultural Research and Farm Structural Change: Bovine Growth Hormone and Beyond," *Agriculture and Human Values* 4 (Fall 1986):88–98; F. H. Buttel and C. C. Geisler, "The Social Impacts of Bovine Somatotropin: Emerging Issues," paper presented at the National Invitational Bovine Somatotropin Workshop, St. Louis, Mo., September 1987.

38. By this I mean, for example, projections of increases in milk production per cow of roughly 25 percent (as opposed to 8–12 percent which many observers now consider more likely). I have placed "optimistic" in quotation marks to note that defenders and critics of the technology will have far different views of what would constitute an optimistic scenario.

39. F. H. Buttel, "Social Relations and the Growth of Modern Agriculture," in *The Ecology of Agriculture*, ed. C. R. Carroll et al. (New York: Macmillan, forthcoming).

40. OTA has projected that total farm numbers are likely to decline by one million, from roughly 2.2 to 1.2 million, by the year 2000. This projection is doubtful, and hence the denominator of this percent figure may well be larger, and the percentage smaller, than the estimate I have given above.

41. Office of Technology Assessment (OTA), *Technology, Public Policy, and the Changing Structure of American Agriculture* (Washington, D.C.: OTA, 1986); R. J. Kalter, "The New Biotech Agriculture: Unforeseen Economic Consequences," *Issues in Science and Technology* (Fall 1985):125–33.

42. I suspect that the OTA (*Technology, Public Policy, and the Changing Structure of American Agriculture* [Washington, D.C.: OTA, 1986]) is correct in projecting that the capital intensity of agricultural biotechnologies will, on the whole, be only slightly greater than that of current petrochemical-based technologies.

43. Office of Technology Assessment (OTA), *Technology, Public Policy, and the Changing Structure of American Agriculture* (Washington, D.C.: OTA, 1986); R. J. Kalter, "The New Biotech Agriculture: Unforeseen Economic Consequences," *Issues in Science and Technology* (Fall 1985):125–33.

44. J. J. Molnar and H. Kinnucan, "Biotechnology and the Small Farm: Implications of an Emerging Trend," in *Strategy for the Survival of Small Farmers,* T. T. Williams (Tuskegee Institute, Ala.: Human Resources Development Center, Tuskegee Institute, 1985).

45. At the same time, I would suggest that redirection of the priorities of public agricultural research institutions, without parallel changes in public policy, is a weak instrument for achieving social change. This is particularly the case now, as private sector agricultural research spending is as much as two-fold that of the public sector (V. W. Ruttan, *Agricultural Research Policy* [Minneapolis: University of Minnesota Press, 1982]). Accordingly, the public agricultural research system, dwarfed as it is by the private research system, is no longer preeminent in shaping the course of technological change in American agriculture.

46. F. H. Buttel, "The Land-grant System: A Sociological Perspective of Value Conflicts and Ethical Issues," *Agriculture and Human Values* 2 (Spring 1985):78–95.

47. F. H. Buttel, M. Kenney, J. Kloppenburg, Jr., J. T. Cowan, and D. Smith, "Industry/ Land-grant University Relationships in Transition," in *The Agricultural Scientific Enterprise,* ed. L. Busch and W. B. Lacy (Boulder, Colo.: Westview Press, 1986), 296–312.

48. K. A. Dahlberg, "Introduction: Changing Contexts and Goals and the Need for New Evaluative Approaches," in *New Directions for Agriculture and Agricultural Research,* ed. K. A. Dahlberg (Totowa, N.J.: Rowman and Allanheld, 1986), 1–27.

18

Democracy in Technological Research: Planning a Future for Family Farming

CHUCK HASSEBROOK

Technological research is a powerful form of socioeconomic planning. Decisions on what research is undertaken have great influence over what technologies are ultimately developed and become cost effective. These technologies in turn profoundly shape our economic and social structures. As wrote University of Minnesota Regents Professor Vernon Ruttan, "Agronomists and other agricultural scientists, along with engineers and health scientists, have been the true revolutionaries of the 20th century. But they are reluctant revolutionaries! They have wanted to revolutionize technology but have preferred to neglect the revolutionary impact of technology on society."[1]

As we recognize technological research as a form of social planning with potentially revolutionary results, we must also recognize that in a democratic society, decisions about what research to do are too important to be made by scientists alone and too important to be made without consideration of basic questions about what kind of a world we want to live in. It is not enough to ask which research—increasing production or reducing cost (as measured by the market)—holds the most potential for solving a particular technical problem. We must ask which research will do those things at the same time that it helps us create the kind of economic opportunities and social structure desired by the citizenry. We must ask how a particular line of research will affect the structure of agriculture, including how many farms we will have and how they will be owned and controlled.

My values on this matter and, I believe, those of most rural people in the Upper Midwest suggest that we ought to strive to maximize opportunities for self-employment in agriculture and seek to the greatest extent possi-

251

ble to build a relatively egalitarian social structure. In other words, we ought to strive for as many owner-operated farms as possible and avoid creation of an industrial-type class structure divided along lines of owners, managers, and labor. I recognize that many people who see themselves as practical and scientific scorn such consideration as subjective and value-laden. It is value-laden but no less so than the assumption that reducing the cost of producing a bushel of corn by five cents is more important than an egalitarian social structure. The importance of such a social structure is something that can be measured as well as felt. Dean MacCannell, University of California at Davis, has noted that

> as farm size and absentee ownership increase, social conditions in the local community deteriorate. We have found depressed median family incomes, high levels of poverty, low education levels, social and economic inequality between ethnic groups, etc., associated with land and capital concentration in agriculture. . . . Communities that are surrounded by farms that are larger than can be operated by a family unit have a bi-modal income distribution with a few wealthy elites, a majority of poor laborers, and virtually no middle class. The absence of a middle class at the community level has a serious negative effect on both the quality and quantity of social and commercial services, public education, local governments, etc.[2]

Many see a fundamental inconsistency between technological progress and a stable family farm population. They suggest that removing people from agriculture is synonymous with efficiency and that developing only technologies that do not undermine family farms would end progress. That view, I believe, underestimates technology and sees in it the potential to do only that which has been done in the past. It also ignores the fact that the efficiencies gained by removing people from agriculture have been trade-offs. We have replaced labor with capital and family farmers with employees of giant agribusiness corporations. Technology might just as well make agriculture more efficient by replacing capital and purchased inputs with the labor and management of more family farmers. For example, a technology which makes it possible for a family farmer to reduce capital costs by two dollars for each additional dollar worth of his/her time spent on labor and management would both increase the efficiency of U.S. agriculture as measured by the market and increase opportunity for people in agriculture.

The assumption that replacing people with capital, including natural resource base inputs, is implicitly efficient is particularly flawed in light of the situation in the world economy today. We face massive crop surpluses, natural resource depletion, and severe underemployment and unemployment. It makes little sense to focus scarce research funds on finding ways for farmers to use more scarce resources to produce more corn and soybeans on every acre with less labor when we already have more corn and soybeans than we can use and more people than can be employed. It would appear that the more efficient approach would be to find ways for family farmers to use their labor and management skills to reduce the use of scarce natural resources in meeting food needs.

Assessing the Impacts of Agricultural Research

If we are to focus the public investment in agricultural research on technologies that advance social and structural goals, we must develop the capacity to anticipate the impacts of the technologies likely to result from a particular line of research. This endeavor applies primarily to applied research and not basic research, by definition. As a starting point, I would propose that the following questions be asked of applied agricultural research projects to help in predicting their impacts on the structure of agriculture:[3]

What are the forces causing the research to be undertaken? — What goals do the parties pursuing the technology have and how do they plan to put it in use?

How does the technology affect the balance between the use of labor and management relative to capital? — In areas where labor and management are provided by owner-operators, technologies which replace labor and management with capital reduce the number of farmers. By contrast, technologies which increase returns to labor and management relative to capital create the need for more people in agriculture. If the technology consumes a high level of limited resources per economic opportunity, it can create opportunity for only a few people.

Will the technology be more conducive to owner operated farms or to industrially structured farms where the functions of ownership, management, and labor are divided among different people? — Will significant skills, responsibilities, and judgment be required of the persons providing the labor input? Or, will skill, responsibility, and judgment be by unskilled, low-paid, and unmotivated employees? Will the management tasks be so complex as to require a trained specialist or will the necessary skills be widely attainable by large numbers of family farmers? Will the technology ultimately result in persons engaged in labor having more or less control over their work?

Will the technology create barriers to entry due to capital, acreage or other requirements that cannot be met by most potential beginning farmers?

Will the risk of adopting the technology discourage widespread early adoption and jeopardize those who adopt the technology, in the event of an economic downturn? — Will it be an all-or-nothing technology, or will a farmer be able to experiment with the technology? If problems arise with the technology, will the farmer be able to fix those problems, or will the farmer be dependent on outside help to correct problems? Will the technology foster diversification or specialization?

How will the technology affect competitive position among different types and sizes of producers? — Will it create economies or diseconomies of size in the production, processing, or input sectors? How will competition between products be affected (i.e., pork versus poultry, on-farm versus off-farm production, United States versus Third World, etc.)?

We must also ask how the technology will affect the natural environment and consumers of food. Will the technology foster genetic diversity or uniformity, monocropping or rotations? Will it contribute to or prevent depletion or contamination of natural resources? What risks of benefits attend its introduction to the environment? How will the technology affect the cost, security, and quality of food? How will it affect diets?

Choices in Emerging Technology

In the pages that follow, I will offer some hypotheses on how these questions would be answered when applied to some of the current thrusts in agricultural research and alternatives.[4]

Biological Control of Pests

Much research is focused on mass production of microorganisms which control insects, fungi, and nematodes. In some cases the biocontrol agents are naturally occurring microorganisms while in other cases they are the result of genetic engineering.

On the positive side, these substances might be environmentally safer than petrochemicals, though that is a matter of some debate. Their manufacture would not deplete nonrenewable petroleum supplies. But from the perspective of farmers, these products would simply replace one purchased product with another to reduce pest constraints, labor, and uncertainty in large-scale monocultures. That has implications for both the environment and farm structure. Since big farms enjoy substantial volume discounts on purchased inputs relative to medium-size farms, purchased inputs create economies of size and an advantage for big farms over family-size farms.[5] Purchased inputs increase the capital cost of farming and thereby create financial barriers to entry. Monocultures reduce labor and management requirements relative to diversified farms and thereby allow the nation's land to be farmed by fewer people. Since the predominant monocultures in this region require more fertilization and result in less soil organic matter than do some alternative rotations, they often result in more nitrate contamination of water and greater soil erosion than is necessary to farm efficiently.

An alternative is to expand farming systems research already underway, to enable farmers to substitute their management, labor, and cultural practices for purchased pesticides, herbicides, and fertilizers. For example, research might focus on developing economically competitive rotations that break insect and weed cycles and replenish nitrogen, and cultural practices that increase the levels of naturally occurring microorganisms which control pests. (University of Nebraska agronomist Chuck Francis has written extensively on management-intensive farming systems.)[6]

The alternative approach would reduce economies of size, barriers to entry, and negative environmental impacts relative to monocultures. It would create a need for more people in agriculture, since the greater number of crops and, in many cases, forages would increase labor and management requirements.

Low-input farming systems require a different set of management skills, skills more conducive to owner-operated farms, than do monocultures. High-input monocultures require excellent financial management and skill in selecting and purchasing the proper combination of inputs. Low-input systems require continuous monitoring and understanding of natural systems as well as responsiveness to what is learned and understood. They require the exercise of a greater degree of hands-on judgment by persons working on the farm. Consequently they are most conducive to owner-operated farms, where the person in the field is a highly motivated manager capable of exercising judgment. These systems would not work so well on industrially structured farms where the manager is in the office and the workers in the field are unskilled and not highly motivated. That is not to say that such systems do not tax the management abilities of many owner operators. Extension programs to help family farmers develop the skills to manage these systems might be needed.

Nitrogen Fixation

Researchers are attempting to increase the nitrogen production of existing legumes by modifying existing rhizobium and engineering new strains of rhizobium which produce more nitrogen than those which currently predominate in most soils. Although more productive strains do currently exist in the soil, they are not very competitive with less productive strains. Researchers are also looking into "free living" nitrogen fixers which occur naturally in the soil free of a host plant. These do not appear to have the potential to fix large amounts of nitrogen, but they could produce some. Finally, some researchers are attempting to develop nitrogen-fixing corn, though many doubt the feasibility of that undertaking.

In general, increased biological nitrogen fixation would reduce pressure on scarce petroleum supplies, reduce the cost advantage gained by big farms through volume purchases of fertilizer, and lower an entry barrier by reducing input costs. Research focused on improving the nitrogen-fixing capacity of forage legumes to be grown in rotation with major crops could lead to enhancement of low-input farming systems relative to high-input monocultures. By contrast, an increase in the nitrogen-fixing capability of soybeans and the creation of that capability in corn could remove a constraint to large-scale monocultures.

Weed Control

The major thrust in weed control is the development of herbicide-resistant crop varieties. Some researchers, however, are exploring other options such as use of cover crops to suppress weed growth (allelopathy) and better intermediate tillage systems which kill weeds while leaving residue on the surface for conservation.

Herbicide resistance is likely to foster a system of fewer farms by reducing the need for labor and management (people) in weed control. It could also increase purchased input costs, which would grant an advantage to volume purchasers over medium-size farms, and lead to the introduction of more toxic substances to the environment. However, I offer a caveat. Crop resistance to effective post emergence herbicides could cause some farmers to reduce herbicide use because it would provide an effective backup to mechanical cultivation and cultural practices. Currently, there is no practical backup system when rain keeps cultivators out of the corn field.

The adoption of allelopathy and soil conserving tillage and weed control systems which don't require herbicides would have beneficial effects on the environment and farm structure, since less purchased inputs/toxics and more labor and management would be required. One new area worth exploring is the development of crop varieties which germinate and rise to a height of six inches more rapidly to allow for more effective mechanical weed control.

New Crops and New Uses for Crops

Much emphasis is being placed on finding new food and industrial uses for existing crops. Researchers are exploring new oil crops and looking at production of pharmaceuticals by easily genetically engineered crops such as tobacco.

The critical question in this regard is whether we focus our research dollars on developing uses for crops that fit our social and environmental goals or simply focus on the crops that are most widely grown today and have the strongest constituencies. In Iowa, for example, should research focus on finding new uses for corn and soybeans so they can continue to be grown on the majority of the planted acres in Iowa, or should research focus on developing new uses for soil-conserving and input-reducing crops to be grown in rotation with corn and soybeans? For example, could we create new uses for oats, clover, and alfalfa which would increase their price and make it profitable to grow them in rotation with corn and soybeans?

One prospect in this regard is to work to develop bacteria capable of making more of the energy in forages available so that forages become more competitive with corn as energy sources for ruminants. Also promising is USDA research to increase the digestibility of forages by treating them with hydrogen peroxide. Researchers claim that the latter allows the same digestible energy production per corn belt acre with native grasses as with corn.[7] However, if that technology is used to remove, treat, and feed residues of grain crops, it could also contribute to increased soil erosion, water contamination, and input use. Another possible alternative would be to develop tree crops for fuel and feed, especially for planting on highly erodible land in the Conservation Reserve Program.

Computers

Improvements in computer software, hardware, and networks offer greater access to data and information from off of the farm as well as the capability to monitor conditions on the farm, such as moisture in grain or soil and temperature, feed intake, and milk production of cows. In some cases, sensors feed into a computer which triggers a reaction such as turning on an irrigation system or feeding a ration mixed specifically for a particular animal. In other cases (e.g., a cow running a fever or refusing to eat), the computer alerts the manager. Researchers are also using computers to develop expert problem-solving systems for agriculture along the lines of computer systems used by mechanics to pinpoint the problem in a car engine.

It would appear that the use of the computer as an information source could improve the competitive position of small farmers by equalizing their access to information, relative to a trained specialist employed on an industrialized farm. Such information could aid management-intensive systems, relative to capital-intensive systems. The use of computers to monitor events and conditions on the farm and automatically respond or give instructions could make up for the inability of unskilled labor to do these things on industrialized farms. It could be a great help in monitoring herd health on farms with employees who are not sensitive to animal health. It would allow one person to manage and control much greater amounts of activity and thereby help large farms avoid management diseconomies of size.

Hog Production

Growth hormones have been the major focus of discussion on emerging technologies for hog production. Researchers are also looking at hormones to speed rebreeding and synchronize breeding and birthing. Biotechnology is allowing the development of new, more effective vaccines with less risk of infecting animals. Researchers are exploring means of strengthening the animals' immune systems and breeding disease resistance into animals. Bioengineered proteins are being used to develop kits to diagnose diseases, which some say could be used by farmers. Researchers are studying the impact of various stresses on growth rate and immunity; they are also studying means to eliminate the suppression of the immune response system which can result from stress. Some work is being done to develop soybeans free of tripson inhibitor, which would allow farmers to feed their own soybeans without processing. Much work is being done on improving embryo transfer techniques, including sexing of embryos.

Much of the disease- and stress-related research seeks to undo constraints on particular production systems. For example, research to speed rebreeding of swine seeks to address the high costs of confinement systems by reducing the amount of time an animal must be housed to produce a litter. Likewise, research to synchronize breeding aims at getting the most efficient use possible for expensive facilities. Research directed at removing constraints to total-confinement hog production is most conducive to large employee-operated farms because the high capital cost of confinement creates economies of size and barriers to entry. Total confinement routinizes the labor associated with hog production and reduces the need for judgment on the part of labor, thereby adapting hog production to industrialized systems.

An alternative would be a research program with the goal of making sure of farmers' labor and management to reduce the investment requirements of raising hogs efficiently. It would include development of low-cost facilities as well as management systems to fit those facilities. Researchers might seek to adapt hogs and remove constraints to such systems. Low-investment hog production systems would reduce barriers to entry, reduce economies of size, create a greater role for people in agriculture, and fit owner-operated farms better than employee-operated farms.

The impact of porcine growth hormone (PGH) is not easily predicted. I will only pose some questions and suggest some possible impacts, as follows. If the delivery system requires a monthly injection, that would increase the need for labor in hog production. If PGH increases pork consumption by lowering the price of pork and reducing its fat content, additional farmers might be required to produce the additional hogs. However, monthly injections might be much easier in total-confinement systems where hogs have less room to move than in less-intensive systems. The reduced time to market resulting from growth hormones would reduce the need for labor.

PGH will undoubtedly be cheaper for large operations than for small; this will create an economy of size. If PGH requires higher-protein diets for hogs, as appears likely, that would increase the share of the ration purchased from off the farm, creating another cost advantage for volume purchasers. However, if researchers were to develop a tripson-inhibitor–free soybean which could be fed to livestock without processing, that factor would be eliminated.

Will the lower backfat of hogs given PGH affect their tolerance of less-intensive facilities in cold climates? Cornell University research indicates that sows receiving PGH increase their milk production substantially late in lactation.[8] Will this grant an advantage to less-intensive systems, which can afford to wean later than total-confinement systems due to the high facility costs of keeping sows and piglets together? Will management differences among farms of different sizes and types affect response to PGH?[9]

PGH would probably improve human health (by reducing the fat content of pork) unless there are unknown health effects.

Application to the Land-Grant System

As revolutionary social planners in a democratic society, land-grant researchers need to begin to predict the impacts of alternative lines of applied research and compare those impacts with the social goals of the citizenry. That will take a substantial commitment on the part of the land-grant colleges. They must develop the capacity to better predict the ultimate impacts of research and engage the public in discussions about what kind of communities they want to live in. These discussions must focus on fundamental and, yes, value-laden issues such as what structure of agriculture we want and what social structure we want agriculture and agriculture technology to foster. These discussions must involve a broad cross section of rural Americans and not be limited to people of particular ideologies or social circumstances. One promising move in this direction is the bioethics program at Iowa State University.

The public must do its part in turn by participating in these discussions and supporting the land-grant system. The land-grant system is our one hope to have social goals drive agricultural research priorities. The system must be adequately funded so it can respond to those considerations, regardless of whether supplementary private funding is available. It must be adequately funded to do not just basic research but applied agricultural research, which does so much to shape our lives and communities. Finally, the system may need additional funds to develop innovative education and extension programs to develop the skills to compete in people who might otherwise be left behind by rapidly changing technology.

Notes

1. V. W. Ruttan, "Agricultural Scientists as Reluctant Revolutionaries," *Choices: The Magazine of Food, Farm, and Resource Issues* (Third Quarter 1987):3.

2. Dean MacConnell, "Agribusiness and the Small Community," (Department of Sociology Seminar, University of California, Davis, 1985).

3. Some, but not all, of these questions reflect the input of University of Kentucky rural sociologist Larry Busch, though he is not responsible for their final forms. (Larry Busch, interview by Chuck Hassebrook, 1987.)

4. U.S. Congress, Office of Technology Assessment, *Technology, Public Policy, and the Changing Structure of American Agriculture* (Washington, D.C.: OTA, 1986).

5. Roy N. Van Arsdall and Kenneth E. Nelson, "Economics of Size in Hog Production," Economic Research Service Technical Bulletin #1712 (USDA, December 1985):34.

6. See, for example, Charles A. Francis and James W. King, "Overview of Sustainable Agriculture," *1988 Proceedings: Sustainable Agriculture in the Midwest* (Lincoln: Cooperative Extension Service, University of Nebraska, 1988).

7. J. Michael Gould and Lee B. Dexter, "Warm Season Grasses—New Agricultural Crop for the Cornbelt," in *Prairie: Past, Present, and Future,* eds. G. K. Clambey and R. H. Pemble (Fargo, N.D.: Tri-College University Center for Environmental Studies, 1986).

8. Dean R. Boyd et al., "The Effect and Practical Implications of Recombinant Porcine Growth Hormone on Lactation Performance of Sows," (Department of Animal Science Seminar, Cornell University, November 1985).

9. Some, but not all, of these questions regarding PGH reflect input from University of Nebraska Extension swine specialist Bill Alschwede, though he is not responsible for their final forms. (William Alschwede, interview by Chuck Hassebrook, Lincoln, Nebraska, 1988.)

19

Perceptions of the Role of University Research in Biotechnology: Town, Gown, and Industry

MACK C. SHELLEY II, WILLIAM F. WOODMAN,
BRIAN J. REICHEL, and PAUL LASLEY

Universities, as large, complex organizations, have at least a three-fold mission of teaching, research, and service. Academic institutions and units within those institutions have various degrees of emphasis on these areas of their mission. As new areas of research open up, usually because of the availability of funding to support such efforts, key decision-makers within universities must consider whether and to what extent they wish the institution to invest its resources in an attempt to capture portions of the external funding that has become available. They must also decide to what extent a redistribution of current and expected future resources within the institution is necessary in order to sustain the revised research agenda. There are potential consequences growing out of these kinds of decisions about resource allocation and research agendas for a number of groups within the university and for groups external to but related to and interested in the university's decisions. The highly permeable nature of university agenda-setting decisions is particularly clear with respect to the impact that private industry may have on university resources and research goals through offers of financial support and cooperative research and the feedback received through "constituent" reactions to pending decisions or to decisions already made by university officials.

Land-grant universities are perhaps peculiarly sensitive within American academia to the interplay of pressures from internal groups, industry, and relevant external publics. They must be responsive to the needs and demands of state legislatures and state executives; to the wishes of their internal administrators, their faculty, and their students; to the private corporate sector on which they depend increasingly for research funding; and to the citizens for whom they provide extension services and other support. Political constituents, in particular, are often inclined to insist on overly precise measures of the job-creating outcomes of their investment of public funds in universities for economic development.[1] The expectations, reactions, and attitudes of all such groups, whether reasonable and well informed or not, need to be taken into consideration in the process of formulating broad research agendas, calculating probable trade-offs among various choices, and monitoring the progress achieved from the decisions that have been made. The argument for measuring organizational effectiveness in a multiconstituency context is made, for example, by Connolly, Conlon, and Deutsch,[2] who, however, eschew a "social justice,"[3] minimum-regret,[4] or otherwise normative criterion.

In 1986, the Iowa state legislature made a commitment to channel seventeen million dollars over four years in funds from the state's lottery into research on agricultural biotechnology (molecular biology) at Iowa State University, the state's land-grant academic institution. A biotechnology council was formed to determine allocations of those funds to selected research projects, in consultation with an industrial advisory group. Shortly thereafter, in response to a legislative provision in the funding bill, a bioethics committee was also established. That committee subsequently was assisted by its own external advisory group. One function of the bioethics committee was to look into the possible impacts on the university of this funding for research in biotechnology.

To further the goal of evaluating the internal dynamics of biotechnology funding, a series of surveys were mailed to what were perceived as relevant internal and external constituent groups. The surveyed groups at Iowa State University included virtually all faculty members who were listed in an internal directory of faculty involved in biotechnology research, a random sample of nonbiotechnology faculty members, a random sample of graduate students selected without regard to their involvement or lack of involvement in biotechnology work, and virtually all university administrators with ranks of departmental executive officer through vice president. A companion survey was distributed to all known biotechnology companies in the United States, as listed in *Genetic Engineering News.* In addition, survey data on perceptions among a large random sample of Iowa farmers were made available through the Iowa Farm and Rural Life Poll, conducted through the College of Agriculture and the Department of Sociology by Paul Lasley. Our results, then, transcend the "internal dominant coalition" of decision-makers and are inclusive of a much wider range of constituency views.[5]

The purpose of this chapter is to present some findings from this diverse data base regarding two major issues of university research. First, to what extent is there a congruence or a divergence among these groups' attitudes toward the roles that could be adopted by universities in conducting, disseminating, and evaluating the burgeoning research in the biotechnology area? Second, precisely what are the current expectations of each of these groups for university activities in the emerging locus of research on biotechnology? Both of these questions relate to the fundamental issue of whether what Gurwitz has called "the new faith in high tech" is a correctly directed faith in this instance.[6]

Although the results for our respondents from academe are drawn from a single university, Iowa State is sufficiently similar in scope and mission to other land-grant universities in the nation so that the attitudes measured by our internal surveys are quite likely to be representative of those that might be found at comparable academic institutions. Similarly, although the Iowa farm operators' survey is limited to one state and to one occupational group, it is the farm sector that is likely to feel the greatest effect from work in agricultural biotechnology both at universities and in industry, and there probably is no obvious reason why Iowa farmers would be expected to have markedly different attitudes toward biotechnology than would farmers in the nation as a whole.

In reviewing the results of these surveys and in attempting to make sense of their meaning for the ethics and practicality of university research in biotechnology, it is instructive to keep in mind that academic institutions have been said by two leading organizational theorists to be characterized by "organized anarchy."[7] Further, it is well to recall the views of March and Olson that institutions of higher education are "complex 'garbage cans' into which a striking variety of problems, solutions, and participants may be dumped."[8] Thus, our results should be seen in light of the normal creatively chaotic state of academia and the consequent lack of specific knowledge on the part of many of its institutional actors regarding policy.

In our four original surveys, and in Paul Lasley's Farm and Rural Life Poll, there are a total of eleven questions that were asked of all sets of respondents. A statistical summary of the results of the responses to these questions is shown in Table 19.1.

These eleven questions are divided into two groups. The first cluster of six items is focused on matters pertaining to the consequences for economic development which may eventuate from a research emphasis on biotechnology. The latter group of five questions pertains to various aspects of university research agendas which might be affected by the push for biotechnology applications. In general, the first set of questions is related to the impact on the rural economy from university-based biotechnology re-

Table 19.1. Descriptive statistics on questions common to the surveys

Economic Development

Item 1: *Biotechnology will help solve the problem of farm surpluses by finding new uses for crops and livestock.*

	SA	A	U	D	SD	Mean	s	n
Biotechnology faculty	8.3%	42.4%	24.2%	22.0%	3.0%	2.689	1.005	132
Other faculty	3.5	31.0	32.7	29.2	3.5	2.982	0.945	113
Graduate students	10.2	31.8	31.8	25.0	1.1	2.750	0.986	88
Administrators	4.4	33.3	36.8	22.8	2.6	2.860	0.911	114
Biotechnology companies	6.7	37.8	39.5	16.0	0.0	2.647	0.829	119
	VD	SD	U	SU	VU			
Farmers	35.0%	28.0%	28.0%	5.0%	5.0%	2.20	1.20	1884

Item 2: *Advances in biotechnology will probably benefit persons with large farm operations more than persons on middle-sized and small farms.*

	SA	A	U	D	SD	Mean	s	n
Biotechnology faculty	6.1%	34.8%	33.3%	22.0%	3.8%	2.826	0.969	132
Other faculty	13.3	41.6	31.9	11.5	1.8	2.469	0.927	113
Graduate students	10.1	34.8	30.3	19.1	5.6	2.753	1.058	89
Administrators	13.2	43.0	24.6	17.5	1.8	2.518	0.989	114
Biotechnology companies	2.5	24.4	31.9	33.6	7.6	3.193	0.977	119
	VD	SD	U	SU	VU			
Farmers	4.0%	8.0%	25.0%	30.0%	34.0%	3.85	1.09	1895

Item 3: *Through biotechnology, scientists will be able to develop new species of animals.*

	SA	A	U	D	SD	Mean	s	n
Biotechnology faculty	4.6%	31.3%	17.6%	35.1%	11.5%	3.176	1.133	131
Other faculty	5.3	27.4	28.3	31.9	7.1	3.080	1.045	113
Graduate students	9.0	23.6	27.0	27.0	13.5	3.124	1.185	89
Administrators	8.0	52.2	19.5	16.8	3.5	2.558	0.981	113
Biotechnology companies	5.1	61.9	19.5	13.6	0.0	2.415	0.788	118
	VD	SD	U	SU	VU			
Farmers	7.0%	17.0%	44.0%	15.0%	17.0%	3.18	1.25	1896

Item 4: *Research in biotechnology will increase the efficiency of feed conversion in livestock production.*

	SA	A	U	D	SD	Mean	s	n
Biotechnology faculty	17.4%	68.2%	12.1%	2.3%	0.0%	1.992	0.624	132
Other faculty	15.0	54.9	26.5	3.5	0.0	2.186	0.726	113
Graduate students	12.4	58.4	28.1	1.1	0.0	2.180	0.650	89
Administrators	19.3	64.0	15.8	0.9	0.0	1.982	0.624	114
Biotechnology companies	16.8	69.7	12.6	0.8	0.0	1.975	0.574	119
	VD	SD	U	SU	VU			
Farmers	27.0%	45.0%	18.0%	6.0%	3.0%	2.10	0.99	1894

Note: SA = strongly agree; A = agree; U = uncertain; D = disagree; SD = strongly disagree; VD = very desirable; SD = somewhat desirable; U = uncertain; SU = somewhat undesirable; VU = very undesirable.

Table 19.1. (*cont.*)

Economic Development

Item 5: *Biotechnology will lead farmers to become more dependent upon large corporations for many of their inputs, such as seeds, growth hormones, and feed additives.*

	SA	A	U	D	SD	Mean	s	n
Biotechnology faculty	14.3%	45.9%	17.3%	20.4%	2.3%	2.504	1.042	133
Other faculty	18.6	38.9	27.4	12.4	2.7	2.416	1.015	113
Graduate students	14.6	46.1	27.0	11.2	1.1	2.382	0.911	89
Administrators	12.3	50.9	24.6	10.5	1.8	2.386	0.897	114
Biotechnology companies	5.9	29.4	31.9	25.2	7.6	2.992	1.046	119
	VD	SD	U	SU	VU			
Farmers	3.0%	7.0%	19.0%	35.0%	36.0%	3.94	1.10	1896

Item 6: *Greater quantities of crops and livestock products will be available for sale as a result of biotechnology.*

	SA	A	U	D	SD	Mean	s	n
Biotechnology faculty	12.1%	47.0%	27.3%	12.1%	1.5%	2.439	0.910	132
Other faculty	8.0	41.6	41.6	7.1	1.8	2.531	0.814	113
Graduate students	15.7	43.8	29.2	9.0	2.2	2.382	0.935	89
Administrators	18.6	50.4	24.8	6.2	0.0	2.186	0.808	113
Biotechnology companies	10.1	63.9	19.3	6.7	0.0	2.227	0.718	119
	VD	SD	U	SU	VU			
Farmers	5.0%	19.0%	31.0%	27.0%	18.0%	3.34	1.26	1886

University Research

Item 7: *Universities should work closely with private businesses and industry, including the agri-business sector.*

	SA	A	U	D	SD	Mean	s	n
Biotechnology faculty	24.6%	44.8%	15.7%	10.4%	4.5%	2.254	1.081	134
Other faculty	15.0	42.5	18.6	14.2	9.7	2.611	1.191	113
Graduate students	23.3	36.7	22.2	11.1	6.7	2.411	1.160	90
Administrators	21.1	49.1	17.5	7.0	5.3	2.263	1.040	114
Biotechnology companies	51.2	38.0	4.1	4.1	2.5	1.686	0.922	121
Farmers	27.0	49.0	14.0	7.0	3.0	2.10	0.91	1911

Item 8: *Scientists, rather than the agri-business community, should determine what types of problems need to be investigated.*

	SA	A	U	D	SD	Mean	s	n
Biotechnology faculty	20.5%	31.8%	18.9%	23.5%	5.3%	2.614	1.202	132
Other faculty	15.0	33.6	16.8	32.7	1.8	2.726	1.128	113
Graduate students	10.0	24.4	20.0	31.1	14.4	3.156	1.235	90
Administrators	12.3	29.8	22.8	31.6	3.5	2.842	1.110	114
Biotechnology companies	2.5	18.3	12.5	45.8	20.8	3.642	1.083	120
Farmers	6.0	17.0	22.0	38.0	18.0	3.48	1.13	1917

Table 19.1. *(cont.)*

University Research

Item 9: New discoveries by university scientists should be patented by the university and sold to the highest bidder, who would then make these products commercially available.

	SA	A	U	D	SD	Mean	s	n
Biotechnology faculty	19.7%	45.5%	15.2%	11.4%	8.3%	2.432	1.173	132
Other faculty	11.6	41.1	28.6	13.4	5.4	2.598	1.035	112
Graduate students	13.2	34.1	28.6	14.3	9.9	2.736	1.163	91
Administrators	14.2	52.2	22.1	9.7	1.8	2.327	0.901	113
Biotechnology companies	6.7	36.7	25.0	17.5	14.2	2.958	1.177	120
Farmers	12.0	27.0	28.0	20.0	14.0	3.00	1.42	1913

Item 10: The amount of private consulting by university faculty should be curtailed.

	SA	A	U	D	SD	Mean	s	n
Biotechnology faculty	3.0%	9.8%	19.5%	37.6%	30.1%	3.820	1.065	133
Other faculty	5.4	9.8	20.5	46.4	17.9	3.616	1.059	112
Graduate students	3.3	14.3	26.4	44.0	12.1	3.473	0.993	91
Administrators	4.3	10.4	20.9	44.3	20.0	3.652	1.052	115
Biotechnology companies	3.3	13.9	13.9	48.4	20.5	3.689	1.053	122
Farmers	5.0	20.0	44.0	25.0	6.0	3.07	0.89	1902

Item 11: More public funds should be used to support the development of new uses for agricultural commodities.

	SA	A	U	D	SD	Mean	s	n
Biotechnology faculty	26.1%	50.0%	10.4%	9.7%	3.7%	2.149	1.037	134
Other faculty	16.1	54.5	13.4	11.6	4.5	2.339	1.027	112
Graduate students	18.9	35.6	31.1	14.4	0.0	2.411	0.959	90
Administrators	27.0	49.6	7.8	12.2	3.5	2.157	1.065	115
Biotechnology companies	24.6	48.4	13.1	9.0	4.9	2.213	1.070	122
Farmers	30.0	45.0	16.0	6.0	3.0	2.07	0.97	1912

Note: SA = strongly agree; A = agree; U = uncertain; D = disagree; SD = strongly disagree; VD = very desirable; SD = somewhat desirable; U = uncertain; SU = somewhat undesirable; VU = very undesirable.

search, and the second set of items addresses the sometimes uneasy multipartite research partnership among universities, industry, and government. For all eleven items, the text (there were some slight variations from survey to survey) of the question is given together with a number of summary measures of the responses received from the six different groups of respondents.

The percentage distribution of responses is given for each point along a traditional Likert five-point, fixed-response range. For all respondents except the farm operators, the choices for every question were: *strongly agree* (SA), *agree* (A), *uncertain* (U), *disagree* (D), and *strongly disagree* (SD). In the Iowa Farm and Rural Life Poll, respondents were asked to indicate their opinions about the desirability of the impacts addressed in the six economic development questions, with equivalent responses of *very desirable* (VD), *somewhat desirable* (SD), *uncertain* (U), *somewhat undesirable* (SU), and *very undesirable* (VU). The remaining five questions were coded along the agree-disagree continuum for farm operators. Also shown are the sample mean score averaged over all responses, the sample standard deviation of the responses (s), and the number of valid responses (n) for each item. Means and standard deviations were calculated by assigning numeric values of 1 to a *strongly agree* response, 2 to an *agree* response, 3 to an *uncertain* response, 4 to a *disagree* response, and 5 to a *strongly disagree* response. A larger value for the mean, then, corresponds to a greater tendency on average to disagree with a statement, and a smaller mean value indicates that respondents on average were more likely to be in agreement with a statement. A visual summary of the comparative magnitudes of the item means is shown in Figure 19.1.

A number of fairly dramatic differences are evident from an examina-

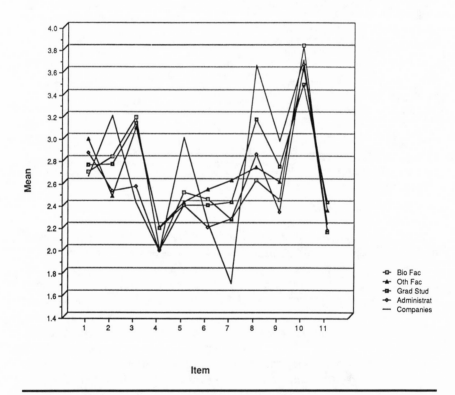

Fig. 19.1.   Means for eleven items from biotechnology survey.

tion of the data. For Item 1 ("Biotechnology will help solve the problem of farm surpluses by finding new uses for crops and livestock"), there are very distinctive patterns in the responses from farmers as opposed to the responses from all other groups. The strength with which farmers believe that biotechnology research will or should alleviate the problem of surplus production by developing alternate uses for crops and livestock is clearly different from the responses of any other group. A large majority of farmers agree or agree strongly with the proposition that biotechnology will or should take care of farm surpluses, compared to just a bare majority of biotechnology faculty and much smaller proportions of the other respondent groups. As measured by the mean level of their responses, the distinctiveness of the farmers' attitudes is quite readily apparent.

An even larger gap between farmers' perceptions and the perceptions of other groups is demonstrated in Item 2 ("Advances in biotechnology will probably benefit persons with large farm operations more than persons on middle-sized and small farms"). Vastly larger proportions of farmers disagree with the correctness or desirability of this statement than do other respondents. This result, combined with the previous finding regarding solutions to the farm surplus problem, suggests that farmers' perceptions of the role of biotechnology are quite different, and perhaps less realistic, than other relevant groups' beliefs about the economic development role to be played by biotechnology. It is worth noting, however, that attitudes among respondents from biotechnology companies are more likely than the attitudes of other groups to reflect the farmers' viewpoint regarding differential benefits as a function of farm size. Note, too, that differences among the university respondent groups (biotechnology faculty, other faculty, graduate students, and administrators) do not vary greatly.

Quite different response patterns are characteristic of Items 3 and 4 than were true of the first two items. As measured by mean values of their responses, the administrators and biotechnology company respondents were of generally like mind in tending to agree that new animal species are likely to result from biotechnology (Item 3). With regard to Item 3 ("Through biotechnology, scientists will be able to develop new species of animals"), the four other groups (biotechnology faculty, other faculty, graduate students, and farmers) showed very similar mean responses and differed from the administrator/biotechnology company pattern in their greater inclination to disagree with the likelihood or desirability of this proposition.

Item 4 ("Research in biotechnology will increase the efficiency of feed conversion in livestock production") produces a virtually uniform set of responses from all six groups. All groups agree or strongly agree with the accuracy or desirability of the proposition that more efficient feed conversion in livestock production is likely to result from biotechnology research. The absence of any clear differences on this issue does not necessarily mean that all groups were responding out of the same underlying thought process. Farmers may well perceive this item (and likely other items, too) in terms of a wonderful innovation that will make their individual farms more profitable or at least better able to withstand the stresses of the rural economy. University respondents may well think of feed conversion as something that may (except for the biotechnology faculty) be of little immediate relevance to their lives or research agendas but as something that would be nice if it would happen. Respondents from biotechnology companies may envision potential profits or possible dollar losses, depending on whether they can "get ahead of the curve" and use more efficient feed conversion processes in their own product and service line, but may feel overall that this is one of the most likely changes to be brought about by biotechnology research. That is, with this and other items, the patterns of responses are undoubtedly conditioned by the dominant realities of each group's portion of the social world. Nonetheless, it is fairly remarkable to see in this case that groups with multiple and perhaps contradictory motivations have converged in agreement on the issue of feed conversion.

Item 5 ("Biotechnology will lead farmers to become more dependent upon large corporations for many of their inputs, such as seeds, growth hormones, and feed additives") seems at first glance, to have results essentially the same as those shown for Items 1 and 2. However, this would be a misleading assessment. The distinctive feature of the response patterns for Item 5 is that practically a whole category worth of difference (that is, a magnitude of nearly one for the values of the means) separates farmers' average perceptions of the extent to which biotechnology will or should result in further concentration within the farm sector from the perceptions of any other group. This large difference is further reflected in the virtual unanimity (71 percent disagreement or strong disagreement) with which farmers reject the notion of more concentrated agriculture, as compared with nearly 33 percent of biotechnology company respondents, almost 23 percent of biotechnology faculty, just 15 percent of other faculty, slightly over 12 percent of administrators, and barely 12 percent of graduate students who hold with the farm respondents that biotechnology will not or should not cause farmers to become more dependent upon large corporations. The dramatic gap between farmers' and other groups' perceptions here may reflect wishful thinking among family farmers based on the belief that biotechnology will be or should be sufficiently "portable" that even smaller farms will find its applications profitable or embedded in the belief that political forces will come to the rescue of family farmers who could wield their legislative and lobbying influence to shake off the grip of agribusiness. Clearly, whatever the mix of motivations behind the farm operators' responses on this item may be, their outlooks are not shared by any other relevant group. The closest of the five other groups to the farm respondents on the dependency issue are the biotechnology company respondents, but even they are far less inclined than the farm survey sample to believe that further dependency will not or should not eventuate.

Item 6 ("Greater quantities of crops and livestock products will be available for sale as a result of biotechnology") produces a general similarity of views between university administrators and the respondents from biotechnology companies. The three other groups of respondents from academia are less likely to agree with the university administrator/biotechnology company view in this respect, and farmers by a very wide margin differ from any of the other groups. Here, farmers are far more likely to believe that biotechnology will or should result in the marketing of greater quantities of agricultural products.

The five questions forming the second portion of Table 1 elicit variable patterns of responses on several aspects of the role of university research in the field of biotechnology.

Item 7 ("Universities should work closely with private businesses and industry, including the agri-business sector") is a straightforward assertion of expanded university-industry common research agendas. The biotechnology company respondents were nearly unanimous (over 89 percent agreeing) in accepting this statement. In fact, all groups of respondents were highly supportive of the general proposition of closer university-industry research links. It is interesting, though, that the "other" faculty, that is to say those who are not directly in the biotechnology area, representing a very wide diversity of departments and research emphases, were least supportive of the principle of cooperative research. This may simply reflect the lesser chance that faculty members in, for instance, the social sciences or humanities would ever be involved directly in such research endeavors; or, again, this result may represent an alternative view in the nonbiotechnology portion of the faculty about the propriety of or the need for such cooperative research programs.

The next item (Item 8, "Scientists, rather than the agri-business community, should determine what types of problems need to be investigated") produces an altogether different pattern of responses. This proposition, which states essentially that university researchers and other scientists should govern biotechnology research directions, is soundly rejected by the biotechnology company respondents and by the farm survey respondents and rejected only slightly less by graduate students. Agreement with this proposition is strongest among the biotechnology faculty, who presumably would be affected most directly by whether they or the agribusiness firms had the primary say over research agendas. Other faculty and university administrators give at least moderate support to the notion of relatively free scientific research. One interesting aspect of these results is that large proportions of university respondents were not willing to rise to the support of "pure" research, versus research driven by the needs of the business community. That is a potentially telling point as universities debate internally the relative merits of basic versus applied research.

The sensitive issue of patent rights is addressed in Item 9 ("New discoveries by university scientists should be patented by the university and sold to the highest bidder, who would then make these products commercially available"). The strongest opposition to university control over patent rights comes from farmers, followed closely by the respondents from biotechnology companies. One can understand rather easily why biotechnology company representatives would want to limit university control over patents for new processes or organisms developed by academic researchers, particularly if the research were underwritten at least in part by the business community. However, the responses from farmers are a bit curious. It is difficult to understand immediately what difference it would make to farmers where their commercial contacts are made. Perhaps most farmers have become used to dealing directly with corporate entities rather than working through universities for the rights to use products. Or perhaps the responses suggest that farmers are predisposed to working with what they may regard as fellow businesspeople who might be perceived as more familiar, more available, and more efficiently businesslike than university representatives. It is equally interesting, although not particularly surprising, that university administrators were most inclined to agree with the notion of university-patented products being made available to the commercial sector. Presumably, this reflects both a reluctance to "let go" of products that were developed internally through the use of increasingly scarce university resources and a desire to retain control over the output from intellectual and human capital. This, too, may indicate that university administrators are disinclined to deal directly or through intermediaries with commercial outlets that may be seen as having little appreciation of the complexities of university research agendas and budgeting needs.

The most consistent disagreement with any of the eleven items common to all of the surveys occurred on Item 10 ("The amount of private consulting by university faculty should be curtailed"). Precious few faculty members (in biotechnology or otherwise), graduate students, or administrators were willing to agree with that proposition. Although it remains a sensitive and controversial topic, consulting by university faculty (as a means to enhance academic incomes or as an additional outlet for research interests) does not draw widespread criticism in our results. It is worth noting, however, that the farm survey respondents were appreciably more likely than university respondents or than respondents from biotechnology companies to favor some curtailment of private consulting. This may reflect a popular image of university faculty as being relatively well paid and as leading comparatively "softer" work lives than do farmers or other nonacademic workers. It may be, too, that farm survey respondents generally have had relatively little contact with academic consultants and may therefore not fully appreciate the virtues of such interaction.

Finally, Item 11 ("More public funds should be used to support the development of new uses for agricultural commodities") shows relatively small differences among the different respondent groups. All sets of respondents are heavily in agreement with the proposition of increased public funding to support research in new agricultural commodities. Farm respondents, however, are more likely to support this proposal than are the other groups. This may be due to the fact that the farm survey respondents react to this item principally as interested parties who expect to benefit directly by the expenditures of money (much of it their own) on new applications of technology to utilize the commodities that they produce in abundance. Another possible interpretation is that hope springs eternal for the emergence of a viable "third crop" or another partial solution to the recurrent difficulties of the farm economy.

On the whole, it is quite clear that respondents to the farm survey were by and large of a different mind with regard to both economic development and university biotechnology research than were the various internal university constituencies or the respondents from biotechnology companies. How does this inform our answers to the questions posed at the beginning of this chapter: (1) What may be the "proper" roles that could be adopted by universities in conducting, disseminating, and evaluating the burgeoning research in the biotechnology area? and (2) What is the range of current expectations of the activities by academic institutions in that research area?

The first, normative question can be addressed in the following way. Given that both consensus and disagreement are present along various axes of perceptions about university roles in biotechnology research, there is a general commonality of views within the academy. In a very general sense, biotechnology faculty, other (i.e., nonbiotechnology) faculty, graduate students, and university administrators do not in general have strongly different points of view about what are the "proper" roles for university actors in biotechnology research. Working closely with private businesses and industry is generally acceptable; scientists should have latitude, within reason, in determining the directions of research (opinions of graduate students are an exception to this generalization); patents should be marketable by universities to commercial firms; private consulting by university faculty should not be curtailed; and public funds should be made available in greater amounts to develop new uses for agricultural commodities. Farmers and biotechnology companies resist the direction of research agendas drawn up principally by scientists; they both resist, though less intensely, the control of patent rights by universities; and farmers are more inclined than other respondents to favor limits on private faculty consulting and to support increased public funding for developing new agricultural commodities.

While the fact that internal disagreement on the "proper" directions for university research in biotechnology is fairly minimal among academic constituent groups does not in and of itself provide sufficient justification for doing what is agreed upon, future university actions in this area are not likely, based on these empirical results, to be the source of intense internal battles over ethics. It is much more likely, though, that debates would center on the merits of questions such as, "What will this research do to farmers?" or "What will it mean for the farm economy?" This is adumbrated by discussions that have already been held in legislative bodies, within the university, and between citizens and university actors about the impact of biotechnology research on the family farm. Beyond the consensus point that the family farm should not be harmed by the outcomes of biotechnology research is the fundamental reality that not all rural residents are likely to benefit equally from university-based research in biotechnology. In fact, differential benefits of any new technology are an article of faith among students of the process of technological innovation and diffusion. It is through these differential benefits that the ethical issues are likely to be played out.

As for the second, and more easily handled, empirical question, the expectations of university-based biotechnology research with respect to economic development can be summarized. Respondents generally believe that research in biotechnology will or should produce more efficient feed conversion and help solve the problem of farm surpluses by finding new uses for crops and livestock. Beyond those perhaps fairly straightforward expectations, however, little consensus is available. Academics believe that research in biotechnology probably will benefit large farm operators more than it will benefit persons on middle-size and small farms. Farmers emphatically do not agree with that notion, and the respondents from biotechnology corporations are split on the issue but lean toward the farm respondents' viewpoint. University administrators and biotechnology company respondents anticipate that biotechnology will allow the development of new species of animals. The other groups are at least moderately inclined to take an opposite view. Farm respondents stand alone in the degree to which they hold that biotechnology will not or should not make farmers more dependent upon large corporations for many of their inputs. Here there is a vast and difficult-to-bridge gap between citizen views and the perceptions of academia and of the biotechnology company respondents. A similar, though less intense, difference separates farm respondents from all other groups in their rejection of the belief that biotechnology will or should produce greater quantities of agricultural products for sale. Who is shown to be correct on these issues of disagreement will be determined by events over the next few years. In the meantime, a number of questions remain: Do citizens, and perhaps farm respondents particularly, need to be "educated" on the issues? Or, is it the academics who need the "education" provided by the experiences and perceptions of other groups? Is this an issue that ought to be addressed through public opinion, or is it so technical that only experts can provide policy guidance? Will events move so rapidly that attempts at regulation (and attempts to circumvent regulations) may prove inadequate to the task at hand?

These queries, which are not directly answerable from the data we have available now, are likely to be among those which will have to be considered in very short order. A fertile merger of ethics with empirical investigation of these points promises to provide effective guidance for informed policy judgments regarding these and related issues.

Notes

1. Scott Jaschik, "States Trying to Assess the Effectiveness of Highly Touted Economic Programs," *The Chronicle of Higher Education* (June 3, 1987):19,26.

2. Terry Connolly, Edward J. Conlon, and Stuart Jay Deutsch, "Organizational Effectiveness: A Multiple-Constituency Approach," *Academy of Management Review* 5 (1980):211–17.

3. John Rawls, *A Theory of Justice* (Cambridge, Mass: Harvard University Press, 1971).

4. Michael Keeley, "A Social-Justice Approach to Organizational Evaluation," *Administrative Science Quarterly* 23(1978):272–92.

5. James D. Thompson, *Organizations in Action* (New York: McGraw-Hill, 1967); Kim Cameron, "Measuring Organizational Effectiveness in Institutions of Higher Education," *Administrative Science Quarterly* 23(1978):604–32.

6. Aaron S. Gurwitz, "The New Faith in High Tech," *The Wall Street Journal* (October 27, 1982).

7. Michael Cohen and James G. March, *Leadership and Ambiguity: The American College President* (New York: McGraw-Hill, Carnegie Commission for the Future of Higher Education, 1974).

8. James G. March and Johan P. Olson, *Ambiguity and Choice in Organizations* (Oslo: Universitetsforlaget, 1976):176.

PART VI

Ethical Dilemmas

20 | Moral Responsibility, Values, and Making Decisions about Biotechnology

RACHELLE D. HOLLANDER

This chapter uses an illustration from nineteenth-century literature and some conceptual distinctions made by a contemporary philosopher to clarify important aspects of moral responsibility: foresight and nonexclusivity. Examples from the current scientific literature demonstrate how value presumptions and differences often underlie uses of evidence in policy disputes. Current position statements from federal agencies and scientific organizations continue in this tradition. When these value assumptions go unnoticed, the disputants cannot address issues of grave importance to them as well as to everyone else. Exercising only limited foresight, these individuals and institutions cannot behave responsibly. Incorporating views and addressing concerns outside of this mainstream is required for morally responsible behavior.

Moral Responsibility and *An Enemy of the People*

Henrik Ibsen wrote the play *An Enemy of the People* in 1882.[1] *An Enemy of the People* concerns the possibility of contamination in the water supply that feeds a town's new baths. It is those baths that attract the summer visitors who have rejuvenated the community. Dr. Thomas Stockmann has investigated and discovered the problem, he has documented it, and he is gleeful to have made the discovery. After all, he had given the town the idea. He had warned the town fathers about the problem when they designed the water supply, and they didn't listen. He presents the truth

279

as he sees it—and he sees it in the worst possible light—to his brother Peter, the mayor, who had organized the efforts to bring the baths to fruition. The mayor is much less concerned and convinced than is the doctor about threats to people's health—typhoid, gastric troubles—and much more concerned than is the doctor about the costs of reworking the water supply. Mayor Peter Stockmann does not believe the threat can be significant, and he believes the doctor's report should be kept confidential. When the mayor tells representatives of the press and small trade and business interests that the gentlemen who own the baths don't intend to pay for the expensive remodeling that will be necessary and that the tax assessment will have to rise to accommodate it, the town turns against the doctor. The newspaper, which had been about to publish the report of possible water contamination, suppresses it. In the end, the doctor is dismissed from his position. His home is vandalized; his sons are attacked at school; he and his family are nearly run out of town. But he decides to stay—out of spite and conviction—and carry on his campaign for truth, morality, and justice. He is going to start by educating his sons, along with street urchins and guttersnipes, at home. The play concludes with his statement: "The strongest man in the world is he who stands most alone."

A film version of this play was on the agenda of a science, technology, and public policy seminar at the Western Executive Management Center in 1983. The center is a training facility for managers in the federal government. The audience's response to the film was to blame the doctor. *Nobody* saw him as the man in the white hat. Now it's true the doctor was obnoxious. But all the major characters in the opposition suffered from at least as many and as bad character flaws and shortcomings as he. And the doctor was concerned, wasn't he? And wasn't he supposed to be? Wasn't that his professional responsibility—public health? The opposition was solely, or at least primarily, concerned about their pocketbooks. Is it astonishing when an audience of bureaucrats *all* blame the doctor for not presenting his message better, for hiding his investigation and then springing it on his brother, and so forth and so on? This audience of persons supposedly committed to the public interest thought the doctor was the "bad guy." Everyone identified with the mayor.

What was Dr. Stockmann? He was a whistleblower. And the response that Dr. Stockmann got from the audience, viewed as an example of what we know happens to whistleblowers, is not at all remarkable. In April 1987, *Recreation News,* the official publication of the League of Federal Recreation Associations, Inc., ran a cover article on whistleblowing.[2] The authors viewed whistleblowers as providing an important public service at enormous personal cost. I recall seeing more letters—pro and con—about having that article in the paper than I can remember seeing about any other article published in it. Perhaps this shows how controversial a subject it is. The authors had a list of "painful lessons of experience" at the end. The last sentence was, "Remember the words of Hugh Kaufman, a resister who still works for the Environmental Protection Agency: 'If you have God, the law, the press, and the courts on your side, you have a 50–50 chance of winning.' "

Interestingly, when the film of *An Enemy of the People* was accompanied by a discussion of this article in a presentation at the Center in 1987, the audience's responses were much more positive toward the doctor. Furthermore, at the statement that many whistleblowers are obnoxious, several members of the audience protested vigorously. They insisted that the kind of perseverance and strongmindedness whistleblowers needed, in the face of organizational constraints, meant that they were almost inevitably going to be perceived as difficult.

Now, it might be that the samples were skewed on these two occasions, and bureaucrats haven't changed. But perhaps the difference illustrates an awakening to the positive societal function whistleblowers serve. In no small part, this may be due to new laws and regulations that institutionalize the whistleblowing function and provide protection for whistleblowers. At the very least, the audience's response in 1987 shows that many bureaucrats are well aware of the difficulties whistleblowers face and are positively inclined toward them.

The example from *An Enemy of the People* is intrinsically interesting. It shows that these problems have long been with us. When it was presented at another conference in the summer of 1986, one participant mentioned that a similar episode had recently occurred in a European country. With the increased attention to—some would say obsession about—risks nowadays, societies are unlikely to remain unaware of these incidents. But this example is important in the context of this chapter because it helps to illustrate an interesting point about moral responsibility, a point that shows that even the doctor doesn't wear the white hat.

An article by Professor John Ladd that appeared in the June 1982 *IEEE Technology and Society* helps us to think about the implications of this story for moral responsibility. The article was called "Collective and Individual Moral Responsibility in Engineering: Some Questions."[3] In it, Ladd points to several qualities that he says distinguish moral responsibilities from other kinds of responsibilities. Two will be discussed here. First, he emphasizes that moral responsibility is forward-looking and thus can be distinguished from legal responsibility. The latter wants to know where to fix blame when something has gone wrong. The former wants to exercise moral foresight, to prevent things from going wrong. The second quality is Ladd's insistence that moral responsibility is nonexclusive. Just because someone is morally responsible for something, it does not follow that others are not responsible. In this way, moral responsibility contrasts with the notion of role or task responsibility. If something is part of my job, it can be said to be exclusively my responsibility. But my moral responsibilities, on and off the job, are not exclusively mine. In fact they mostly are and can be shared. They may vary in degrees, says Ladd, "only to the extent that one person is better able to do something about [them] than others."

To quote from the Ladd article, "If many people can be responsible for the same thing, then there can be such a thing as . . . collective responsibility. . . . Thus, the whole family is responsible for seeing that the baby does not get hurt. The whole community is responsible for the health and safety of its citizens. And all the engineers, as well as others, working on a project are responsible for its safety"(p. 9).

In this light, some of the resentment that first audience expressed about and toward Dr. Stockmann in its discussion of *An Enemy of the People* may become understandable and even morally justifiable. That is, Stockmann, for all the better part of his intentions, was not able to help the others in his community exercise their moral responsibilities. This is his tragedy. The tragedy for the others was that they were not able to help him exercise his. Moral responsibility puts each of us on the hook. We each must exercise foresight. We must deliberate about the effects of our actions on ourselves and others. And the more we come to know about these effects, the more the job of exercising our moral responsibilities entails. We cannot assign the job to others. It is inalienably ours. Yes, we cannot exercise it alone. It requires cooperation and collaborative effort. If enough people exercise it, we can collectively succeed in behaving morally. We can achieve the desired moral results.

Value Differences in Policy Disputes

This is the context within which to consider how the ostensibly scientific literature and certainly the policy statements of some governmental organizations and national scientific groups incorporate unnoticed and unremarked value statements that presuppose a limited and inadequate sense of moral responsibility. Several papers that appeared in *Science* early in 1985 will be used to illustrate. Brief quotations from related policy statements also make the point. The discussion of the *Science* articles is drawn substantially from my article in *Agriculture and Human Values,* Summer 1986.[4]

That article adopts the values typology William Aiken developed to describe normative views he heard in discussions about agricultural research.[5] Aiken identifies four views: top-priority, net-gain, constraints, and the holistic or compatabilistic views. He differentiates them as follows: The top-priority view sees values as goals. This view decides on an appropriate set of goals for agricultural research and the correct ranking of those goals. The net-gain view sees values as benefits or harms. Unlike the top-priority view, the net-gain view does not pursue the top-ranked goal, despite its effects on others. Rather, it attempts to balance incompatible or conflicting benefits and associated harms and pursue the course of greatest net gain. The constraints view is again different; it divides values into goods and constraints. It may prohibit types of acts that violate constraints—interference with human rights for instance, whether or not they would interfere with attaining the top-ranked goal or the greatest net benefit. Finally, holism differs from these others in determining what is valuable by examining the functioning of the whole system. Effects that other views might identify as negative need not be thought undesirable; rather they may be viewed as important means of promoting the whole. Effects other views might identify as positive may be viewed unfavorably for their bad effects on the system.

Briefly, with respect to agricultural research, the four views have had typical adherents. The top-priority and net-gain views have focused on productivity, although the latter looks for "productivity on balance." The constraints view has emphasized human health and safety. While the holistic view could condone economic or ecological Darwinism, it has traditionally emphasized preservation of small rural communities, organic or traditional farming practices, and ecosystems.

Many problems come with the use of these views. People often use more than one to justify policies or practices they want, even when a careful analysis may reveal inconsistencies. What each view means or implies in application is often vague or ambiguous, and the views are often used to justify policies or practices where it is not clear that they can fulfill claims made. There is need both to clarify these views and to assess and be prepared to revise policies and practices that are in effect because it is believed they will satisfy one or more of the claims.

I shall not do that here. Rather I shall try to identify the moral difficulties that arise when disputants do not recognize that they are proceeding from different conceptions of what is important. A *Science* article on a current controversy in agricultural research, which is the subject of a lawsuit, illustrates this kind of problem.

"The Agricultural Mechanization Controversy"

California Rural Legal Assistance (CRLA) has brought suit against the University of California (UC), alleging that mechanization research displaces farmworkers. In "The Agricultural Mechanization Controversy," published in *Science*, February 8, 1985, Philip L. Martin and Alan L. Olmstead review the circumstances of the case and attempt to refute the CRLA allegation.[6]

The two disputants, CRLA and the UC system, may appear to disagree with the facts of farmworker displacement. In the summary of their article, Martin and Olmstead seem to think so; they conclude that they have reviewed the evidence and it does not support the charge.

But their approach to the dispute—their use of evidence—involves certain value assumptions that their opponents needn't share. Because of this, their review cannot definitively refute the CRLA charges. For instance, CRLA claims that mechanization research displaced farm workers. Martin and Olmstead respond that mechanization may have kept the tomato processing industry in the United States rather than allowing it to be lost to Mexico. They point out that, "Since 1960, job losses due to mechanization have been more than off-set by the expansion of labor-intensive agriculture in California."

This claim, even if true, does not refute a claim that workers were displaced. Both factual claims may be true, so one cannot refute the other. To understand the dispute, we need a different level of analysis—an analysis of the value assumptions being used to support conflicting normative positions.

At this level, we see that Martin and Olmstead's response relies on a net-gain or trade-offs view. The CRLA criticism can be based in a constraints or human rights view, however. From this latter perspective, it is the fact that workers were displaced that is wrong, because individuals should not be treated in this way. Martin and Olmstead, viewing worker displacement in the aggregate, justify loss of individual jobs by pointing to net gain in jobs—the greatest good for the greatest number.

In the article, Martin and Olmstead attack and attempt to refute five CRLA allegations against the university. Often, they assume the net-gain view as justification. Against the CRLA claim that mechanization research eliminates small farms, they note, "While farm production has become increasingly concentrated, the number of small farms continues to increase." But this CRLA claim can also be founded in the constraints or human-rights view. It could call on the top-priority view for justification. If increasing the number of small farms is top priority, then the best policies would be those that retain all extant small farms in addition to encouraging others. Alternately, the CRLA claims could draw on a vision of what holism requires. Sometimes, Martin and Olmstead's statements can be viewed as a contrasting vision of holism. For example, "Mechanization plays a role in this evolving structure of the U.S. farming industry, permitting some farmers to manage large units efficiently while allowing others to operate small farms as part-time or hobby operations."

Furthermore, the authors call on the constraints view to serve their ends in several places. They note a goal of public policy, to eliminate dangerous and undesirable jobs, and assert that mechanization research and the enactment of health and safety standards fulfill this goal by making workers more expensive relative to machines. They maintain that this protects workers, keeps the American agricultural industry competitive, and eases the nation's immigration problems. Concern for workers is a hallmark of the constraints view, so its use, when the result is loss of jobs, is somewhat ironic. More direct acknowledgement of the constraints view is found in their conclusion, where they recommend establishing "a program to assist displaced workers" as part of a proposal for an integrated and rational policy.

By now, Martin and Olmstead's concerns have moved from net-gain into constraints territory. And their concern for limiting immigration does not appear to be easily defensible on either view. This concern may require calling upon the holistic view for justification. When a letter to *Science* took them to task for disloyalty to the net-gain and the constraints views, the authors cited the universally recognized rights of nation-states to exclusionary borders.[7] Perhaps the holistic view can justify this as required for peaceful and prosperous human societies.

This illustration demonstrates how policy disputes can masquerade as factual disputes and confuse even the participants. It also exhibits a laudable if implicit discomfort about relying on just one value-laden perspective in making policy decisions. Disputes about policies and practices in agricultural research often incorporate disputes about values. It is important to consider what the world is likely to be like as a result of these policy decisions, which will affect productivity, net gain, constraints, and the interactions and balances in and between components of our ecosystems and communities. So it is important to identify the value-laden perspectives that are germane to the decisions at hand and incorporate, to the extent feasible, an understanding of how the decisions will affect those values in discussions of how to proceed.

This analysis helps to demonstrate how "facts" serve as "evidence" in policy disputes only in conjunction with values. This is unavoidable. To be used to justify or oppose human endeavors, factual assertions must be combined with value judgements about the facts. Policies and practices are put in place to attempt to assure certain valued results. Facts become "evidence" in policy disputes as they bear on whether or not those results are achieved. To be used correctly, evidence must be weighed in a balance containing values before there can be a "correct" finding or decision.

Current Debates over Biotechnology Research

If an area such as research on new machines for agriculture, where the results of new discoveries on the workforce may be foreseen with considerable reliability, can generate such controversy, we can expect much more with the developments and applications of new recombinant-DNA techniques.

Martin and Olmstead's article, published in February 1985, generated the letter to the editor mentioned above, which was published with the authors' reply in an October issue. In contrast, an article about biotechnology by Winston J. Brill, published in January 1985, provoked two letters and an author's reply in the issue of July 12.[8] The argument continued with letters in the October 18 and January 10, 1986, issues.[9] Brill's article provoked four times the response of Martin and Olmsteads'.

"Safety Concerns and Genetic Engineering in Agriculture" takes the position that "Problems caused by introduction of organisms such as kudzu and the gypsy moth into a foreign environment do not imply problems for an organism, currently considered safe in its habitat, with characterized recombinant genes added to its genome." Brill takes the net gain view: "The economic and environmental benefits expected to accrue from agricultural use of recombinant organisms are great . . . and should be considered in relation to the potential risks." In his figuring, the potential risks are not great. He believes that there is no need to treat genetically engineered organisms "differently from conventionally altered organisms with regard to safety evaluation."

Not all ecologists share his views of the potential risks. They stress the need to develop protocols for testing, on a case-by-case basis, proposed introductions of organisms modified by recombinant-DNA technique.[10] Nonetheless, this debate in Science is basically limited to a discussion of benefits and risks. One does not hear these scientists' voices charging that the proposed experiments or field tests should not occur at all unless they can satisfy human-rights requirements. Nor are they raising questions about the effects of these new developments on the balance among elements in the food production system in the United States. Where constraints perspectives arise, the notion of human rights that prevails is limited to mortality and morbidity risks. Ecologists' concerns for the "balance of nature" do exhibit some holistic elements—although here the notion of ecosystems seems artificially limited by leaving humans and human societies out. Putting them in, of course, poses great difficulties for a scientific notion of "balance of nature." Furthermore, the notions of risk in the top-priority, net-gain, and constraints views leave out many that the holistic view might find important. Some social scientists are beginning to raise holistic concerns explicitly. For example, Frederick H. Buttel describes new developments in biotechnology and accompanying institutional changes that could be detrimental to relationships among experiment stations and land-grant universities and could exacerbate conflicts between them and state legislators, other state-level groups, and national interests.

Buttel and the research team of Lawrence Busch, Michael Hansen, Jeffrey Burkhardt, William B. Lacy, and Laura R. Lacy believe that current directions in biotechnology research are likely to favor seed and chemical companies and processors over farmers as "the primary clients for plant breeding research."[11] In an article titled "Plant Breeding and Biotechnology," that group of researchers writes that "neither scientists nor administrators appear aware of the potential for conflict between the interests of farmers and those of agribusiness." They note other changes in system functions that they believe may be undesirable—for instance, inattention to the need for training the next generation of plant breeders and inhibitions on open flow of information among biotechnology scientists.

These changes, among others, come with the increasing influence of certain private sector institutions in setting research agendas for agriculture. Buttel and the Burkhardt, Busch, Hansen, Lacy, and Lacy research team do not believe the changes are good for the system; the problems they identify are problems from a holistic point of view.

Taking a broad perspective on human rights, these problems are relevant for the constraints view as well, at least insofar as they project further consolidation and displacement of small farming operations. Concern for the safety of farm operators and workers handling the new products and for community members and consumers exposed to them are also salient to this view.

The federal research and regulatory establishments take the scientific traditionalists' limited view of values and, thereby, of their responsibilities in this area. For instance, the National Academy of Sciences (NAS) issued a report this year titled "Introduction of Recombinant DNA-Engineered Organisms into the Environment: Key Issues.[12] The NAS bills itself as a "private, nonprofit, self-perpetuating society of distinguished scholars engaged in science and engineering research, dedicated to the furtherance of science and technology and their uses for the general welfare." But the booklet limits the view of the general welfare primarily to the top priority or trade-offs perspectives. It states, "Mounting concerns about environmental degradation, together with the pressing problems of ensuring adequate food and health care for a rapidly expanding global population, provide a compelling rationale for the accelerated study and development of biological organisms for use in agriculture, health care, and biosphere management. The committee concludes that R-DNA techniques constitute a powerful and safe new means for the modification of organisms" (p. 6). There is a tip of the hat there to environmental concerns, but it seems to be in the context that these powerful new tools will allow us to increase productivity or use less land.

Similarly, in its report on the results of the deliberations of the interagency working groups on biotechnology, the *Federal Register* notice of June 26, 1986, states, "The working group sought to achieve a balance between regulation adequate to ensure health and environmental safety while maintaining sufficient regulatory flexibility to avoid impeding the growth of an infant industry."[13] This is a trade-offs view, within which a very narrow range of effects is deemed appropriate for consideration.

In its report to the National Science Foundation (NSF) in 1984 on policy issues for biotechnology, Arthur D. Little, Inc. summarizes its "suggested research agenda, selected with regard to the objectives of the Division of Policy Research and Analysis" in NSF.[14] This agenda

> includes the following high priority areas: 1) university/industry relationships (the commercial potential of biotechnology calls for new policies regarding relationships between academia and commercial enterprises); 2) commercialization of biotechnology (a number of questions associated with the commercial use of basic science have emerged, focussing on the nature of proprietary rights and the flow of scientific information); and 3) risks and biotechnology (an accurate assessment of the various risks potentially associated with this new technology and its products is needed and requires new methods and detailed study).

This agenda is oriented to the top-priority view, with the top goal being "commercialization." But whose commercialization?

The recent announcement from NSF for proposals for plant science centers also seems to take a top-priority view, with a tip of the hat to trade-offs: "The goal of the Program is to encourage the best in basic research in plant science. Growth in imaginative basic research is believed necessary to insure the future competitive position of U.S. agriculture and production of renewable resources. The program is founded on the needs to improve the quality of food and fiber, to increase the efficiency of their production, to develop and use new products, and to sustain a renewable resource base."[15]

Note the use of the phrase *is believed necessary.* This phrase may well demonstrate the commendable restraint of the scientists who put this announcement together. They may have doubts as to whether there is a direct causal connection between imaginative basic research and the future competitive position of U.S. agriculture. Their top priority—their goal —is to foster imaginative basic research. The goal may be laudable, but this statement is also a narrow and a value-laden point of view.

By often limiting their perspectives on the values at issue to those of priorities or trade-offs, the research and regulatory establishments demonstrate a correspondingly limited view of moral responsibility on these matters. Foresight needn't extend to encompass the constraints or holistic perspectives. It would be unfair, however, to overlook other sectors of these establishments for whom the constraints and holistic views seem more important. The Biotechnology Study Group of the EPA Science Advisory Board has called for more research on environmental and health effects, remedial action, and destruction of environmental pollutants.[16] A recent report from the Office of Technology Assessment concludes that unless the U.S. government helps small and medium-size farmers get access to new technologies, they and their communities will suffer.[17] The report also expresses a concern about private firms' inequitable capture of benefits from publicly supported research and the effects of that on competition. In another example, the National Agricultural Chemicals Association is cooperating with the Campaign for Pesticide Reform, developing an unusual agreement to reform the federal pesticide law, an agreement that pays attention to all four views.[18] And a recent report from the Board on Agriculture of the National Research Council also calls for a systems or holistic approach to *managing* pesticide resistance.[19]

But of course, all of these establishments aren't the only ones with responsibilities here. Because they are nonexclusive, moral responsibilities are shared by everyone. People have to be prepared to raise these perspectives in forums and ways which can make a difference.

Of fundamental importance is the recognition that all these positions and decisions are not value-free. In fact, they often downgrade or overlook certain values and advance others. In doing so, they downgrade certain interests and interest groups and advance others. Couched often in what appears to be value-free language, and cloaked often by a presumption that science is value-free, positions articulated by scientists and science advisors can encourage the authors and others to limit moral responsibilities inappropriately.

To exercise foresight in these circumstances requires going beyond the narrow confines of justifications in terms of top priorities and trade-offs and thinking more carefully about the implications of the constraints and holistic views. The effects of research and development in these new biotechnologies on human rights and on the organizations, groups, and institutions that compose local communities, states, nations, and their relationships to each other and to the rest of the world must be considered. The question must be posted: What should the earth look like and be like in the coming centuries? It will take careful planning, careful thinking, care to evaluate and overcome mistakes. But first, it will take the recognition that these moral responsibilities cannot be assigned away to any particular individuals or groups as only their job. This is a distinctively human endeavor in which all must participate. Until individuals and organizations institutionalize ways in which to do so, they—like Dr. Stockmann and the individuals and organizations in his community—cannot fulfill their moral responsibilities.

Notes

1. H. Ibsen, *An Enemy of the People,* in H. Ibsen, *Four Major Plays, Volume II* (N.Y.: The New American Library, 1970).

2. M. P. Glazer and P. M. Glazer, "Whistleblowers Who Risked It All," *Recreation News* 5, no. 4 (1987):1, 10–12.

3. J. Ladd, "Collective and Individual Moral Responsibility in Engineering: Some Questions," *IEEE Technology and Society* (1982):3–10.

4. R. D. Hollander, "Values and Making Decisions about Agricultural Research," *Agriculture and Human Values* 3, no. 3 (1986):33–40.

5. W. Aiken, "On Evaluating Agricultural Research," in *New Directions for Agriculture and Agricultural Research, Neglected Dimensions and Emerging Alternatives,* ed. Kenneth Dalberg (Totowa, N.J.: Rowman and Allenheld, 1986), 31–41.

6. P. L. Martin and A. L. Olmstead, "The Agricultural Mechanization Controversy," *Science* 227 (1985):601–6.

7. D. H. Gieringer, P. L. Martin, and A. L. Olmstead, "Letters: Restricting Immigrant Labor," *Science* 230 (1985):237–38.

8. W. J. Brill, "Safety Concerns and Genetic Engineering in Agriculture," *Science* 227 (1985):381–84; W. J. Brill, "Letters: Genetic Engineering in Agriculture," *Science* 229 (1985):111–17.

9. "Letters: Genetic 'Engineering'?" *Science* 230 (1985):237–38; "Letters: Genetic Engineering," *Science* 231 (1986):103.

10. G. Kolata, "How Safe Are Engineered Organisms," *Science* 229 (1985):34–35; G. Kolata, "Letters: Biotechnology and the Biosphere," *Science* 229 (1985):1338; G. Kolata, "Letters: Ice Nucleating Bacteria," *Science* 231 (1986):536.

11. F. H. Buttel, "Biotechnology and Agricultural Research Policy: Emergent Issues," *New Directions for Agriculture and Agricultural Research,* 312–47. See Aiken reference in note 5. The point was originally made in his paper "Biotechnology and the Struggle over Agricultural Research Policy," presented at the Workshop on Ethical and Value Choices in the Selection of Agricultural Research Goals, Washington, D.C., April 12–13, 1984. See also

Lawrence Busch, Michael Hansen, Jeffrey Burkhardt, William B. Lacy, "The Impact of Biotechnology on Public Agricultural Research: The Case of Plant Breeding," presented at the AAAS Annual Meeting in New York, May 1984; Michael Hansen, Lawrence Busch, Jeffrey Burkhardt, William B. Lacy, and Laura R. Lacy, "Plant Breeding and Biotechnology," *Bioscience* 36 (January 1986):29–39.

12. Committee on the Introduction of Genetically Engineered Organisms into the Environment, *Introduction of Recombinant DNA-Engineered Organisms into the Environment: Key Issues* (Washington, D.C.: National Academy Press, 1987).

13. Office of Science and Technology Policy, "Coordinated Framework for Regulation of Biotechnology: Announcement of Policy and Notice for Public Comment," *Federal Register* (June 26, 1986), 23302–3.

14. A. D. Little, Inc., *Study of Federal Biotechnology Policy Issues Report,* prepared for the Division of Policy Research and Analysis, National Science Foundation, Washington, D.C., BOA no. PRA-8400692, ADL reference 50362, 1984, v.

15. *Planet Science Centers Program,* Interagency Program Announcement, OMB 3145–0058, NSF 87–71.

16. "Biotechnology Products: EPA Finds Gaps in Risk Assessment," *CEN* (February 17, 1986):6–7.

17. U.S. Congress, Office of Technology Assessment, *Technology, Public Policy, and the Changing Structure of American Agriculture,* OTA-F-285 (Washington, D.C.: U.S. Government Printing Office, March 1986).

18. "Agreement Ends Impasse over Pesticide Law," *CEN* 14 (March 24, 1986). Marjory Sun, "Antagonists Agree on Pesticide Law Reform," *Science* 231 (1986):16–17.

19. "Pesticide Resistance Calls for New Strategies," *CEN* 6 (April 14, 1986).

The "Frankenstein Thing": The Moral Impact of Genetic Engineering of Agricultural Animals on Society and Future Science

21

B. E. ROLLIN

Shortly after I had accepted the invitation to address the Iowa State University Agricultural Bioethics Symposium, I remarked to a friend of mine (a nonscientist) that I was going to address a conference on genetic engineering of animals. "Ah," he said, "the Frankenstein thing!" I didn't pay much mind to his remark until perhaps a week later, when, while perusing the new acquisitions in our library, I encountered an extraordinary, newly published, five-hundred-page volume entitled *The Frankenstein Catalog: Being a Comprehensive History of Novels, Translations, Adaptations, Stories, Critical Works, Popular Articles, Series, Fumetti, Verse, Stage Plays, Films, Cartoons, Puppetry, Radio and Television Programs, Comics, Satire and Humor, Spoken and Musical Recordings, Tapes and Sheet Music Featuring Frankenstein's Monster and/or Descended from Mary Shelley's Novel.*[1] The entire book is precisely a descriptive catalogue, a list and very brief descriptions of the works mentioned in the title. Amazing though it is that anyone would publish such a book, its content is even more incredible, for it in fact lists 2,666 such works (including 145 editions of Shelley's novel), the vast majority of which date from the mid-twentieth century. All of this obviously indicates that in the Frankenstein story is an archetypal myth or category which somehow speaks to or for twentieth-century concerns and which could perhaps be used to shed light on the social and moral issues raised by genetic engineering of animals. My intuition was confirmed while visiting Australia and discussing with an Australian agricultural researcher the, to him, surprising public hostility and protest that his research into teratology in animals had provoked. "I can't understand it," he told me. "There was absolutely no pain or suffering endured by any of the animals. All I can think of," he said, "is that it must have been the Frankenstein thing." And in its cover story on the fortieth anniversary of the Hiroshima bombing, *Time* magazine again invoked the Frankenstein theme as a major voice in post–World War II popular culture, indicating that it was society's way of expressing its fear and horror of a science and technology that had unleashed the atomic bomb.[2]

Given this pervasive reaction, it seems valuable to explore the social and moral concerns about research into genetic engineering of agricultural animals using the Frankenstein myth as a framework for our discussion. As I shall try to show, the social concerns and the genuine moral concerns are not always identical and are, in fact, sometimes confounded and not clearly separated in the public mind and, indeed, in the minds of many scientists. Furthermore, some of the deepest and most genuine moral concerns encapsulated in the Frankenstein story are undoubtedly least discussed and explored either by the scientific community or the public.

Before pursuing this inquiry, it is worth pausing to stress that, in general and not just in the case of genetic engineering, both the scientific community and the general public often miss the mark in their attention to the ethical issues growing out of scientific activity. My good friend, the late Dr. Bernard Schoenberg, associate dean of the Columbia University College of Physicians and Surgeons, used to remark that while the public and the medical community alike were spilling a great deal of ink on issues such as disconnecting the respirator from Karen Ann Quinlan, almost no one was discussing the far more fundamental moral issue of fee for service in medicine! By the same token, when the Baby Fae affair occurred, generating much debate, neither scientists nor the public seemed to realize that there was little moral difference between this case and any case of killing an animal for possible human benefit or for research. For that matter, it was hard to see the moral difference between harvesting hearts from baboons and harvesting heart valves from pigs – something which has been standard practice for some time and yet something which no one had raised as an ethical question. The practical rather than moral difference, of course, was in the sensational nature of the story – transplanting hearts from animals plucks at primordial emotions. As I told the press, this issue was not from a conceptual point of view worth discussing in isolation from the general question of whether science has the right, in far less dramatic cases, to expend animal lives, sometimes with far more suffering, anxiety, and fear on the part of the animal than that experienced by the anesthetized baboon.

Again, and in the same vein, I recently gave the keynote speech at an Australian conference on the moral issues in animal experimentation. In the course of my talk, I pointed out that merely citing a list of human benefits engendered by research on animals does not in itself logically serve to morally justify that invasive use of animals, anymore than a listing of benefits which emerged from medical research on political prisoners, concentration camp inmates, slaves, criminals, and the like would justify doing such research without obtaining noncoerced, informed consent. Despite this obvious point, many researchers, in their talks, continued to base their defense of animal use solely on benefits to man, as indeed the U.S. medical research community has tended to do.

Unfortunately, the general public is usually too ignorant about science to be able to sort out the genuine moral issues emerging from scientific activity and in practice tends to rely on the media to do the job for it. The media, in turn, is of course less interested in conceptual or factual accuracy than in selling papers, as one reporter candidly told me during the Baby Fae case. So that, as we shall see in the case of genetic engineering, what gets presented to the public as major moral issues are often not moral issues at all. At the same time, scientists are themselves often unable to discriminate the ethical issues implicit in or arising out of their own activity, and essentially wait to have the issues defined for them by the public, or by the same people who define them for the public, so that the issues do not get adequately dealt with from the scientific side either. The failure of scientists to discriminate moral issues in science in turn raises doubt about what I have called "the ideology of science"—in essence, the set of philosophical principles, positions, assumptions, presuppositions, and values that scientists tend to acquire unconsciously along with their scientific knowledge in the course of their training. This pervasive ideology is rooted in the logical positivism and behaviorism of the 1920s, and suggests that science deals only with what is observable and verifiable—with "facts." Since statements about values, including moral values, are not verifiable, they are alleged not to fall within the scientist's purview, at least in his or her capacity as scientist. This is often codified as the slogan that science is "value-free," and is accompanied by the claim that, although values perhaps enter into the use to which science is put by society, values never enter into science itself. Value judgements, including ethical ones, are often viewed by scientists as emotive responses and matters of individual preference or taste, and hence not as rationally adjudicable; after all, *de gustibus non disputandum est*. Thus, philosophically, many scientists see nothing wrong with ignoring moral issues or even with being emotional about moral issues, since their unspoken philosophical training leads them to believe that moral issues are nothing but emotional issues.

In actual fact, as I have taken pains to demonstrate elsewhere, science is not value-free and includes ethical values.[3] Indeed, all science is permeated with valuational presuppositions. Surprisingly, perhaps, the very notion of what will count as a fact, as a legitimate object of investigation, or as data relevant to a given question, rests squarely upon valuational presuppositions. Consider, for example, the Scientific Revolution, during which the commonsense, sense-experience–based physics and cosmology of Aristotle were replaced by the rationalistic, mathematical, geometrical physics of Galileo and Newton. The discovery of new data or new facts is not what forced the rejection of Aristotelianism—on the contrary, empirical observations all buttressed Aristotle's idea of a world of qualitative differences! What led to the rejection of Aristotelianism was essentially a change in *value*—a discrediting of information provided by the senses, as Descartes does so well in his *Meditations,* and a correlative valuing of the rational and mathematically expressible over the empirical, of Plato's philosophy over Aristotle's. This was nicely expressed in Galileo's claim that, in essence, an omniscient deity would have to be a mathematician and create a mathematical unity underlying apparent diversity.

Few of you would go along with one of my acquaintances, an accomplished medical researcher and Rhodes scholar, who heatedly informed me that the question of the use of animals in science is simply a scientific, not a moral, question, and that, indeed, science has nothing to do with ethics. In an attempt to show him that he had not thought out the logic of his position, I pointed out that if science is indeed constrained only by scientific concerns, why don't we use children for research, since they are better models for humans than are animals. His reply, amazingly enough, was, "Because they won't let us." And none of you who have watched the obviously morally based changes in scientific opinion on whether race differences and intelligence are legitimate objects of study, on whether homosexuality is a disease or an alternative lifestyle, and on whether alcoholism or wife-beating is sickness or badness can truly deny that science is rooted in moral valuational assumptions.[4]

In any event, my main concern thus far has been with showing that our understanding of moral issues does not usually keep pace with the scientific progress that generates these issues. And if I should stress to you any urgent message at all, let it be that scientists ignore or shunt off these issues until they assume crisis proportions at the scientists' own peril. In the final analysis, public money pays for science and ever-increasingly demands accountability. A failure on the part of any area of science to clearly define the moral issues growing out of its activity and to deal with them puts that area's very existence in peril, as the case of animal research around the world dramatically illustrates. Furthermore, in a moral variation on Gresham's Law, bad moral thinking can drive good moral thinking out of circulation. Thus a failure on the part of scientists to articulate the genuine moral issues in genetic engineering or any area leaves open the very real possibility of false and irrelevant, but sensationalistic, issues occupying the public mind and being used as a basis for social policy. And, as we shall now see as we return to the "Frankenstein thing" as a basis for discussing genetic engineering of animals, the same sort of thing can happen here.

A nice illustration of my moral Gresham's Law may be found in the fact that probably the most socially pervasive component of the Frankenstein metaphor as it applies to genetic engineering of animals is also the least interesting morally. This component may be characterized in terms of the classic line from old Frankenstein genre movies that "there are certain things man was not meant to know" (or to do or to explore). In other words, there is certain scientific knowledge or activity, or application of scientific knowledge, that in and of itself is taboo, irrespective of its consequences. In the case of genetic engineering of animals, this would most likely be attributed by those who hold such a view to the creation of chimeras or crossing of species lines; to major modifications within a species which are phenotypically apparent (such as genetically manipulating for leglessness in farm animals); or even, as press and public reaction to the Fox-Rifkin lawsuit against the U.S. Department of Agriculture (USDA) indicates, to introduction of genetic material derived from humans into animals, or, presumably, to the introduction of animal-derived genetic material into humans. (As I suggested earlier, a similar strain of thought arose during the Baby Fae case; numerous people seemed to have perceived unspecified ethical difficulties in a human having an animal part.)

The pattern of thinking represented in this sort of version of the "Frankenstein thing," though widespread, does not represent a genuine moral issue and does not raise moral questions requiring social adjudication. It appears to me to have a variety of nonmoral sources which are typically confused with moral concerns.

One such source is most certainly theological: the Judeo-Christian notion that God created living things "each according to its own kind," with the clear implication, expressed both in nineteenth-century and contemporary opposition to Darwin, that species are fixed, clearly separated from one another, and immutable—and furthermore, ought to be. A nontheological, historically influential, philosophical vector buttressing this view in Western thought is Platonized Aristotelianism, which again postulates fixed natural kinds, again immutable and clearly demarcated from one another. Indeed, Aristotle defends this view on the grounds that its contrary would make knowledge impossible. (An opposite tendency also found in Aristotle, which suggests an infinite continuum and graduation in species, has been all but ignored.) But, of course, such theological and philosophical prejudices are not in themselves legitimate bases for moral questioning of genetic engineering, though they help explain certain people's knee-jerk bias against it. And, of course, to a religious person, anything that violates any of his or her religious tenets must be seen as morally problematic.

But reservations against "meddling with species" stem from sources beyond theology and Aristotle. They stem also from a common but scientifically unsophisticated and rather muddled understanding by a virtually scientifically illiterate public of *species* as being, as it were, the building blocks or atoms of the biological world, out of which the biological world is built and upon which it rests. To tinker with species is, in this view, to tinker with the stability of nature, to (in some unarticulated way) shake the entire Great Chain of Being, as Coleridge's Ancient Mariner did when he killed the albatross. The fact that species are, in current biological theory, dynamic rather than static, stop-action views of a continuing evolutionary process is ignored by such critics. These critics also ignore the fact that the notion of (genetic) species is highly complex and problematic, and that it has been rejected by some biologists such as Kensch in favor of notions like subspecies, races, *Rassenkreis,* or *Formemkreis* as not being the fundamental taxonomic unit.[5] (On the other hand, the fact that most biologists do treat species as the fundamental taxonomic unit and as being "more real" than other such units, as Michael Ruse puts it, lends support to such critics.[6] If subspecies is the fundamental unit, incidentally, then we have been genetically engineering biological reality with no fuss for thousands of years. For that matter, if one takes seriously the currently standard definition of a species as a naturally interbreeding population, then one could argue that certain subspecies we have genetically engineered by breeding, such as the Great Dane and Chihuahua, in fact constitute separate species.

Incidentally, as I have argued elsewhere, much of the debate about the reality or nonreality of species rests upon a deep and ancient philosophical mistake, the attempt to classify all phenomena as being either *nomos* or *physis,* nature or convention.[7] In actual fact, it appears that species represent something of both; what species we find in the world depend on the scientific-theoretical lenses with which we examine the world. Given current theories of evolution and molecular genetics, such procedures as DNA matching and serological evidence from protein matching give us an objective method of species classification. But, at the same time, we must recognize that these objective tests are based on accepted biological theories and that, given an alternative biological theory — one, for instance, oriented far more to whole-organism function or ecological place than to the molecular basis of life — we would probably generate a completely different taxonomy, complete with a totally different set of objective tests.[8]

Another factor which appears to me to foster the belief in inviolability and sacredness of species is the environmental movement. It is a psychologically small, albeit conceptually untenable and logically vast, step to go from concern that species not be allowed to become extinct to the idea that we ought not change them. Or perhaps, a bit more reasonably, the movement of thought is rather from the idea that species ought not be allowed to vanish as a result of what humans do to the idea that they ought also not change at our hands. Built into the environmental movement is, in short, a "nature knows best, hands off nature" mindset, but that is more an attitude than a reasoned position.

In my view, as I have argued elsewhere, species are not the sorts of things which are legitimate objects of moral concern.[9] It makes little sense to me to assert that it would be permissible to shoot ten Siberian tigers as long as there were plenty of Siberian tigers or to suggest, as one of my environmental ethicist colleagues has written, that if a species of endangered moss is in the migratory path of a species of plentiful elk, it is not only permissible but obligatory to save the moss by shooting the elk.[10] In my view, as I shall discuss later in detail, only sentient individuals are legitimate objects of moral concern; species count morally only insofar as they represent a group of individuals, and the last ten Siberian tigers are no different as a moral issue than are any other ten Siberian tigers.[11] There is certainly a great loss in species becoming extinct, but it is fundamentally, perhaps, an aesthetic one, analogous perhaps to our repulsion at trampling a flower. Ethics is relevant only insofar as one is morally obligated not to destroy aesthetic objects, or to deprive future generations of having them in their *umwelt*.

In any case, I think we can conclude from all of the above that the first aspect of the "Frankenstein thing," namely that "there are certain things we simply ought not do, and species modification by genetic engineering is one of them," does not represent a defensible moral claim even if it may be so perceived by large numbers of people. To respond to this pervasive idea, however, the research community needs to do a great deal of public education, necessarily preceded by self-education in ethical issues.

Any rational attempt to extract a genuine moral issue from the first aspect of the "Frankenstein thing" we have discussed must be based in a second aspect of the "Frankenstein thing," to which we now turn. Crucial to most versions of the Frankenstein myth is the danger to humans that grows out of unbridled scientific curiosity. Thus, the dictum that "there are certain things that are just wrong to do" becomes replaced in this aspect of the myth by the dictum that "there are certain things that are wrong to do because they must or will inevitably lead to great harm to human beings." The archetypal image of this is Dr. Frankenstein's monster on a rampage — terrorizing, hurting, killing, and harming the innocent. Despite the scientist's noble intentions (Dr. Frankenstein's purpose, in the novel, was to help humanity), his activity was morally wrong not (or not merely) because of hybris but because of his unjustifiable failure to foresee the dangerous consequences of his actions or even to consider the possibility of such consequences and take steps and precautions to limit them. And to this objection, of course, twentieth-century science and technology is quite vulnerable. We have tended to believe that if we can do something, we should, and we forge ahead as quickly as possible, damn the torpedoes. And we have also tended to believe, as part of the ideology of science discussed earlier, that scientists are not morally responsible for the pernicious uses to which their explorations are put; the responsibility for these consequences allegedly belongs to politicians, governments, military agencies, or corporations. There are, of course, notable exceptions to this claim, as Asilomar nobly illustrates,[12] but in the main, scientists are vulnerable to this criticism, as any of us who have served on university biosafety or surveillance committees knows all too well. The recent discovery of killer bees on the loose in California represents another example of unjustifiable negligence on the part of scientists, who of course imported and bred these insects apparently without proper regard for the dangers involved.

What, if any, are the potential dangers inherent in genetic manipulation of animals in agriculture? This is certainly a legitimate issue which should be addressed by all of those working in the area. Even a cursory examination of the area suggests a number of possibilities that should be raised, explored, and assessed in terms of likely risk, and for which mechanisms of minimizing the risk should be devised before embarking upon genetic engineering of animals utilizing new principles of biotechnology.

I would suggest that any country contemplating such work establish formal mechanisms to ensure that the social questions associated with potential risks growing out of genetic engineering of animals be fully evaluated and made known to the public, much in the way recombinant DNA work has been dealt with in the United States. I have recommended to the USDA the establishment of something analogous to the National Institutes of Health's (NIH's) Recombinant DNA Advisory Committee (RAC) to assess potential risks and other ethical and social issues associated with genetic engineering of agricultural animals. This ought to proceed in a number of stages. First, a fairly large committee consisting of scientists, attorneys, public policy people, ethics people, and members of the public should delineate the issues and suggest broad guidelines for assessing and minimizing risk. If possible, levels of risk should be identified and broad characterizations of types of research and applications thereof delineated. Subsequently, local committees analogous to human research committees, animal research committees, and biosafety committees, with significant public membership, should be appointed at institutions engaged in research or application of genetic engineering of animals. As much accurate publicity as possible should accompany all aspects of this process, both to dispel irrational components of the "Frankenstein thing" and to show responsiveness to legitimate concerns. Such committees should also engage an entirely different set of ethical questions which we will outline shortly in discussing the last component of the Frankenstein myth.

The sorts of hazards, risks, and potential dangers associated with genetic engineering of agricultural animals appear to be the following (doubtless, most of you can supplement my list significantly). At this stage, I believe that it is vital to err on the side of caution, to look at and consider every possible danger, however apparently unlikely. It is usually far easier to prevent than to amend, especially in an area like agriculture in which vast amounts of money or food are at stake when a technological tool or procedure becomes integral to an operation and is later found questionable or unsuccessful. The use of antibiotics in feeds provides a clear example, as do overly intensified and overly capitalized systems in pork production and crop decimation growing out of unanticipated disease and genetic uniformity.

The first set of potential dangers emerging from the new forms of genetic engineering of agricultural animals obviously stems from the rapidity with which such activity can introduce wholesale change in organisms. Traditional genetic engineering, of course, was done by selective breeding over long periods of time, during which time one had ample opportunity to observe the untoward effects of one's narrow selection for isolated characteristics. But with the techniques we are discussing here, we are doing our selection "in the fast lane." This leads to two sorts of potential danger.

First of all, there may be untoward consequences affecting the organism which one is rapidly changing. The characteristic consequence is that genetic engineering may have implications that are unsuspected. Thus, for example, when wheat was genetically engineered for resistance to blast, that characteristic was looked at in isolation, and the genetic basis for this resistance was encoded into the organism. The back-up gene for general resistance was, however, ignored. As a result, the new organism was very susceptible to all sorts of viruses which, in one generation, mutated sufficiently to devastate the crop.

Second, the isolated characteristic being engineered into the organism may have unsuspected harmful consequences to humans who consume the resultant animal. Thus one can imagine genetically engineering, for example, faster growth in beef cattle in such a way as to increase certain levels of hormones, which, when increased in concentration, turn out to be either carcinogens for human beings when ingested over a thirty-year period or teratogens, in a manner similar to diethylstilbestrol. The deep issue here is that one can of course genetically engineer traits in animals without a full understanding of the mechanisms involved in phenotypic expression of the traits, with resulting disaster. This in turn suggests that it would be prudent to be cautious in one's engineering until one has at least a reasonable sense of the physiological mechanisms affected.

A second set of risks growing out of genetic engineering of the sort we have been discussing replicates and amplifies problems already inherent in selection by breeding—namely the narrowing of a gene pool; the tendency towards genetic uniformity; the emergence of harmful recessives; the loss of hybrid vigor; and, of course, the greater susceptibility of organisms to devastation by pathogens, as has been shown in genetic engineering in crops. (On the other hand, genetic engineering can have the opposite effect in making available to the gene pool greater variety than ever before, as in the case of artificial insemination making new genetic material available to beef breeders.)

A third set of risks arises out of the fact that in certain cases when one changes animals, one can thereby change the pathogens to which they are host. This can occur in two conceivable ways. First, if one were genetically engineering for resistance to a given pathogen in an animal, one could thereby unwittingly be selecting for new variants among the natural mutations of that microbe to which the modified animal would not be resistant. These new organisms could then be infectious to these animals, other animals, or humans. Second, even if one were changing the animal in nonimmunological ways, one could be changing the pathogens to which it is host by changing the microenvironment where they live. This in turn could result in these pathogens becoming dangerous to humans or to other animals. Thus, in changing agricultural animals by accelerated genetic means, one runs the risk of affecting the pathogenicity of the microorganisms that inhabit the organism in unknown and unpredictable ways. And the more precipitous the change, the more inestimable the effects on the pathogens are likely to be.

A fourth set of risks is environmental and ecological and is associated with the possibility of radically altering an animal and then having it get loose in an environment which was not anticipated. While this certainly seems like a minimal danger when dealing with intensively maintained cattle or chicken, it could surely pose a real problem with extensively managed swine, or with rabbits, or even with extensively managed cattle. Bitter experience teaches us that such dangers cannot be estimated, even with species whose characteristics are well known (witness what happened with rabbits and cats in Australia and with the mongoose in Hawaii); a fortiori, an ignorance of what would happen with newly engineered creatures, is even more certain.

We have talked briefly of the potential risks of genetically engineering agricultural animals involving the animals themselves, the general human population, other animals, and the environment. A fifth set of risks is relevant to a special subgroup of the human population—namely, those who will actually be doing the experimentation on and genetic manipulations of the animals. Common sense tells us (and there is ample evidence to support this claim) that people working directly with dangerous materials are at greater risk than is the general population. The last smallpox death in England resulted from a virus contracted in a laboratory, and standard precautions are taken universally to protect people working with dangerous substances. But the need for extra vigilance in dealing with new situations is well illustrated by the deaths caused twenty years ago by Marburg virus, which did not affect laboratory workers who had been handling live monkeys, but which killed those people who had been collecting cells from dead animals for cell culture. In the case of genetic engineering, people handling the vectors used to introduce the genes could conceivably be at risk.

For any significant risks which we have discussed, or for others I may have omitted and which might pose real danger, the imperative for risk management can be generated without recourse to ethical considerations; rational self-interest and prudence would dictate that one not be cavalier about them. Thus even if a person has absolutely no concern for anyone but himself and his loved ones, he would wish to see anything that might do massive harm controlled, since he and his might just as easily as anyone else fall victim to its effects. Therefore, in my view, following up an insight of Kant's, it is difficult to separate moral from prudential reckoning in such areas. Only when we consider the third and final aspect of the "Frankenstein thing" do we in fact encounter something that requires purely moral deliberation and decision, because morality and self-interest are very unlikely to coincide in these cases. In other words, one is unlikely to do the right thing for prudential reasons and, in actual fact, moral behavior in this area is likely to exact costs in self-interest. It is to these questions we now turn.

The final aspect of the Frankenstein myth is more difficult than the others to find in many of the popular renditions of the myth but was in fact a central theme in Mary Shelley's novel. This dimension concerns the plight of the creature engendered by abuse of science. In the novel, the creature is innocent yet isolated: shunned, mocked, abused, and persecuted in a plight not of its own making. Seeking love and companionship, it finds only hatred and rejection. One can find traces of this concern for the monster in the classic *Frankenstein* movie, and it is in fact a central theme in the recent remake of *King Kong*. Translated into the arena of genetic engineering of agricultural animals, this aspect of the myth, in essence, raises the question of the moral status of animals, of the rights of these animals — certainly the most difficult of the moral questions we have examined in our discussion. And it is so difficult for a complex of reasons worth briefly detailing.

In the first place, when it comes to trying to get a purchase on our obligations to other creatures, we get little help from common sense, our intuitions, ordinary practice, the law, or even traditional moral philosophy. Common sense and ordinary practice say little about our obligations to animals other than enjoining us to avoid cruelty, hardly a great help since most animal suffering and death is not the result of cruelty. (The great emphasis on cruelty to animals and love for animals is the major failing in the traditional animal welfare movement. Most scientists and agriculturists are not cruel, yet they invasively use countless numbers of animals. On the other hand, loving something is neither necessary nor sufficient for treating it morally. I certainly don't love most of the human surgeons I know; I don't even like them; yet, I am bound to treat them morally. By the same token, many people who love their pets mistreat them in countless ways, from providing improper diets to denying them exercise.) Our intuitions on animals are incoherent; the same people who condemn branding of cattle may dock the ears and tails of their dogs. The law is of no help — in the eyes of the law animals are property, either private property or community property. The Animal Welfare Act, reflecting irrational social prejudice, does not consider rats, mice, or domestic farm animals to be animals; for purposes of the act, a dead dog used in research is an animal, a live mouse is not. And traditional moral philosophy is of no help either, since for most of its history it was virtually mute on the subject of our obligations to other creatures. More has been written on this subject, in fact, in the past ten years than in the previous three thousand.

All of this is further complicated by a major component of the same ideology of science that we discussed earlier and that I have explored in detail elsewhere.[13] From about 1920 until the mid-1970s, behaviorism was a major component of scientific ideology, and it was dogma to assert that we could not scientifically know that animals were conscious or even that they felt pain. Indeed, this is still dogma in many quarters — a USDA inspector recently told me that a medical researcher had informed him that dogs lack a sufficiently highly developed cerebral cortex to experience pain, and I have heard variations on this theme over and over. A leading veterinary pain expert told me that the majority of veterinarians still view anesthesia as a way of restraining the animal. (This was confirmed for me when I was lecturing at a leading veterinary school early in my involvement with this issue and naively remarked that at least veterinarians could not doubt that animals felt pain, or else why would they study anesthesia and analgesia?

Up jumped the associate dean, livid with rage. "Anesthesia and analgesia have nothing to do with pain," he shouted, "They are methods of chemical restraint.") Analgesia is virtually never used on laboratory animals and very rarely used in clinical veterinary practice. Ironically, rodents are the most infrequent recipients of analgesia, yet most pain and analgesic research is done on rodents, so the dose response curves are well known.

Obviously, much scientific research, agricultural practice, and, indeed, ordinary activity rest on exploiting animals, so that it is far easier and more comfortable not to think about animals in moral categories. Nonetheless, common sense has never denied that what we do to animals matters to them; that they have needs and interests, physical and psychological, and that they can suffer, physically and psychologically, when those needs and interests are thwarted and infringed upon. During the past decade, society has just begun to realize the implications of its own assumptions about animals and has begun to be aware that we do have moral obligations to them. Hence the rise of the animal rights movement, a massive international stirring which cannot be ignored, which questions much of our traditional treatment of animals, and which has been called "the Vietnam of the 80s."

At all events, a growing number of people in the scientific community are beginning to think seriously about the moral status of animals. In the past eight years, I have lectured to over thirty veterinary schools all over the world on these issues, as well as to biomedical scientists of all sorts, attorneys, agriculturalists, psychologists, government officials, farmers, ranchers, and scores of other groups. I have testified before Congress and state legislatures and have served as a consultant to various agencies of three national governments. In the course of this decade, I have tried to develop an ethic to guide us on the uses of animals, one that I believe follows logically from moral assumptions we all share by virtue of living in democratic societies. In other words, rather than generate my own ethic and attempt to force it on others, I have attempted, following Socrates, to extract from others (though they may not and often do not realize it) what their own moral assumptions entail about animals. Such an ethic is necessary, not as a blueprint for instant social change in all areas but, as Aristotle put it, as a target to aim at, and as a yardstick to measure our current conduct. Without such an ethic, as my colleague Dr. Harry Gorman, surgeon and researcher, has beautifully put it to me, we tend to confuse what we are doing with what we ought to be doing.

Given the constraints of time, I can only present the briefest sketch of this ideal for animals. For those of you who wish to pursue the topic more deeply, I would refer you to my book, *Animal Rights and Human Morality*.[14] Stated boldly, I ask people to consider whether they can present any rationally defensible grounds for excluding animals from the moral arena or from the scope of moral concern and deliberation. Surely animals are more like children than like wheelbarrows in that they can be hurt and that what we do to them matters to them. None of the standard, historically pervasive differences that have been cited to exclude animals from the moral arena will bear rational scrutiny. The claims that man has a soul and animals do not, that man is evolutionarily superior to animals, that man is superior in force to animals, that man is rational and animals are not, do not suffice to exclude animals from the moral arena and from falling under the purview of our socially pervasive moral concepts. In other words, given the logic of our moral ideas, there is no way to preclude extending them to animals. And this is not difficult to do.

In democratic societies, we accept the notion that individual humans — not the state, the Reich, the *Volk,* or some other abstract entity — are the basic objects of moral concern. We attempt to cash out this insight, in part, by generally making many of our social decisions in terms that would benefit the majority, the preponderance of individuals (i.e., in utilitarian terms of greatest benefit to the greatest number). In such calculations, each individual is counted as one, and thus no one's interests are ignored. But such decision making presents the risk of riding roughshod over the minority in any given case. So democratic societies have developed the notion of individual rights, protective fences built around the individual which guard him or her in certain ways from encroachment by the interests of the majority. These rights are based upon plausible hypotheses about human nature — about the interests or needs of human beings that are central to people, and whose infringement or thwarting *matters most* to people (or, we feel, *ought* to matter). So, for example, we protect freedom of speech, even when virtually no one wishes to hear the speaker's ideas. Similarly, we protect the right of assembly; the right to choose one's own companions and one's own beliefs; and also the individual's right not to be tortured even if it is in the general interest to torture, as in the case of a criminal who has stolen and hidden vast amounts of public money. And all of these rights are not simply abstract moral notions but are built into the legal system.

The extension of this logic to animals is clear. Animals too have natures (i.e., fundamental interests central to their existences, whose thwarting or infringement matters to them). This set of needs and interests — physical and psychological, genetically encoded and environmentally expressed — which make up the animal's nature I call the animal's *telos,* following Aristotle. It is the pigness of the pig, the dogness of the dog. Such a notion is not mystical; it follows, in fact, from modern biology. Thus, it ill serves the issue at hand when scientists sneer at this notion, as one person at the NIH did recently, by suggesting that an animal's only nature is to serve us and die. According to the logic of our position, animals' basic interests as determined by their *telos* ought also to be morally and legally protected; this is the cash value of talking about rights. This, then, is what I take to be the logical extension of our socially sanctioned, moral notions when applied to animals, and when one cannot cite a morally relevant difference between people and animals which would forestall such application. Obviously animals do not have the same rights as humans, even ideally, since they do not have the same natures. So it will not do to ridicule the position by saying that I am urging that turtles have the right to vote or dogs have freedom of bark.

I have devoted much of my recent activity to attempting to actualize this ethic as far as is practically possible into veterinary medicine and research uses of animals, where its relevance is evident. But what does it tell us about genetic engineering of animals? Let me first of all clear up a misconception which has arisen about my notion of *telos*. It has been asserted by some opponents of genetic engineering that in my view *telos* is inviolable, and it is immoral to change it. I have never said that. What I argue is that, *given* an animal's *telos,* certain interests which are part of that *telos* ought to be inviolable. Thus, given a burrowing animal, it is wrong to cage it so that it can't burrow. But I have never asserted that there is anything wrong with changing the *telos* of a burrowing animal so that burrowing no longer matters to it.

The proper application of these ideas to genetic engineering of farm animals is made quite interesting by the fact that so much of our current intensive agricultural use of animals involves forcing animals into environmental contexts for which their natures are not suited. As a result, we must perpetually depend on highly artificial devices like debeaking in chickens and extensive uses of drugs and chemicals as well as contend with "production diseases." While extensive agriculture has its own problems, at least the problems are, as it were, natural to the animals rather than being created by the humanly devised management system. Ideally, from the point of view of the animals' welfare, I would like to see society back off from ever-increasing intensification. (This would, I think, have certain social and economic benefits as well.) But in all likelihood, increasing intensification is here to stay. So the main *moral* challenge to those involved in genetic engineering of agricultural animals is to avoid modifying the animal for the sake of efficiency and productivity at the expense of the animal's happiness or satisfaction of its nature. Economic pressures, of course, in the main, militate against my recommendation. This is why I asserted earlier that this is truly a moral challenge. Also militating against this is the fact that hitherto the animal's welfare (except insofar as it affects economic productivity of an entire operation) has entered neither into intensive agricultural decision making nor into research serving that decision making. (This was freely admitted to me by a group of high-ranking USDA officials.)

Nonetheless, given the increasing public concern about the welfare of all animals including agricultural animals, as well as the strong moral arguments in favor of concern for animals, it is imperative that this moral vector enter into agriculture. And certainly genetic engineering is an excellent place for this vector to be felt. The basic principles that should guide thinking in this area are not hard to see. Obviously, as a minimal principle, the animals should suffer no more as a result of genetic intervention than they would have without it. Ideally, they should suffer less and be happier. Thus, in my view, it would be grossly immoral (as has actually been suggested) to use genetic engineering to change chickens into wingless, legless, and featherless creatures who could be hooked to food pumps and not waste energy. Similarly (as has also been suggested), it would be wrong to manipulate the genome of pigs to produce leglessness, with the animals after all still having all the psychological urges to move. On the other hand, if genetic engineering is used to genuinely suit the animal to its stipulated environment and therefore eliminate the friction between *telos* and environment which clearly results in suffering, boredom, pain, stress, and disease, and this conduces to the animal's happiness, it does not appear morally problematic. Thus, if one were to genetically alter chickens' physical and psychological needs so that all evidence (such as results of preference testing, physiological signs of stress, behavioral signs of stress, individual animal productivity, and health) indicated that the animals were happy, this would be morally acceptable according to the theory I have been expounding, though many people, myself included, would certainly not be quite comfortable with it, probably on aesthetic grounds.

Obviously, therefore, these considerations of the animal's welfare, independently of the effect on humans, should be weighed and considered before a piece of genetic engineering is undertaken. And such consideration should be part of the formal charge of the committee we discussed earlier. Thus, if someone were to suggest using genetic engineering to create larger beef cattle, the researcher should be required to show that there is good reason to believe that the animal's joints could withstand the extra stress and that no new suffering would be engendered by such genetic manipulation. In this way, we can at least begin to assure that the animals' interests are weighed along with ours. In the case of totally virgin territory, as in the creation of chimeras, an even stronger burden of proof should be put on the proposer to demonstrate that his manipulation would not lead to suffering.

In sum, in my view, the genetic engineering of animals in and of itself is morally neutral, very much like the traditional breeding of animals or, indeed, like any tool. If it is used judiciously to benefit humans and animals, with foreseeable risks controlled and the welfare of the animals kept clearly in mind as a goal and a governor, it is certainly morally nonproblematic and can provide great benefits. On the other hand, if it is used simply because it is there, in a manner guided at most only by considerations of economic expediency and "efficiency" or by quest for knowledge for its own sake with no moral thinking tempering its development, it could well instantiate the worst rational fears encapsulated in the "Frankenstein thing." To those of you upon whom the primary responsibility for this choice rests, let me conclude by reminding you that, though Frankenstein was in fact the name of the scientist, virtually everyone thinks it is the name of the monster.[15]

Notes

1. D. F. Glut, *The Frankenstein Catalog* (Jefferson, N.C.: McFarland, 1984).
2. *Time* (July 29, 1985):54–59.
3. B. E. Rollin, *The Teaching of Responsibility* (Hertfordshire, England: Universities Federation for Animal Welfare, Potters Bar, 1983); "The Moral Status of Research Animals in Psychology," *American Psychologist* 40, no. 8 (1985):920–26.
4. B. E. Rollin, "On the Nature of Illness," *Man and Medicine* 4, no. 3 (1979):157ff.
5. J. R. Baker, *Race* (London: Oxford University Press, 1984).
6. M. Ruse, *The Philosophy of Biology* (London: Hutchinson, 1973).
7. B. E. Rollin, "Nature, Convention, and Genre Theory," *Poetics* 10 (1981):127–43.
8. Ibid.
9. B. E. Rollin, *Animal Rights and Human Morality* (Buffalo: Prometheus, 1981).
10. H. Rolston, "Duties to Endangered Species," *BioScience* (1984).
11. Rollin, *Animal Rights and Human Morality.*
12. See the foreword of this volume for more on Asilomar.
13. B. E. Rollin, "Animal Consciousness and Scientific Change," *New Ideas in Psychology* (in press); R. E. Rollin, "Animal Pain," *Advances in Animal Welfare Science* (in press).
14. Rollin, *Animal Rights and Human Morality.*
15. I wish to thank Linda Rollin, M. Lynn Kesel, David Neil, Murray Nabors, Robert Ellis, George Seidel, and Dan Lyons for dialogue and criticisms.

22 | The Case against bGH

GARY COMSTOCK

> There are times when the drive [for technological progress] needs moral encouragement, when hope and daring rather than fear and caution should lead. Ours is not one of them.
>
> —Hans Jonas[1]

Bovine growth hormone (bGH) is a protein that occurs naturally in cattle. Produced by the pituitary gland, it regulates the cow's lactational cycle; generally speaking, the more bGH a cow has, the more milk she gives. Using the techniques of genetic engineering, researchers at Monsanto Company have devised low-cost techniques to manufacture the drug artificially. The plan is to sell the product to farmers who will administer it in daily doses to their animals. Monsanto's motivation is not hard to discern: a single dose of bGH may cost them ten cents to make and yet be sold to farmers for fifty cents; a worldwide market of one billion dollars a year is predicted by Monsanto's vice president, Lee Miller; and a profit ratio of two dollars returned for every dollar invested is foreseen.[2] The first agricultural biotechnology to hit the market, bGH will be commercially available as soon as the Food and Drug Administration finds it safe for consumers. Approval is expected before the end of 1990.

The product works. Daily injections cause dairy cows to increase production of milk from 10 to 15 percent.[3] (A study funded jointly by Cornell University and Monsanto in 1984 showed some cows increasing production by as much as 41 percent, but these results are not expected in the field.) In addition to Monsanto, at least three pharmaceutical companies—Eli Lilly,

Upjohn, and American Cyanamid — believe that they can produce and sell bGH so cheaply that large, efficient dairy farmers will not be able to do without it.

The social benefits seem clear: some farmers will be able to produce more milk from fewer cows using less labor. Dairy operations with large herds are expected to cull their less productive cows, put more feed into the remaining ones, and get the same amount of milk. All this, presumably, while farmers reduce their working hours. As the senior vice president for research and development at Monsanto exclaims, "In the future, a farmer using BST will be able to produce as much milk with 70 or 80 cows as can be produced with 100 cows today, use 15 percent less feed to produce that milk, and finally have a chance to be more profitable!"[4] Consumers are also supposed to benefit: as dairy farmers save money, their decreased costs will be passed along to shoppers in the form of lower milk prices. And the nation's poor, who need the same amount of calories and calcium as others but who have to pay a larger share of their personal income to get it, will benefit even more than the middle-class consumer from slightly lowered food costs.

With so many benefits promised, why has bGH become anathema to some farm and consumer groups? In April of 1986 a coalition of farm and environmental organizations asked the FDA to prepare an environmental impact statement on bGH. Led by Jeremy Rifkin and his organization, the Foundation on Economic Trends, the group included the Humane Society, the Wisconsin Family Farm Defense Fund, Wisconsin's secretary of state, and was later to be joined by the Audubon Society, the Wisconsin Farmers Union, and the Farmers Union Milk Marketing Cooperative.[5] The FDA turned down the request. Without examining the philosophical merits of the farmers' case, the regulatory affairs commissioner of the FDA claimed that the requested assessments had already been performed and, in fact, were "submitted to the FDA with investigational new animal drug (INAD) applications pending before the agency."[6] While the FDA has not expressed interest in the wider ramifications of the farmers' case, those interested in the future of rural America, agriculture, and farm animals have. As a result, a voluminous literature on the subject of bGH has quickly developed.[7] Unfortunately, no attempt has been made to frame the issue in specifically moral terms or to address systematically its ethical implications.[8]

The farmers' opposition is based on three claims: that bGH is harmful to the environment, constitutes inhumane treatment of cows, and will displace farmers from already-distressed rural communities. Since any "environmental" damage caused by the drug would be linked to its effect on dairy cows or humans, the farmers' case effectively consists of two claims:

that bGH represents an inhumane method of treating animals, and that bGH threatens to dislocate an unacceptably high number of disadvantaged farmers. Predicting that the use of bGH will drive as many as 30 percent of all dairy farmers out of business, Rifkin claimed that bGH usage would lead to "the single most devastating economic dislocation in U.S. agricultural history."[9] Nor will the farmers affected be randomly selected; arguably, they will be primarily small- and medium-size farmers in later stages of their careers with small herds or high debt loads and lacking highly mechanized and intensively managed operations.

I consider the farmers' two claims below.

Humane Treatment of Cows

Several contemporary philosophers including Tom Regan and Ned Hettinger have argued that higher mammals such as cows possess all of the characteristics needed to be bearers of moral rights: sentience, purpose, interests, social life, intelligence, emotions, and so forth.[10] To possess moral rights is to be entitled to fully equal treatment; we do not countenance discrimination against children with Downs syndrome even though they are not as sentient, purposive, or rational as we are. Since they have moral value, they have it fully and are entitled to equal treatment.

If adult higher mammals possess moral rights, then we must treat them the same way we treat humans who, like animals, lack certain characteristics of normal humans. It is permissible for us to act paternalistically toward them insofar as they need extra care. But we may not exploit those beings who lack a certain measure of linguistic ability or emotional security or physical autonomy. If Regan and Hettinger are right, we ought not to do to cows anything that we would not do to mentally enfeebled human beings; the differences between cows and the "marginal human" cases are morally irrelevant.[11] On the animal rights view, allowing scientists to administer bGH to cows simply to observe its effects would be similar to allowing scientists to administer it to brain-damaged adults for the same purpose. We would not allow this to be done to any human who was not capable of giving (or withholding) informed consent; consequently, we ought not to allow it to be done to other beings in the identical position.

The strictness of the animal rights view has been criticized as failing to make relevant moral discriminations. For example, moral value is not like a light switch that is either off or on. It comes in gradations, as our ability to acquire more of it (through education) and to lose some of it (by entering an irreversibly comatose state) shows. The quality, intensity, and complexity of different animals' mental and social lives make them bearers of different gradations of moral value. In addition, it is sometimes appropriate to use another as a means to our own ends even if the other possesses the full complement of moral value. We do this often, as when we allow attendants to fill our gas tanks or ask our hosts to provide us with a glass of seltzer. It is not always morally objectionable to use another as a means to our own ends even if that other is the possessor of supreme moral worth. Each of these considerations points to a morally relevant distinction that Regan fails to make in his either/or case (either adult mammals have moral rights in the same sense that humans do, or they do not).

A less controversial stance is that animals have gradations of inherent value determined in part by the complexity and intensity of their social and mental life, and that we must act toward them in ways that respect this value.[12] Supposing that we could successfully defend the "humane treatment" of animals view, would the use of bGH be acceptable?

An answer to this question relies on our being able to assess the degree to which bGH use diminishes the quality of the animals' physical and psychological health and, if it does, whether this harm is justified by the benefits it confers. Accurate data about the long-term effects of bGH are not available, but studies have been completed of the effects of using bGH during one lactational cycle.

bGH works by stimulating the division of muscle and liver cells and, apparently, inhibiting the growth of fat cells. (This is the reason for its attractiveness beyond the dairy industry; beef and swine producers expect it to lead to leaner meat.) Evaluations of the effect of the drug on the overall health of the animal are divided between those who see few, if any, adverse effects and those who are more skeptical. Don Beitz, animal scientist at Iowa State University, notes that while use of bGH leads to increased feed consumption, bone growth (in young animals), muscle growth (in adults), and milk production, the efficiency of the digestive tract and reproductive system seems to be unaffected; the birth rate of calves is the same for treated and untreated mothers.[13] Beitz acknowledges that treated animals do require more intensive management since their nutrient requirements are greater, but he does not anticipate deleterious effects from proper usage of the protein.

Others are more concerned. bGH will put the cows' body metabolism under greater physiological stress. David Kronfeld of the University of Pennsylvania claims that high levels of bGH result in "subclinical hypermetabolic ketosis, a condition associated with reduced reproductive efficiency, mastitis, decreased immune function and the full gamut of other diseases typical of early lactation."[14] Research at the University of Missouri, according to Kronfeld, also supports the view that the drug negatively effects the reproductive efficiency and health of many animals. It is worth pointing out, however, that mastitis—a painful infection of the udder—is a very common problem for dairy cows even without bGH, and that the dangers associated with decreased immune function—lowered resistance to infectious and contagious diseases—may be minimized with good veterinary care.

Both the proponents and critics of bGH are relying on scientific data taken from experiments lasting only a short term. Until we have studies that look at the longer-term effects of bGH, studies covering several lactations, we will not be able to say with much confidence whether the drug seriously impoverishes the lives of the cows or not. But on the basis of what we do know, it seems reasonable to conclude that bGH is relatively safe for the cows if carefully administered: that is, given for one lactational cycle and in low doses. Under such conditions, the treatment seems no more inhumane than do many other practices typical of modern dairy operations.

This judgment must be set in the context of current conditions, however. We face a period of shrinking profits for farmers; if bGH is readily accessible for as little as fifty cents a dose, it will be used by farmers in many ways. Will it always be used prudently? Having invested in the accessories needed to use the drug, will a farmer have the incentive to restrict use to one cycle at low levels? Knowing that animals are destined for slaughter eventually, will farmers be tempted to overstimulate older cows whose lactational cycle is in the latter stages? Even if scientific research shows that increased use of the drug does *not* lead to increased returns on capital, will farmers in the field know this? What is to keep them from administering doses above the "safe" levels? Unless they are legally required to do so, it is difficult to imagine all farmers using bGH uniformly in the moderate style of the university trials; the financial rewards associated with more intensive usage seem too attractive. In the real world of dairy farming, it is improbable that bGH would be restricted to one cycle at low doses. And, we may safely assume that the risk of mistreating animals grows with each level of drug usage.

Further, even if effective regulations controlled abuses, we may ask about the management techniques encouraged by the drug: will dairy cows on bGH be more likely to have access to pasture and open land, or is it more likely that they will be located on intensive operations where almost all of the inputs are purchased and the space allowed for the cows is minimal? Factory farming is already objectionable on the modest philosophical grounds suggested above; will bGH help to alleviate the trend toward treating dairy cattle as nothing more than milk machines on legs?[15] Conceivably, the use of bGH could lead to a decrease in the number of cows in the national herd, and that might appear to ease the need for confinement. But, again, we must match reality against theory: dairy cows are not confined because of the number of them in the national herd. They are confined because the intensive style of modern dairy farming requires easy access to them. The prospect of having to inject the cows with a daily dose of bGH does not hold out much promise of turning us toward more sustainable agricultural practices or toward a kind of dairying in which animals are treated more humanely.

This objection from humane treatment might lose force if other considerations outweighed it. Do current economic conditions justify the risks associated with bGH usage? If we were at war and milk supplies were endangered, if extreme shortages were anticipated in the short or long term, if our children were calcium deficient because our cows were such poor milkers, then our need to exploit the cows' ability to produce might outweigh the risks to the animals' health. Few would argue, however, that this is currently the case. In developed countries, there is too much milk, not too little; the U.S. Congress is trying to decrease milk production by 8.7 percent by paying producers $1.2 billion to get out of dairy production. Human need for more cow milk does not outweigh the risks associated with the drug's use.

One might argue that bGH is needed in developing countries. Here it is necessary to examine the broader problems of hunger and poverty in nations such as Guatemala, Ethiopia, and Bangladesh. Do such countries need more milk? In tropical climates, milk production from cows is at a minimum; the weather, for one thing, discourages milk production by, for example, making the growing of hay and forage almost impossible. Moreover, many of the people in such cultures would not consume more milk even if it were abundant since they have a natural biological intolerance for it. And finally, infants in these countries ought not to be nourished on cows' milk at all, but on their mothers' milk. So even in the Third World— where one might think that milk production needs a boost—bGH turns out to be a bad answer to an irrelevant question.

We ought also to consider the wider economic dimensions of agriculture in developing countries. Is a capital- and management-intensive technology an appropriate solution to these countries' complex food problems? The style of farming associated with bGH usage is more adaptable by *latifundios,* large plantation-like farms, than by smaller independent farms. Yet the smaller, independently owned farms hold the most hope for relieving widespread hunger and poverty in the long run. So, even if more milk *were* needed in the Third World, the system of large-scale dairying likely to be required (or induced) by bovine somatotropin may not be the answer.

If other considerations justify the risks to dairy cattle associated with intense bGH usage, we have not been shown what they are. Lacking any persuasive arguments to that effect, one would have to argue for bGH on ever more general grounds; a boycott against bGH would send the wrong signal to industries investing in biotechnology and using animals in their research. Isn't opposition to bGH on humane grounds equivalent to opposition to all recombinant DNA research on animal rights grounds? Clearly not. Unlike most animal rightists, "humane-ists" see potential benefits in genetic engineering of animals. When genes for disease resistance are microinjected into mice eggs, many of the offspring possess altered immune response systems which give them a genetic ability to survive diseases that would kill the parent.[16] Few genuinely interested in the physical health of animals may object to this result of rDNA research. Biotechnology may enhance an animal's capacity to flourish and may enable successive generations to be less susceptible to disease.

But, again, the abstract potential of biotechnology must be weighed against the realities of the scientific world. Of the many lists stating the explicit goals of biotech research, not one that I have seen includes the claim that biotech is aimed at improving the quality of animal life for the sake of animals. Indeed, we want to improve animals not for their sake but for ours. It is only the bodies of animals we are interested in, not their spirits. We want the bodies to be better suited to our interests in milking them, eating them, carving them, chasing them, riding them, hunting them, or using them as factories. Indeed, a good part of the scientific community is presently most interested in producing mice that are naturally susceptible to diseases such as diabetes, cancer, and AIDS, since such animals make better models for understanding the function that genes might play in eliminating disease from humans. These genetically engineered animals are intentionally programmed to lead lives of protracted deprivation or suffering. We are a long way here from engineering animals for increased capacity to flourish. So, while biotechnology has the potential to improve animal life simply for the sake of animal life—and ought to be encouraged to do so for this reason—this sort of research is unlikely to attract the dollars necessary for significant research projects.

The conclusion suggested by this discussion is not one favorable to the marketing and use of bGH. The drug itself is a potential threat to the well-being of the animals as it is likely to be administered to them in doses that have deleterious or unknown effects. It is also likely to exacerbate the problems involved in the treatment of animals on factory farms. The Wisconsin farmers' first claim—that bGH represents an inhumane method of treating animals—is not without merit for anyone taking seriously the inherent value of animals.[17]

Social and Economic Effects

The Wisconsin farmers also called for a boycott against the use of bGH on the grounds that it would dislocate too many producers. The argument here cannot be that the technology will put *some* workers out of business; if we were to object to inventions on those grounds we would have had to oppose railroads, electricity, and electronic printing presses. With the introduction of each of these technologies, blacksmiths, lamp-makers, and typesetters were put out of business. Yet society is better off for having the advantages brought by computers, rapid transit, and widespread literacy. We should not underestimate the pain involved when workers must move into new professions and neighborhoods. But most of us are willing to accept some costs, as long as they are outweighed by social gains.

Our concerns are raised not when new inventions displace labor, but when new inventions displace labor in ways that seem unnecessary, unfair, arbitrary, or completely unaccompanied by redemptive benefits. People are not infinitely plastic: attachment to place, profession, and way of life is part of human nature. So, even in a market economy in which inventiveness and entrepreneurial independence is valued, it is rational to try to minimize the pain associated with rapid social change, and actively to oppose those changes that benefit only those already most advantaged. Is the new invention needed? If so, how can it be introduced with the least amount of suffering? If not, why is it being promoted and who stands to gain from it? These questions force us to look more carefully at the data about bGH's predicted effects.

Robert Kalter himself has taken pains to point out that his study has been misused by Rifkin. He does not predict that bGH will drive 30 percent of all dairy farmers out of business.[18] He claims that many technical changes—including bGH, but not limited to it—combined with the removal of dairy price supports, could cause a 25 to 30 percent increase in the nation's milk supply. Since the demand for milk is relatively static, however, this extra milk would not be consumed. Market equilibrium, then, would require a 25 to 30 percent reduction in the number of cows and farms in order to bring supply in line with demand. Since not all farms going out of dairy production would go out of farming and since bGH is only part of the broader technical change expected in the future dairy industry, Kalter expects that the above scenario might send between 23.3 and 46.0 percent of dairy farmers out of milking.[19]

But this decrease must be compared to what we can expect for dairying without bGH in its future. If the drug is kept off the market, not all dairy farmers will stay in operation; between 17.2 and 20.4 percent of them are expected to go out of business even if there is no technical change. So the technology itself cannot be held responsible for all of the 23 to 46 percent reduction foreseen by Kalter. How much could be blamed on bGH? If my reckoning is correct, the figures would be between 15.9 and 25.6 percent.[20]

In New York, there were 17,500 dairy farms in 1984. If price supports are removed, Kalter predicts that the number will fall to somewhere between 12,600 and 15,800 depending on the rate of adoption. This decline of 2200 to 4900 is too conservative by the estimates of Magrath and Tauer.[21] They predict that as many as 5400 farms will fail in New York. But they also point out that over the last ten years "conventional technological changes and ongoing structural change has resulted in the exit of 4000 dairy farms." Of course, this means that bGH would still take down more dairy operations than have been lost in the last ten years.

We must also put this reduction in the broader history of declining farm numbers. In the years between 1964 and 1984, the United States saw a decrease of 77 percent of dairy farms and, Kalter points out, "this happened without hormone technology."[22] The decrease is due to a number of factors, but the improved efficiency brought about by artificial insemination, embryo transfer, and computerized record keeping plays a large role. Since the current "farm crisis" has between 9 and 24 percent of all dairy farmers getting out of the business over three-year periods, bGH will only add to the total. This leads Kalter to conclude that bGH will simply "speed up the process a little."[23]

While Kalter's estimates are more conservative than Rifkin's rhetoric, the figures command attention. And, since so much rests on the accuracy of these figures, a brief consideration of Kalter's methodology is required. Buttel and Geisler have questioned whether bGH will actually be adopted as quickly and widely as Kalter assumes. They note that he used a mail survey in which information about bGH was included and farmers were asked whether they would use the product described. Kalter's response rate with dairy farmers was only 13 percent, "well below the typical rate of 65 to 70 percent in mail surveys among the general public."[24] A different method was used by the O.T.A.; experts were asked for their judgments about who would adopt the technology. Based on this "consensor" method, sharp disparities were predicted between adoption rates of large and small farmers, with an 80 to 90 percent adoption rate predicted among farmers with over a half million dollars in annual sales, but only 10 to 20 percent among those with less than twenty thousand.

Buttel and Geisler are skeptical about Kalter's results because mail surveys, like personal interviews, do not always give objective results. Respondents "have a tendency to provide responses that they feel the . . . survey authors want."[25] Consequently, among those few who chose to respond to Kalter's survey, most probably did not want "to admit that they [were] inattentive [or would be slow to adopt] new technologies."[26]

The point is well taken. Surveys using more reliable methods are desirable. But the force of the argument is not clear. Is it supposed to show that Kalter's study is methodologically suspect? This might have academic interest, but it would not necessarily invalidate the results. Or is the point that the results are necessarily skewed, and that adoption rates will clearly be much lower? This would have greater relevance for those concerned about bGH, but the authors do not ask us to draw this conclusion. As the authors themselves admit, we have good reasons for suspecting Kalter's method but not necessarily his results. Beyond the academic point that Buttel and Geisler have scored, nothing seems to be changed.

The authors have a second argument: Kalter's estimates were skewed upward by the sort of information about bGH given in the mailing. "If . . . hypothetical fact sheets and advertisements" given to respondents "paint a rosy picture of a technology, adoption rates will be biased upwards."[27] Again, this sounds reasonable. But did the information in this case actually affect the results? On the authors' own admission, it appears not. They write, "In contrast to arguments that the rapid adoption rate estimated by Kalter et al. is due to the high return-over-cost margin that farmers would enjoy with BST, studies in Alabama, which gave farmers far less encouraging data, yielded virtually identical rates."[28] Unless I misunderstand the sentence, this bit of evidence confirms rather than repudiates Kalter's study. If a much less rosy picture of bGH does not affect Alabama's farmers' estimates of their adoption rate, why should we believe Buttel and Geisler that "the methodologies used to estimate the configuration of adoption curves have led to some exaggeration of the rate of adoption"?[29]

Buttel and Geisler offer a third argument against the rapid adoption assumption. Smaller, part-time farmers have different criteria for making decisions. Whereas large farmers on the whole try to achieve the highest average rate of profit, smaller farmers often place more emphasis on preserving a certain rotation of crops or on holding onto the family farm for their sons or daughters. "Agricultural census data on smaller, part-time farmers demonstrate well that these farmers' decision-making criteria are not primarily those of maximizing returns to equity capital."[30] Since decisions are made on different bases, we should not assume that all dairy farmers will value equally the increased efficiencies found in bGH.

This is another valid point. It is a too little appreciated fact that Amish farmers refused to adopt no-till farming techniques not only on religious or environmental grounds, but because it would mean fewer Amish sons would need to be employed in farming.[31] Few commercial farmers would consider such a factor to be relevant to their business decisions, but the fact that some Amish did indicates the diversity of decision-making criteria among farmers.

If the general point is right, it is again difficult to see its relevance to Kalter's estimates. Small farmers may gather in Wisconsin and state their opposition to bGH. But what these small farmers say they would like to do as a collective group is not the same thing they will do as individual operators. The Wisconsin farmers do not want bGH because it will drive some of them out of business. But they know as well as we do that if the coalition fails in its goal to keep bGH off the market, most of them will be forced to use it if they want to stay competitive. These are well-informed and politically astute farmers. But they are also financially stressed. So even if they have different decision-making criteria and express these as a group, when it comes to saving their operations, they will act just like the big producers. A survey in Wisconsin showed that while two-thirds of farmers said that they did not want bGH, most of those same farmers said that they would use it if it were made available.

It is not only rational for dairy farmers to oppose bGH as a group and yet to use it as individuals; it is very likely to occur. So, again, the Buttel-Geisler reservations about the high rate of adoption assumed by Kalter fail to convince.

There seem to be few good philosophical reasons to doubt Kalter's estimates. Even on the most modest of assumptions, then, technical change (of which bGH will be a large part) seems likely to increase the expected rate of farmers leaving dairying. Without bGH we can expect at least 17.2 percent of farmers to go out of business. With it, that figure rises to at least 23.3 percent. Notice, however, that this is an increase of some 33 percent in the number of farm failures. (The number could go as high as 120 percent. Using Kalter's figures for a low inelasticity of demand and a high rate of technical change, farm failures could go from 20.4 to 46.0 percent, an increase of over 100 percent.) If Kalter's numbers seem reliable, then, we might wonder at his judgment. Is a 33 percent increase in the number of dairy farmers forced out of dairying to be interpreted simply as "speeding up the process *a little*" [emphasis added]? In a time of milk surpluses and rural crises, are these additional failures needed? Is it fair to ask a very small percentage of society to bear all of the costs for a marginal increase in the efficiency of milk production?

Now, someone might respond by admitting that even though we do not need bGH, we should not try to keep it off the market; doing so would put a chill on future industry investment that could cause our country to fall behind others in the biotech race. For the vice president of Monsanto, "the choice is clear: Either be an innovative farmer or compete with one." And Howard Schneiderman wants American agriculture to have the "innovative edge" on other farmers around the world, claiming that "if we do not continue to innovate, we will be forced out of business."[32]

The answer to this worry is painfully obvious; it is most improbable that American businesses will pull out of a race for $40 billion in prizes just because they see an obstacle on the track. Corporate America is much better at jumping hurdles than the vice president of Monsanto gives it credit for being. Its lawyers are very good at anticipating and responding to legislative and judicial constraints. Its marketing analysts are very good at figuring out what sorts of products consumers want and do not want. Its scientists are very good at identifying and locating various genes of agronomic and economic importance. So those wishing to stop bGH are no threat to the Monsantos and Eli Lillys of the world. The argument that stopping bGH will put a chill on research is not only unpersuasive, it demonstrates a lack of confidence in a large and resourceful industry.

Part of the problem with bGH is that it discriminates against small- and medium-size farmers, the same farmers who helped to pay for research on it. The genetic engineering techniques that industry will use in making the protein were perfected at universities like Cornell using public monies. And, in research funded jointly by Monsanto, dairy scientists at that land-grant school tested the validity of the drug while agricultural economists at the same university devised econometric models to gauge its market viability and impact. In both indirect and direct fashion, the potentially displaced farmers paid monies for public research which, in turn, led to private sector developments that promise to put the farmers out of business. Many of these farmers have families that have been in the dairy business for generations. Prima facie, they are justified in believing they have been treated unfairly.

Assessing the deeper merits of this belief, however, is no simple matter. There are several problems here, touching on issues of fundamental disagreement between social philosophers. What is distributive justice in economic matters? What does it require in this case? Don't the greater benefits brought by the free operation of markets outweigh the social costs incurred in the constant shifting of labor resources in capitalism? If so, isn't bGH really for our common good, even if it displaces one-fifth of all our dairy farmers?

Before taking up these questions, I want to lay my cards on the table. It is my intuition that the Wisconsin farmers are right: something about bGH's social and economic effects is objectionable. On examination, however, I have found it very difficult to say exactly what that is. No laws have been broken, no contracts circumvented, no federal regulations ignored. Not even Jeremy Rifkin claims that any legal damages have yet been done to any party. So the "injustice"—if we are to call it that—is taking a very strange form. None of it has happened: the 15.9 to 25.6 percent of dairy failures due to bGH are hypothetical (even if probable) *future* events.

If the oddness of this case tempts us to throw up our hands, we will have to resist; if we ever needed a language in which to discuss "potential future injustices," it is now. The skill of social scientists to make sensitive ex-ante studies about the likelihood of various consequences of new technologies grows. As it does, their sophistication in predicting the future quickly surpasses our ability to assess the results of their studies morally. And yet— if it is in our power to do so—it is surely better to prevent an injustice before the fact than to try to remedy one after. So the urgency of trying to assess the farmers' second charge is as great as the conceptual difficulties involved in doing it.

If bGH is unfair to farmers, it is not yet clear how or why. We might begin by specifying the group that, at some future point, is supposed to be the one offended. According to Kalter, bGH is size-neutral: it can be used by farmers whether they have "ten cows or a thousand."[33] Contrary to the claims of bGH's proponents, however, many studies have contested the claim that bGH is size-neutral because winners and losers will not be evenly distributed throughout the farming population.

Even though bGH may be marketed at a low cost per dosage, successful use of the product will require significant managerial expertise and access to capital. "These constraints," write Barnes and Nowak, "will be most problematic for smaller and less efficient farms that have operators that are less knowledgeable and older."[34] Nowak, Kloppenburg, and Barnes point out that their research found "substantial differences among [dairy] operations" in ability to use bGH:

> For example, the managerial constraints (forage testing, ration balancing, and DHI records) became more problematic as the age of the respondent increased, and the average pounds of milk (cwt) sold, number of cows in the operation, and rolling herd average all decreased. To the extent that the use of bGH will require these managerial inputs, then operators or operations with these characteristics will either not use it or use it in an inefficient manner.[35]

A new technology is not size neutral when its cost-effectiveness improves as the number of cows and the quality of managerial skills increase. And yet, even if individual doses are priced low, larger and younger and better-educated farmers will reap disproportionately greater benefits than older, less "aggressive" farmers. bGH is not size-neutral.

Fred Buttel goes even further, asserting that it is difficult to imagine any agricultural biotechnology that would be scale-neutral:

> Scale neutrality of a technology is often taken to mean that the technology will have no impact on the size distribution of farms when, in fact, few new agricultural technologies are neutral in their impacts on farm size distribution. . . . The essence of technological change is the substitution of relatively cheap, abundant factors or means of production (e.g., bGH) for relatively expensive, scarce factors (e.g., land, cows), along with new management or organizational means for altering the mix of input factors. . . . A new technology such as bGH *requires* [my emphasis] substantial managerial expertise for its successful adoption by farmers, and if farmers vary substantially according to scale of production in their management expertise, the technology is, *a priori, nonscale-neutral* [Buttel's emphasis].³⁶

Traditional patterns of technology adoption suggest that larger, more "progressive" producers take earlier advantage of innovations, reaping whatever rewards there might be in increased efficiency. When the rest of the group catches on, these comparative advantages fade. In the case of bGH, early adopters will probably be those dairy farmers with large pedigreed herds who have significant investments in management and labor, access to capital, and low debt loads. They will be the winners. The losers will be those with high debt loads or poor soils or small herds or so-called bad management techniques, the producers that the agricultural establishment sometimes calls "inefficient." These are likely to be subsistence farmers in Appalachia, black farmers in the South, and medium-size farms with high debt throughout the country. They will be the losers.

Have the losers been treated unjustly by the agricultural research establishment? An answer to this question requires us to define *justice,* no easy task. Many definitions have enjoyed favor throughout the centuries of reflection on the matter, but three considerations seem to recur in all of the discussions: equality, contribution, and need. Following contemporary philosophical practice, I will discuss these issues under the headings of *distributive justice* and the *common good.*

Distributive Justice and bGH

The argument from unequal treatment assumes that there is an unwritten contract between agricultural research institutions and the farmers who support them. The farmers pay taxes that go for salaries and equipment; the institutions are supposed to deliver seeds, machines, and techniques that will make farming more productive and profitable for all kinds of farmers. Now, if institutions do research that speaks only to the needs of a certain class of farmers and thereby gives them a comparative advantage over others, then the contract has been broken. The institutions have unfairly privileged one class, and put another at a disadvantage.

There is strong evidence for thinking that operators of smaller-size farms have been treated differently from operators of larger-size farms. Jim Hightower's book *Hard Tomatoes, Hard Times* popularized the case of the mechanical tomato harvester in California, and the ongoing California court case that resulted from it is adding the weight of legal opinion to Hightower's charge.[37] Of course, some benefits from the university research in question have accrued to farmers with small- and medium-size operations and these need to be added into the calculus. Nonetheless, when one considers the kinds of technologies that have come out of agricultural research institutions since the Second World War—including, but not limited to, chemical herbicides and pesticides, large tractors and implements, automated milking parlors, artificial insemination, petroleum-fueled machines, embryo transfer, and hybrid seeds—a presumption in favor of Hightower's charge appears. Even farmers themselves tend to think that their own farms always need to be a little bigger; there is an ideology of growth in farming that has been caused by, and in turn helps to fuel, institutional research biased toward large-scale, capital-intensive, mechanized agriculture. So the ball is in the opponent's court; the burden of proof is on those who believe that operators of small- and medium-size farms have *not* been discriminated against.[38]

One might argue that the skewing of research was justified because large farmers assumed a larger share of the tax burden. If the more aggressive operators had paid substantially larger sums, wouldn't they be entitled to the increased attention they received? Even if it were true that big farmers had shouldered most of the burden, this would not justify an unbroken legacy of hard tomatoes and hard tomato harvesters. Which innovations favored smaller producers? Which hybrid seeds, which machines, which chemicals gave assurance that farmers could remain competitive while retaining their present size? Even Extension has focused on the "progressive" farmers in a community in the apparent hope that benefits would trickle down to others. Thus, even if large farmers had paid the largest taxes, this would not justify the extent of research bias. And it is still to be shown that big operators did in fact contribute more.

The severity of this research bias would be of one magnitude if operators of small- and medium-size farms had simply not been able to increase

their profits. But the situation is much worse; these farmers have not remained where they were; they have gone through years of financial and
emotional upheaval. Many have ended in bankruptcy. As the farm crisis
drags on, successive groups of farmers are moved toward the end of a
conveyor belt and dumped over the edge. The machine is not broken; it is
moving. With each new jerk of the belt, the status quo is changed. Those
farmers with the most comparative advantage in the first round are quite
safe; they are not on the treadmill and continue to prosper from each new
round of innovations. Meanwhile, medium-size farmers struggle to get bigger. A few succeed; the rest are dumped. Insight into this chronic cycle may
have caused Earl Butz to tell farmers to get big or get out, but it also lies
behind the O.T.A.'s much-cited prediction that medium-size farms will
have disappeared by the year 2000.[39]

Doesn't the rapid growth of small farms also predicted by the O.T.A.
give evidence that the research establishment has not favored large operations? No. Small farms are flourishing not because they are efficient at
producing food, but precisely for an opposite reason: their efficiency at
producing food does not particularly matter to them since this is not their
primary source of income. Indeed, the growth of small farms may contribute to the demise of traditional family farms because their willingness to
accept lower prices for their products depresses the prices received by those
trying to make a living primarily from their farms.

The extent of the unfairness cannot be seen if one takes a snapshot of
the conveyor. The belt is turning, and with each turn, a new group of
farmers is dumped off the end. When, as David Braybrooke puts it, "the
game begins again," the terms are different. If the results of the last exchange "were unjust, enriching some people at the expense of others, and
there are no compensating changes, they bring about a distribution of resources (in private property and in other resources like influence) that raises
the prospects of injustice" in the next round of exchanges.[40] As large
farmers increase and consolidate their hold on the industry, the universities
become even more responsive to their needs and to the needs of the private
sector food processors who prefer to deal with a few large producers.
Meanwhile, governmental programs also become increasingly biased toward the larger producers: the amount of governmental assistance provided
to large farms increased tenfold between 1980 and 1985, while the assistance given to medium farms increased only fivefold.[41]

The consequences of such unfair exchanges may be even more troubling than the initial injustices. Not only have the farmers with medium-
size operations lost the value of their tax dollars, but they have also given
up what Braybrooke calls "increments of power and advantage"[42] that they
would have had if the first round had been fair. Their ability to educate

themselves about new farming methods, their incentive to organize into effective political units, their skill in bargaining collectively, their capacity to market their goods strategically—all of these skills may suffer serious erosion as a result of the group's having been mistreated in earlier stages.

Whether my theoretical analysis offers a sufficiently nuanced explanation of the history of America's medium-size farms is arguable. It is admittedly schematic and general. But studies have given us good reasons to believe, more specifically, that (1) prices received by hog and beef farmers in certain portions of the country are artificially lowered because of lack of competition among meat packers in those regions;[43] (2) a concentration in the number of firms in breakfast cereals has artificially inflated prices paid by consumers;[44] (3) tax laws, like rapid depreciation schedules and investment tax credits, have favored large producers over small producers;[45] and (4) the land-grant university system has not taken care to make sure its research is equally beneficial for all sizes of farms.[46]

This illustrative list of problems may or may not add up to a long-standing pattern of discrimination by powerful, tax-funded organizations against the majority of farmers. But the case against bGH does not stand or fall with the answer to that question. Suppose that the process of allocating tax monies for research is judged, as Luther Tweeten argues, *not* to have been biased against family producers.[47] We must still ask ourselves whether the general pattern of the demise of moderate-size farms is socially desirable. In 1986, 6 percent of all farmers went out of business; one farm every four minutes. In 1985, the figure was 5 percent. If those figures seem small compared to the general rate of failures of small businesses, consider that most small businesses have only very recently started up; the farms in question often go back generations. These farms do represent, in the often maligned rhetoric of farm activists, a "way of life" whose value is not measurable in economic terms.[48]

The loss of farmland owned by minorities plays a disproportionately large role in this story. Half a million acres of farmland per year are currently being lost by black owners. The story started, of course, with blacks clearly behind the eight ball; while they constituted approximately 15 percent of the U.S. population, blacks owned almost no farmland at the beginning of the twentieth century. Currently they own 1.4 percent of the farms. Whatever progress black farmers have made, however, is being rapidly eroded. At the current rate, these farmers will be completely landless again by the end of the century.

What is happening to the land? Patterns of land use vary across the country, but in places where conversion to nonagricultural uses is least problematic, the number of absentee landowners is increasing dramatically. In 1981, the number of acres managed by professional farm management companies was 48 million; in 1986, it was 59 million, an area the size of Colorado.[49] While it is not clear from the data which farms in particular are under the most pressure, it is clear that 66 percent of total farm debt was held in 1986 by medium-size farms, those usually owned and operated by families who are dependent on them as their major source of income. These are the farms currently closest to the end of the conveyor belt.[50]

What does this story about publicly funded agricultural research and its effects on rural America have to do with bGH? It helps us to see the broader pattern of which bGH is a continuation. If hardships were distributed evenly, if large and small and medium-size farms — those owned or worked primarily by blacks, whites, and hispanics — had all suffered equally in this tale, then we would have little basis for talking about injustice. But gross discrepancies have been with us for a long time, through several turnings of the belt, and those dumped off the end have not been compensated.

In terms of disparity in income levels and access to power, the situation in agriculture is little different from the wider pattern in the United States. In 1970, the top 20 percent of Americans made 41.6 percent of total family income; the bottom 20 percent made 5.5 percent.[51] By 1985, the top 20 percent were capturing an additional 5.5 percent — up to 47 percent of all earned income — while those on the bottom had dropped to 4.7 percent. Of those working for a living, the most successful in our culture make somewhere in the neighborhood of 100 to 200 times the amount of the least successful.[52] What is the annual income of the CEO of Cargill or Beatrice Foods compared to the income of a migrant worker in Muscatine, Iowa? Suppose, conservatively, that CEOs make $500,000 per annum, while muskmelon harvesters garner $10,000. The difference here is not on the order of 200 or even 100, but of 50. And yet we may ask: Is even this discrepancy justified?

Perhaps the CEOs deserve more or need greater incentives to do their jobs well? This is difficult to believe. Are we to suppose that the corporate manager exerts more physical energy or has longer hours than the fieldworker? Or that the CEO must take more risks? Is the CEO worth more to the vitality and growth of our economy? Of all the possible justifications, this last one is the only one that comes close to being credible. And yet, even if we accepted it (which I am not recommending), would it justify the magnitude of difference? Perhaps so, if we could argue that being the leader of a corporation entails greater psychological stress and that people would not take up this line of work unless the incentives were as great as they are. But people go into equally stressful lines of work (such as air traffic control) for only four or five times the income of migrant workers. So the argument from incentive will not work; capable people will flow into stressful jobs for much less than half a million dollars a year. Perhaps agribusiness leaders need extra intelligence that air traffic controllers (for instance) do not need. But equally intelligent people work in factories and universities and laboratories for, again, a tenth of the CEO's salary. The fact is that there are no persuasive moral arguments to justify the current inequities in pay found across the range of jobs in agribusiness. And the inequities are growing worse.

If the discrepancies were temporary abnormalities, we might be able to overlook them. But to the extent that they are deeply entrenched in our history and likely to persist indefinitely, they indicate a troubling problem in our agricultural market system. For it is, in Braybrooke's words, "the continual repetition of the discrepancies, with one set of people always faring well, and another always faring badly" that fixes our attention. "Some people, and their children, [are] living their lives out – very possibly shortened lives – without having any chance to live decently; others [are] surfeited with pleasures."[53]

We might defend the agricultural market system by arguing that discrepancies of some magnitude are inevitable in any system of allocating resources, and that the agricultural market system could alleviate gross discrepancies by redistributing resources downward – toward those on the bottom – through political measures such as progressive income taxes. In this case, income transfers (for example, via a truly progressive income tax system) from rich CEOs and agribusiness corporations to seasonally employed migrants and poverty-level farmers would be justified on the grounds of equality and need.

This would be a step in the right direction, but the poor need more than income: they need autonomy, meaningful employment, jobs in which their skills can be used and honed and which help to give them control of their lives. The poor need jobs and education through which to be able to meet their own needs for food, shelter, clothing, and companionship. Farming in the traditional sense has offered that sort of employment. The farmers being put out of business by technological advances do not need income enhancements in the long run. They need secure employment. Thus the answer suggested by Michael Novak – to give farmers cash – may show compassion, but it is not directed toward establishing an agricultural economy that plans rationally and deliberately for just compensation of its members.[54]

It may be objected here that my analysis assumes too much control over the inventive process. How can we *plan* to come up with innovations that would help smaller full-time farmers? How could anyone hope to direct the scientific imagination in such a direct way? If this sounds difficult, we need only to look again at the history of agriculture. Contrary to popular mythology, new inventions do not come out of thin air or from lonely, wild-eyed geniuses. Inventions are consciously and rationally pursued by those who have the most to gain from them. Research leading to the invention of the milking parlor did not begin because farmers went on a general strike against milking by hand. It began, in part, because milk processors had an interest in buying milk from as few farmers as possible; they funded research and supported public policies that would help their largest suppliers increase productivity and efficiency. Research on hard tomatoes did not begin because a groundswell of consumers decided they would like dull tasting, thick-skinned vegetables. It began because food handlers wanted to be able to purchase tomatoes in large quantities and to ship them long distances to markets. Research on herbicide-resistant corn has begun not because farmers want to put more powerful chemicals on their fields but because, in part, chemical companies have bought seed companies and want to market an integrated all-Dow or all-Monsanto seed-and-chemical package.[55] Research on bGH may have begun, in part, because scientists were interested in the molecular structure of a specific protein, but it has been pushed through to the marketing stage only by the corporations anticipating significant profits. Expensive biotechnologies do not blossom from people's heads as if they were fresh flowers seeking spring air: they are consciously pursued by powerful organizations with specific plans and needs. This proves that we can and do direct the course of technological invention.

Those who say that "the development of technology" is primarily responsible for the decreasing number of dairy farmers may not intend to mislead us, but they do so when they allow their audiences to infer that history could have followed no other course. In fact, we could have pursued other economic, monetary, and fiscal policies; we could have encouraged farm organizations and cooperatives instead of subsidizing production of targeted crops; we could have concentrated on diversifying our own farms instead of concentrating production on a few export crops; we could have invested in other sorts of research in agriculture — perennial crops, sustainable farming methods, nonchemically driven small-scale planters and reapers. Those who have the most to gain from large, intensively managed, petrochemically dependent methods in farming have played a substantial role in the displacement of farmers, and we should not allow fatalistic rhetoric about the past "development of technology" to hide this fact.

Similarly, those who say that "market forces" such as high input costs (for seed, fertilizer, labor, and operating loans) and low market prices have displaced farmers may also mislead us. Input costs are directly affected by fiscal and monetary policy. When a nation runs large budget deficits it is more difficult to make money available to small-scale entrepreneurs; when it pursues policies that make its currency strong, it negatively affects export-sensitive industries such as agriculture. "Market forces" are no more natural than are the paths of technology development. To talk as if they were is to engage in the same sort of deterministic thinking we saw above.[56] We could collectively decide to change our military budget, tax cuts for the wealthy, schemes for union busting, and refusal to enforce antitrust laws. If we did, the climate for small-scale businesses such as family farms and ranches would be much different.

bGH and the Difference Principle

How should we go about distributing the benefits of technology? John Rawls suggests that social goods should be distributed fairly, and that inequities in distribution should be accepted only when such inequities will enable those on the bottom to be better off than they would have been if the inequities were disallowed.[57] This is the difference principle: unequal distribution of material goods and social status is fair if and only if it improves the lot of those on the bottom. Poor farmers in the South might be denied certain tax breaks given to bigger farmers if and only if the poorer farmers would come out ahead in the long run. Black farmers might be denied Extension Service attention if and only if this would result in their farm operations improving over the long haul. A progressive tax system would be justified, even though it appears to treat the wealthy unfairly, if and only if it improves the condition of the worst off.

Knowing what we now know about bGH, could we justify denying industry and large farmers profits on the grounds of distributive justice to smaller farmers? Advocates would say no: keeping bGH off the market is unfair to some farmers because it denies them the choice of using it. But, according to the difference principle, keeping bGH off the market could be justified if it would improve the lives of agriculture's most disadvantaged.

Would a boycott of bGH improve the lot of the worst-off dairy farmers? There are at least two questions here. The first is, Would banning bGH really be good for the marginal farmers? Lester Thurow argues that while there are an excess of farmers, there are also plenty of good jobs into which they could move.[58] Rather than artificially trying to save farmers' jobs, society would be better off retraining the farmers, helping them to make the transition into other lines of work. This argument might make sense if we decided not to try to count the psychological costs involved in moving farmers, farm families, and associated rural workers out of their way of life. It might make sense, too, if we looked at the history of farming through deterministic glasses, for if the labor requirements of agriculture have been reduced by inevitable, inexorable, economic forces, then it would be foolish to try to retain workers in farming today. Too many inefficiencies in the allocation of resources are promoted by trying to keep farmers employed.[59]

Laying aside for the moment questions about the validity of this view of history, we may still ask whether the argument above takes into account all of the external economic costs involved in moving labor out of agriculture. At the environmental level, what is the cost in soils and water when fewer farmers, increasingly dependent on pesticides and herbicides, increase their landholdings? In many areas, profligate use of marginal lands by farmers intent on increasing the size of their fields in order to use large equipment more efficiently has rendered thousands of acres virtually unusable and unrecoverable. At the national security level, what is the cost of having more and more of our corn crop in less and less diversified species? In 1970 the United States lost 15 percent of its corn crop to southern corn leaf blight due to the uniform nature of the seed used across the corn belt.[60] At the aesthetic level, what do we lose by no longer being able to see countrysides of well-kept farmyards, gardens, and animals in pasture?[61] At the strictly economic level, how much does it cost each taxpayer when one displaced farmer moves into an urban area, fails to find a job, goes on unemployment, and eventually loses incentive to look for work? What human resources are lost in the process? How many tax dollars are spent on Medicare, public nursing, pharmaceutical products, and federal programs in order to care for that farmer? What social costs are incurred by the depopulation of rural areas, the overcrowding of cities, and the malaise and disruption that accompany both?

The fact is that we do not have any idea about the extent of the external costs involved in moving labor out of agriculture. We lack accurate accounting methods "that begin from the assumption that social costs are to be computed so that the public has a far more exact understanding of what any particular item or process costs the society as a whole."[62] So I would not presume to be able to confidently assert that the costs of moving farmers out of their way of life outweighs the benefits of doing so; I have no more privileged way of judging this matter at present than does anyone else. What can be asserted, however, is that those who boldly claim that retraining farmers is the only sensible answer to the farm crisis are either naive or privy to divine revelation.

The second question is whether "banning" bGH would be good for the urban poor, many of them grandsons or granddaughters of farmers. A successful boycott against bGH might prevent the lowering of milk prices or even slightly inflate them and, moreover, have a chilling effect on other avenues of research in industry and university, avenues that might lead to cheaper food for the poor. Advocates of bGH claim that the new biotechnologies will cut costs for farmers and that these cost reductions will be passed on to consumers. History, again, is a good antidote for such rhetoric. In recent years farmers have been pressed to cut their input costs while the prices they received on the market for their wheat, corn, and beans dropped steadily. Did the price of corn flakes to consumers drop? During the summer of 1988, many food manufacturers raised prices at the first media stories of the drought. Their costs, of course, had not gone up; they simply used news reports as cover for increasing profits. The facts are that intermediate markets seem to have a way of absorbing whatever profits are made when farmers' prices go down. There is little reason to think that bGH usage would lower milk prices for the urban poor, or any consumers.

bGH and the Common Good

These considerations compel us to think not simply about distributive justice, but about wider considerations such as the sort of people we are and want to be, the qualities of character we want to encourage in our young, and the type of concerns we wish to pursue together. Our society should be one in which no person goes hungry, in which all who wish to work are employed — in jobs promoting individual autonomy and social cooperation — and in which human flourishing in its moral and spiritual sense is possible. We should pursue objectives that are good in an objective, substantial sense; objectives that allow us "to experience the fullness of human life, as opposed to merely existing."[63]

From the perspective of the common good, bGH appears as a technology that not only will fail to promote the common good, but will actively undermine it. It will add to a decline in the number of dairy farmers, exacerbating the crisis currently affecting rural America. It will degrade rather than enhance the internal goods pursued in the practice of farming since it encourages farmers to treat animals as production machines rather than coinhabitants.[64] It promises to assimilate dairy farming fully into an impersonal, industrialized culture that farmers have long resisted. In short, bGH threatens to undermine the common good not simply of the dairy farmers it will displace, but of us all. It promises, in a small way, to undermine our general well-being.

That conclusion is worth pondering, and its qualifications are worth repeating. bGH *promises* (we should not forget that we are dealing with potential injustices, not yet realized) *in a small way* (it is by no means the most pressing problem in America, though it may be for the less than 1 percent of Americans who are small dairy farmers) *to undermine* (not simply fail to promote) *our general well-being* (it is not simply dairy farmers who are affected, but all of us).

After all of this, defenders of the technology would still have the following response open to them. If we prevent bGH from reaching the marketplace, we may be sending a signal to farmers that inefficient farming is acceptable and that society will always protect them from innovations that might displace them. This would be counterproductive for society as a whole, making farming a less attractive line of work for farmers and driving up the cost of food for consumers.

The objection has merit, and it forces us to admit that we walk a fine line when we get in the business of trying to pick and choose between new technologies. We do not want to stifle the imaginative spirit of public or private scientists, nor the independence of farmers for whom farming is attractive precisely because it allows them freedom to try new things. But while bGH is the first agricultural biotechnology, it will not be the last. And discouraging its use in no way commits us to oppose all technologies. We should oppose only those technologies that unfairly advantage one social group over another, that displace workers at unacceptably high costs, or that threaten the stability, beauty, or integrity of the plant or animal kingdom.[65]

Policy Recommendations

In the interests of the common good, we ought to pursue at least two goals in agriculture. One is to keep farming open to a wide number of people. The second is to allow innovations that will contribute to the number of meaningful jobs. Accomplishing these goals means matching supply with demand. The free market has not demonstrated the ability to do this in the dairy industry. When left to market forces, dairy farmers—like all farmers—have, in the words of John Kenneth Galbraith, "a relentless and wholly normal tendency to overproduce, because of extraordinary productivity gains and because farmers, being powerless to influence or control supply and price, harvest more and more as a way of trying to stay financially afloat."[66] As Galbraith argues, the answer is a system of supply management, something that is "taken for granted in all large-scale industry."[67] We need a way to organize dairy farmers so that each can make a decent living in a relatively stable business atmosphere without relying on government subsidies or having to try to outproduce neighbors. That is a tall political order. My contribution here is only to suggest that bGH — and the sort of technological direction represented by it — is of no help in trying to fill it.

Conclusion

To the extent that the research establishment has clearly favored large producers in its development of techniques and technologies; to the extent that fiscal, monetary, and economic policies have disadvantaged small dairy producers; and to the extent that bGH will only exacerbate the unjust consequences of the past — to those extents we ought to oppose this particular biotechnology. Language about "banning" bGH, of course, is just that — a slogan intended to summarize the case against bGH. There is no governmental body with the authority to ban bGH on the grounds of humane treatment of animals. Nor is there any government agency charged with the task of overseeing — much less regulating — technologies by the criteria of their anticipated socioeconomic effects. This shows the need for legislative attention to this matter. But in the meantime, opposition to the marketing of bGH sends a signal to those in public and private decision-making positions. Not all biotechnologies are acceptable. We do not want those that are destabilizing, inhumane, or ugly; we do want those that will preserve the beauty, integrity, and diversity of the Creation.

Jewish folklore tells of the wonderful town of Chelm, whose inhabitants engaged in the most curious behavior. Knowing full well that the rainy season was upon them and that the prayer hall desperately needed a new roof, they spent all their time putting new carpet on the floor. The next fall, when their schoolchildren had no papers, pencils, or workbooks, they spent their entire fortune on another first edition for the rabbi's library. Chelmians, it seems, always did the opposite of what was in their own interests.

However entertaining fiction may be, contemporary agricultural history is more so. Awash in excess dairy products, our government dumps milk in the ocean, hands out surplus cheese to farmers, and pays operators $1.2 billion to slaughter their cows — all of this while publicly funded institutions quietly spend taxpayers' monies on schemes to increase milk production. There are daring scientific projects that are in our own interests and that need our moral encouragement. bGH is not one of them.[68]

Notes

1. Hans Jonas, *The Imperative of Responsibility: In Search of an Ethics for the Technological Age* (Chicago: The University of Chicago Press, 1984), 203.

2. Matthew H. Shulman, "Bovine Growth Hormone: Who Wins? Who Loses? What's At Stake?" in *Agricultural Bioethics,* ed. S. Gendel et al. (Ames: Iowa State University Press, forthcoming), 127–28.

3. "Experimentally, the greatest response on an annual basis has been a milk increase of 25.6 percent," Kalter et al., *Biotechnology and the Dairy Industry: Production Costs, Commercial Potential, and the Economic Impact of the Bovine Growth Hormone.* A. E. Research 85–20. Ithaca: Department of Agricultural Economics, Cornell University, 1985, 108. This increase is not expected over the animal's complete lifetime. See D. E. Bauman, P. J. De Geeter and G. M. Lanza, "Responses of High Producing Dairy Cows to Long Term Treatment with Pituitary- and Recombinant-Somatotropin," *Journal of Dairy Science* 68 (1985): 1352.

4. Howard Schneiderman, "Innovation in Agriculture," *The Bridge* (Spring 1987): 5.

5. See Shulman, "Bovine Growth Hormone;" and Laura Tangley, "Biotechnology on the Farm," *BioScience* 36 (October 1986): 590–93 and (November 1986): 652–55. At its 1986 annual meeting, the New York State Grange decided to oppose the commercialization of bGH until seven areas had been studied. These included the three areas cited by the Wisconsin farmers.

6. Quoted in Shulman, "Bovine Growth Hormone," 120. How the agency could confidently conclude that "the investigational use of (bGH) in dairy cows does not result in significant impacts on the human environment" without having run its own tests is less than clear.

7. In addition to the works already cited, see, for example, Frederick H. Buttel, "Agricultural Research and Farm Structural Change: Bovine Growth Hormone and Beyond," *Agriculture and Human Values* 3 (Fall 1986): 88–98; Buttel, "Biotechnology and Agricultural Research Policy: Emergent Issues," in *New Directions for Agriculture and Agricultural Research*, ed. K. A. Dahlberg (Totowa, N.J.: Rowman and Allenheld, 1986), 312–47; Buttel, "Biotechnology and Public Agricultural Research Policy," in *Agricultural Science Policy in Transition,* ed. V. J. Rhodes (Bethesda, Md.: Agricultural Research Institute, 1986), 123–56; Buttel and Charles C. Geisler, "The Social Impacts of Bovine Somatotropin: Emerging Issues," paper prepared for the National Invitational Bovine Somatotropin Workshop, sponsored by the Extension Service, USDA, St. Louis, September 22, 1987; William P. Browne, "Bovine Growth Hormone and the Politics of Uncertainty: Fear and Loathing in a Transitional Agriculture," *Agriculture and Human Values* (Winter 1987): 75–80; William P. Browne and Larry G. Hamm, "Political Choices, Social Values, and the Economics of Biotechnology: A Lesson from the Dairy Industry," staff paper 88–33, April 1988, Department of Agricultural Economics, Michigan State University; D. L. Heuth and R. E. Just, "Policy Implications of Agricultural Biotechnology," *American Journal of Agricultural Economics* 69 (May 1987): 426–31; R. J. Kalter, "The New Biotech Agriculture: Unforeseen Economic Consequences," *Issues in Science and Technology* 2 (1985): 125–33; Kalter et al., *Biotechnology and the Dairy Industry: Production Costs and Commercial Potential of the Bovine Growth Hormone.* A. E. Research 84–22. Ithaca: Department of Agricultural Economics, Cornell University, 1984; R. J. Kalter and R. A. Milligan, "Emerging Agricultural Technologies: Economic and Policy Implications for Animal Production." Manuscript, Department of Agricultural Economics, Cornell University, 1986; R. J. Kalter and L. W. Tauer, "Potential Economic Impacts of Agricultural Biotechnology," *American Journal of Agricultural Economics* 69 (1987): 420–25; J. Kloppenburg, Jr., "The Social Impacts of Biogenetic Technology: Past and Future," in *Social Consequences and Challenges of New Agricultural Technologies,* ed. G. M. Berardi and C. C. Geisler (Boulder: Westview Press, 1984); W. B. Magrath and L. W. Tauer, "The Eco-

nomic Impact of bGH on the New York Dairy Sector: Comparative Static Results," *Northeastern Journal of Agricultural and Resource Economics* 15 (1986): 6–13; and R. C. Barnes and P. J. Nowak, "Bovine Somatotropin's Scale Neutrality and Constraints to Adoption." Manuscript, Department of Rural Sociology, University of Wisconsin, 1987.

8. Patrick Madden and Paul B. Thompson have addressed the general issue of ethics and agricultural biotechnology in "Ethical Perspectives on Changing Agricultural Technology in the United States," *Notre Dame Journal of Law, Ethics and Public Policy* 3 (Fall 1987): 85–116. But they are not concerned primarily with bGH. I find their argument marked by a puzzling ambiguity. In the first half of their essay, the agricultural market economy is assumed as a given, and never questioned. The second half describes strong ethical criticisms of that assumption, drawing on the work of E. F. Schumacher and Wendell Berry. Unfortunately, the authors do not show how the challenges raised in the second half of the essay might lead one to revise fundamental assumptions in the first.

9. Tangley, "Biotechnology on the Farm," 590.

10. Tom Regan, *The Case for Animal Rights* (Berkeley: University of California Press, 1983).

11. See Ned Hettinger, "Cohen on the Use of Animals in Biomedical Research," manuscript, Department of Philosophy, Charleston College, Charleston, S.C. Hettinger is responding to Carl Cohen, "The Case for the Use of Animals in Biomedical Research," *New England Journal of Medicine* 315 (1986): 865–70.

12. Among others, Frederick Ferre holds such a view. See "Moderation, Morals, and Meat," *Inquiry* 29 (1986): 391–406. My criticism of Regan is indebted to Ferre's discussion.

13. Don Beitz, "Physiology of Growth Hormone," lecture to Animal Science Roundtable, Iowa State University, May 13, 1988.

14. David S. Kronfeld, quoted in Shulman, "Bovine Growth Hormone," 121.

15. For a description of conditions on factory farms, see Peter Singer, "Down on the Factory Farm," in *Animal Rights and Human Obligations,* ed. Tom Regan and Peter Singer (Englewood Cliffs: Prentice-Hall, 1976).

16. In addition to the work of Phil Leder and his lab at Harvard on transgenic mice, see that of Ursula Storb's lab at the University of Chicago and, closer to the production end of agriculture, the work of Carl Pinkert at University of Missouri. Pinkert, "Gene Transfer and the Production of Transgenic Livestock, *Proceedings of the U.S. Animal Health Association* (in press).

17. I want to reemphasize the promise of biotechnology. If, as is predicted, transgenic animals can be developed that possess the capacity to manufacture important medical drugs, then the milk of healthy, flourishing cows kept in humane conditions could be the source of inexpensive, life-saving proteins such as insulin or the blood component known as human factor IX. If the number of animals needed were reduced and careful attention were given to their physical and psychological needs, purposes, and desires, rDNA research could lead to a revitalizing of our interdependent relations with animals. But, again, this possibility needs to be weighed against institutional and economic forces working against it.

18. "[Kalter] estimates that the number of dairy farms may have to be reduced 25–30 percent to restore market equilibrium. . . . These adjustments will almost certainly have dramatic social, economic, and cultural effects." Andrew Kimbrell and Jeremy Rifkin, "Biotechnology—A Proposal for Regulatory Reform," *Notre Dame Journal of Law, Ethics and Public Policy* 3 (Fall 1987): 125. At a conference on "Public Perceptions of Biotechnology" sponsored by the Agricultural Research Service of the USDA at Airlie House in Virginia in 1986, Dr. Kalter objected strongly to a speech by Rifkin in which Rifkin used Kalter's study in the way recounted here.

19. Kalter, "The New Biotech Agriculture," Table 50, "Changes in Price, Output, Em-

ployment and Cow Numbers from bGH and a Free Market Policy by Elasticity of Demand,"
p. 101.

20. I arrived at these figures by subtracting 20.4 (reduction in farm numbers with 0
technical change and a low elasticity of demand for milk) from 46.0 (reduction with 30 percent
technical change) to get 25.6. I then subtracted 17.2 (reduction in farm numbers with 0
technical change and high elasticity of demand) from 33.1 (reduction with 30 percent technical
change) to get 15.9. These, of course, are very rough calculations. Again, they refer to the
percentage decrease in farm numbers due to technical change of many types, not simply bGH.
If farmers started to milk four times a day instead of three, for example, that would lead to a
greater output of milk and a corresponding need for adjustment. These qualifications need to
be figured into the estimates.

21. Magrath and Tauer, "The Economic Impact of bGH on the New York Dairy Sector,"
12.

22. Quoted in Tangley, "Biotechnology on the Farm," 592.

23. Ibid.

24. Buttel and Geisler, "The Social Impacts of Bovine Somatotropin," 7.

25. Ibid.

26. Ibid.

27. Ibid.

28. Ibid., 8.

29. Ibid.

30. Ibid., 10.

31. For an argument to this effect, see David Kline, "No-Till Farming and Its Threat to
the Amish Community," *Festival Quarterly* (Fall 1986): 7–10.

32. See Schneiderman, "Innovation in Agriculture," 5, 7. For another discussion of the
issue, see *Agricultural Biotechnology: Strategies for National Competitiveness* (National
Academy Press, 1987).

33. Quoted in Shulman, "Bovine Growth Hormone," 125.

34. Barnes and Nowak, "Bovine Somatotropin's Scale Neutrality and Constraint to
Adoption," abstract.

35. P. Nowak, J. Kloppenburg, Jr., and R. Barnes, "bGH: A Survey of Wisconsin Dairy
Producers," in *As You Sow* 18 (July 1987), Department of Rural Sociology, University of
Wisconsin: 2.

36. Buttel, "Agricultural Research and Farm Structural Change," p. 98, n. 5. Buttel adds
the following caveat: "To be sure, most biotechnologies, including bGH, will be *divisible*
inputs—that is, they can be used in either large or small amounts with relatively little difference in the purchase price. And divisible inputs such as biotechnologies will be less biased
toward large farmers than "lumpy" technologies such as large tractors. But divisibility and
scale-neutrality are *not* the same thing."

Luther Tweeten also believes that labor-saving technologies generally favor larger operations, which can provide the lowest cost of production per unit. See Tweeten, "Has the Family
Farm Been Treated Unjustly?" in *Is There a Moral Obligation to Save the Family Farm?* ed.
Gary Comstock (Ames: Iowa State University Press, 1987), 225. But note that Tweeten also
believes that "out-put increasing technologies tend to be somewhat scale neutral," and that
"government price and income payment policy has generally been neutral in its effect on farms
of varying sizes producing program commodities," 219. For supporting evidence, Tweeten cites
R. Spitze, D. Ray, A. Walter, and J. West, "Public Agricultural Food Policies and Small Farms
Project," (Washington: National Rural Center, 1980).

37. Jim Hightower, *Hard Tomatoes, Hard Times* (Cambridge: Schenkman, 1973). Cf.

footnote 46, and Don Paarlberg, "The Land Grant Colleges and the Structure Issue," *American Journal of Agricultural Economics* (February 1981):129–33.

38. Cf. the confirming opinion of Don Paarlberg, "The Land Grant Colleges and the Structure Issue," *American Journal of Agricultural Economics* (February 1981), 129–33.

39. Office of Technology Assessment, *Technology, Public Policy, and the Changing Structure of American Agriculture: A Special Report for the 1985 Farm Bill* (Washington, D.C.: U.S. Congress, OTA-F-272, 1985).

40. David Braybrooke, "Justice and Injustice in Business," in *Just Business: New Introductory Essays in Business Ethics* ed. Tom Regan (New York: Random House, 1984), 173.

41. "In 1985 . . . the average governmental payment to farms was $5,193 for farms with sales between $40,000 and $99,000 . . . and $37,499 for farms with sales over $500,000. The same groups of farms received $1,169 and $3,849 respectively, in 1980, which corresponds to a fivefold increment . . . and a tenfold increment," Alessandro Bonanno, "Agricultural Policies and the Capitalist State," *Agriculture and Human Values* 4 (Sp-Sum 1987): 44.

42. Braybrooke, 175.

43. A study by Quail et al. found that if there had been packer competition instead of shared monopoly (four firms) in such regions as Colorado, Nebraska, and Iowa, that the "average price would have been roughly 24 cents per cwt higher and annual returns to feeders in these . . . regions would have been nearly $42 million greater." G. Quail, B. Marion, F. Geithman, and J. Marquardt, "The Impact of Packer Buyer Concentration on Live Cattle Prices," N. C. Project 117, Working Paper Series, WP-89 (May 1986): 55.

44. In 1972, the Federal Trade Commission attempted to prove this, claiming that the ready-to-eat cereal industry was "highly concentrated (four firm market share of 91) and had high entry barriers." The commission was unsuccessful, because "the judge concluded that the defendants had acted like independently behaving rational oligopolists, which is not sufficient to constitute monopolization." As Bruce Marion points out, this case shows that the courts are not prepared to deal with the complex issue of shared monopolies, choosing instead to apply antitrust laws (if at all) only to markets clearly dominated by a single company. See Bruce Marion, *The Organization and Performance of the U.S. Food System* (Lexington: D. C. Heath and Co., 1986), 396–97.

45. Cf. Luther Tweeten's view that "Federal income tax provisions have, relatively, most favored part-time small farmers and "syndicates" financing, for example, large cattle-feeding operations because farm losses can shelter off-farm income and provide large savings per dollar of farm output." He adds that, in order to serve social justice, we ought "(1) to phase out the investment tax credit and rapid depreciation allowance and (2) to target public program transfers more heavily on farm families with low incomes" (Tweeten, "Has the Family Farm Been Treated Unjustly?" 228, 231). The 1986 tax reform bill, it should be pointed out, promises to accomplish Tweeten's first goal. In general, Tweeten believes that "public programs have not favored large farms over family farms or had a major negative impact on social justice" (231).

46. The California Rural Legal Assistance group sued the University of California in 1979, claiming that the Hatch Act of 1887 obligates Experiment Stations to benefit rural constituents, including small farmers, farm workers, and consumers. They charged that the university had a pattern of research that harmed rather than benefited those groups. In March 1986, Superior Court Judge Raymond Marsh formally ruled that "the Hatch Act obligates the Experiment Station to consider all of the beneficiary interests in evaluating and selecting its Hatch-funded research programs, and that primary consideration must be given to the small family farmer." In January 1987, the university formally admitted "as a matter of fact that . . . it has no process designed to ensure consideration of each legislatively expressed interest."

From a letter written by Bill Hoerger, staff attorney, California Rural Legal Assistance office, 15 June 1987.

47. See note 45 above.

48. For an analysis of arguments from emotion, see Comstock, "Conclusion: Moral Arguments for Family Farms," in Comstock, *Is There a Moral Obligation to Save the Family Farm?* 402–5.

49. The data in this paragraph and the one preceding is from *The Continuing Crisis in Rural America: Fact vs. Fiction*, Prairiefire Rural Action, Des Moines, Iowa, May 15, 1987.

50. Cf. Neil Harl, "The Financial Crisis in the United States," in Comstock, *Is There a Moral Obligation to Save the Family Farm?* 112–28.

51. Lester Thurow, "Toward A Definition of Economic Justice," *Public Interest* 31 (1973): 77, quoted in Albert Borgmann, *Technology and the Character of Contemporary Life* (Chicago: The University of Chicago Press, 1985), 111.

52. Braybrooke, 195.

53. Ibid., 179.

54. Michael Novak, "Cash Income and the Family Farm: Reflections on Catholic Theology and the Democratic Capitalist Political Economy of Agriculture," manuscript, pp. 25–27, forthcoming in *Is There a Conspiracy against Family Farmers?* ed. Gary Comstock, USF Monographs in Religion and Public Policy, Department of Religious Studies, University of South Florida, Tampa.

55. For accounts of mergers between seed and chemical companies, see Jack Doyle, *Altered Harvest: Agriculture, Genetics, and the Fate of the World's Food Supply* (New York: Viking, 1985): 104–106, and M. Kenney et al., "Genetic Engineering and Agriculture: Socioeconomic Aspects of Biotechnology R & D in Developed and Developing Countries," in *Biotech 83: Proceedings of the World Conference on the Commercial Applications and Implications of Biotechnology* (Middlesex, U.K.: Online Conferences Ltd., 1983): 475–89.

56. Cf. these remarks from an otherwise very sensible economist about the farm crisis: "Americans [who can] plan for decline . . . do not attempt to fight the inevitable tides of economic change. . . . Agriculture has been marked by decline and in the future it will be characterized by decline . . . [But, unfortunately,] reality never stopped anyone from going to Mexico for laetrile." Lester Thurow, "The Agricultural Institutions and Arrangements Under Fire," paper presented to the Social Science Agricultural Agenda Project, Phase 1 Workshop, Minneapolis, Minnesota, June 9–11, 1987, 125–26. This sort of rhetoric, found in many agricultural economists who speak confidently about "facing the facts," the "inevitable tides of economic change," and "hard realities," would make for an interesting study in ideological discourse.

57. John Rawls, *A Theory of Justice* (Cambridge: Harvard University Press, 1971).

58. Thurow "The Agricultural Institutions and Arrangements under Fire, 118.

59. Ibid, 124.

60. Doyle, *Altered Harvest*, 7.

61. As Thurow reminds us, if we love "to be or see farmers, it may be rational to protect [our] farmers with tariffs and quotas. What is lost in terms of extra consumption utility is more than gained in extra producer's utility." Thurow, *Dangerous Currents* (New York: Random House, 1983), 121. Quoted in James Montmarquet, "Agrarianism, Wealth, and Economics," *Agriculture and Human Values* 4 (Spring/Summer 1987): 49. I wonder whether Thurow forgot this point between 1983, when he wrote it, and 1987, when he implied that trying to retain labor in agriculture would demonstrate the same naivete as going to Mexico for a miracle cure.

62. Marcus G. Raskin, *The Common Good: Its Politics, Policies, and Philosophy* (New York: Routledge & Kegan Paul, 1986), 147.

63. Bruce Douglass, "The Common Good and the Public Interest," *Political Theory* 8 (February 1980): 105.

64. On the notion of a practice as a human activity whose goods are internal rather than external to it, see Alasdair MacIntyre, *After Virtue: A Study in Moral Theory* (Notre Dame: University of Notre Dame Press, 1981). On the idea of humans as "members and citizens" of a shared kingdom rather than as "conquerors" of it, see Aldo Leopold, *A Sand County Almanac* (New York: Oxford University Press, 1949), 204.

65. Environmentalists will recognize this language as that of Leopold, who said that the basic principle of the land ethic is, "A thing is right when it tends to preserve the integrity, stability, and beauty of the biotic community. It is wrong when it tends otherwise." Leopold, *A Sand County Almanac*, 224–25.

66. Galbraith, quoted by George Anthan in the *Des Moines Register,* October 11, 1987.

67. Ibid.

68. An earlier version of this paper was presented at a symposium on agricultural bioethics at Iowa State University in 1987. I received valuable help in revising that draft from Robin Attfield after presenting the paper at the Fourth International Conference on Social Philosophy, Oxford University, August 1988. The second version was presented at Oregon State University in the National Rural Studies Committee seminar series, October 1988, and at the American Academy of Religion annual meeting in Chicago, November 1988.

It is a pleasure to acknowledge the assistance of those who supported the research: the State of Iowa, the Joyce Foundation, the Northwest Area Foundation, the Kellogg Foundation, and the National Rural Studies Committee at Oregon State University's Center for Western Rural Development.

I would also like to thank David Kline and Jeffrey Burkhardt for incisive criticisms of the argument: would that I could answer all of them.

23 Biotechnology and Bioethics

STEVEN M. GENDEL

The developments in molecular genetics that have occurred in the last fifteen years and the biotechnology they have spawned have resulted in a debate that is unique in the history of science and technology. The issues in this debate have evolved almost as fast as the technology itself has evolved, and the papers presented in this book testify to the breadth of interest and thought that has been generated. Throughout, one of the most intriguing aspects of this—as intriguing as the specifics of the questions raised—has been the structure of the debate. This raises a number of interesting points, which I will try to summarize.

First is the fact that, revolutionary though it seems, modern biotechnology is just an extension of current practices and ideas.[1]

It clearly allows us to progress in some areas much more rapidly than previously, but (at least in agriculture) it has not yet defined any significant new approaches or trends. At most, application of this technology throws existing problems into much sharper focus and will force realistic resolution of difficulties which have been ignored for years. As such, many of the criticisms leveled at particular applications of biotechnology are really aimed at the basic system which underlies our agricultural economy, a system in need of reform. One excellent example of this is in the area of genetically engineered herbicide tolerance and the production of targeted biological herbicides.[2] Critics see this research as environmentally unsound and economically unwise, since it is seen as increasing the power of the agrochemical industry. Regardless of whether either criticism is true, they both simply represent extensions of the current situation, which arise out of an extensive reliance on petrochemical inputs in modern agricultural prac-

340

tice. Biotechnology has the potential either to alleviate both of these problems or to continue the present trends. The difference lies not in the technology itself but upon the economic, legal, and social milieu in which it is applied.

Second, the development of risk-assessment and risk-perception analysis in this area is clearly at a very early stage and is likely to be of little practical use for some time to come.[3] Most such analyses, on both sides of any issue, are strikingly incomplete. The major flaws seem to be a concentration on perceived risk and little examination of any data on real risks. This can be most clearly seen in the various discussions on the risk of environmental release of modified organisms. Although it is true that our ability to make detailed ecological predictions is virtually nonexistent, this does not mean that there is no relevant experimental data available. The scientific literature contains thousands of papers describing deliberate release experiments involving plant pathogens, soil microbes, plant and animal symbionts, and animal pathogens. Most of these have been introduced either into new habitats or into areas where the potential for significant harm exists. Yet, no major disasters have occurred. Clearly, much of the concern about the release of recombinant organisms is unwarranted. Few, I suspect, would suggest that we should drop all controls on these experiments, but I think that experience and common sense suggest that we proceed cautiously and optimistically.

Risk assessment is basically a comparative analysis, although this is seldom acknowledged in discussions of biotechnology. Although biotechnology is a powerful tool, it is often not the only means available to accomplish a particular end. As such, it is important to recognize that the alternatives may present equal or greater risks, particularly since they often result in more poorly defined products (compare, for instance, a herbicide-resistant plant produced by classical crossbreeding with a noxious weed to one produced by transfer of a single defined gene segment). This point has been explicitly recognized in the current government guidelines for regulation of release of engineered organisms, where the organism itself and not the process which produced it is regulated.

The current status of risk assessment as it applies to biotechnology also suggests a third point, which seldom receives serious consideration: namely, the risk of *not* proceeding. Each of these experiments, and each test of a new construct, represents an opportunity for learning. It is seldom possible to judge the potential value of a lesson that has not yet been learned. Nevertheless, it is easy to construct scenarios in which the lack of knowledge could be as disastrous as some of the catastrophic scenarios which have been predicted to result from these experiments. The current AIDS epidemic provides a cautionary example. It has been stated by researchers in the field that our knowledge of and ability to deal with the AIDS virus is two to five years behind where it would have been if the original NIH guidelines had been similar to the current ones.[4] Even more disturbing is the realization that if critics had been successful in banning recombinant DNA research altogether during the early 1970s, we would be virtually helpless in the face of this epidemic and would probably not even be able to clearly define the virus responsible. Although it is unlikely that such an extreme case will arise in agriculture, the principle that lost opportunities represent risks is valid. Again, this argues in favor of cautious optimism. It is difficult to see how the risk of lost opportunity can be calculated or measured, so perhaps the most realistic approach is to assume that these risks balance out the unlikely risks of the doomsday scenarios and to confine quantitative risk analysis to areas where hard data can be obtained.

A third interesting facet of the debate on biotechnology is the confusion that has arisen about the relationship between science and technology. Although many new technological advances can clearly point to some underlying science which made them possible, the reverse is not true. Scientific advances do not necessarily turn into new technology, and any one bit of science could potentially become part of a large range of technological changes. As universities, industries, and governments turn to viewing biotechnology as an economic resource, the value of basic science is often obscured. To the extent that both the critics and supporters of biotechnology try to control and limit scientific inquiry to serve economic or social agendas, the future development of the field is impaired, and the public (and humanity in general) is poorly served.

A basic question about the biotechnology debate is, Why has biotechnology become such a focus for ethical, social, and economic debate while other revolutionary technologies are all but ignored? The period of major growth in biotechnology has seen an even greater growth in computer/ electronics/communication technology; the latter has had a larger direct impact on the daily lives of most people in the United States without a concomitant scrutiny on the impact of this technology. In absolute terms, it is likely that the number of appliance repairers who have lost jobs to the new (unrepairable) electronic technology is greater than the projected number of farmers who will lose dairy farms if the worst-case scenarios for the effects of bovine growth hormone hold true. The application of computer technology will have a tremendous impact on our economy, social structure, and even on our basic world view (we are increasingly moving toward a digital rather than analog model of reality), yet there have been no calls for moratoriums on computer sales, no congressional investigations of the "threat," and even no law suits by "public-interest" businesses. Clearly biological issues touch a sensitive aspect of our culture and lead to deeper and more passionate examination of issues than do issues raised by any other technology. The reasons for this difference are not clear and may stem from the interaction of a number of factors, including the current social climate of general pessimism. Regardless, the fact is that arguments about the merit or appropriateness of biotechnology are made within a value system different from that used to judge other technologies.

This last point brings out the most valuable aspect of this debate on the application of biotechnology: the potential to use the process to assess how our society deals with technological change and advances in general. Unless the answers we reach apply to both farmers and assembly-line workers, to telecommunications as well as to bovine growth hormone, we will have failed in the most important aspects of this dialogue.

Notes

1. See Chapter 17 of this volume, Frederick H. Buttel, "Biotechnology, Agriculture, and Rural America: Socioeconomic and Ethical Issues."
2. See Chapter 13 of this volume, Jack Doyle, "Who Will Gain from Biotechnology."
3. See Chapter 10 of this volume, Ammertte Deibert, "Agricultural Biotechnology: Public Perception of Risk"; see also H. Halvorson, D. Pramer, and M. Rogul, eds., Engineered Organisms in the Environment. (Washington, D.C.: ASM Press, 1985.)
4. See Chapter 2 of this volume, David T. Kingsbury, "The Development of the 'Coordinated Framework' for the Regulation of Biotechnology Research and Products."

Contributors

John M. Asplund is a professor of animal science at the University of Missouri, where he teaches animal nutrition and conducts research with numerous animal species, including sheep, cattle, swine, rats, and mice. His primary research interests are related to amino acid metabolism. Dr. Asplund has conducted research at the Rowett Research Institute in Scotland. He has lectured throughout the world.

Roy Barnes is a doctoral candidate in sociology at the University of Wisconsin-Madison. His work focuses on how property relations, and the behavioral constraints that follow, interact with processes of agency. While an undergraduate student at Pomona College, Mr. Barnes studied anthropology and conducted field work (dealing with ancient agroeconomic systems of Ceylon) in Sri Lanka.

Gordon L. Bultena is a professor of rural sociology at Iowa State University. His main area of research relates to new agricultural practices and technologies and how they are adopted and diffused in the farm population. His most recent work examines socioeconomic factors that are important to farmers' decisions about conservation innovations, particularly those designed to lessen cropland soil erosion.

Brian Buhr is a doctoral candidate in agricultural economics at Iowa State University. His research interests are in production, finance, and econometrics. Currently, he is studying the economic impacts of growth-promotant use on the production of lean meat.

Frederick H. Buttel is a professor of rural sociology at Cornell University. He is particularly interested in the socioeconomic aspects of biotechnology and agriculture in university-industry relationships, in agribusiness reorganization, and in the impacts of biotechnology on Third World Agriculture. Dr. Buttel's numerous consultancies include the Ford Foundation, the Congressional Office of Technology Assessment, the Institute for Alternative Agriculture, the National Rural Center, and the National Council of Churches.

Gary Comstock is chair of the Religious Studies Program at Iowa State University. His areas of interest include agricultural ethics and philosophy of religion. In line with these interests, he is editor of a recently published book of readings on ethics and the farm crisis, *Is There a Moral Obligation to Save the Family Farm?*

Ammertte C. Deibert is a doctoral candidate in the Department of Sociology and Anthropology at Iowa State University, with a research focus on risk perception. She has spent ten years as an educator, teaching subjects that include rural sociology and cross-cultural perspectives.

Dennis D. DiPietre is an assistant professor of economics at Iowa State University. His research interests include social justice issues, economic ethics and justice, bioethics, and personhood and mission.

Jack Doyle is the Biotechnology Project director of the Environmental Policy Institute (EPI), a nonprofit, public interest organization involved in energy and natural resources policy. He has been with EPI since 1974, working on a range of energy, agricultural resources, and farm policy issues. For the last five years he has specialized in agricultural biotechnology, and his recent book on the subject, *Altered Harvest,* has been compared to Rachel Carson's *Silent Spring* and Alvin Toffler's *Future Shock.*

Steven M. Gendel is an assistant professor of genetics at Iowa State University. He conducts basic research in biological nitrogen fixation, with the goal of gaining a more thorough understanding of how plants fix nitrogen in the soil. His interest in bioethics couples his work as a practicing biotechnologist with his long-time concern for ethical accountability in scientific endeavors.

Robert S. Grossman is a legislative aide at the Hawaii State Legislature and is pursuing a Ph.D. in political science at the University of Hawaii. He has been an analyst for the Congressional Office of Technology Assessment and for the Food and Agricultural Organization of the United Nations. He also has surveyed intensive agricultural and horticultural systems in England and has studied species differentiation and evolutionary theory in Ecuador/Galapagos Islands.

Neil E. Harl is Charles F. Curtiss Distinguished Professor in Agriculture and a professor of economics at Iowa State University (ISU). He received the Juris Doctor (law) from The University of Iowa, and the Ph.D. in economics from ISU. Among his most recently conferred awards are the Henry A. Wallace Award for Distinguished Service to Agriculture, ISU, and the Superior Service Award, U.S. Department of Agriculture. His research interests include organization of the farm firm, taxation, estate planning, and legal and economic aspects of farm finance. He has spoken widely on debtor/creditor relations, estate planning, and organization of the farm business.

Chuck Hassebrook is a field organizer and policy analyst at the Center for Rural Affairs at Walthill, Nebraska—a family-farm research and advocacy organization. He works with family farm groups to develop new rural leadership, conducts research on the impact of biotechnology on family farms, and identifies alternative directions for agricultural research. Mr. Hassebrook has been successful in exposing and challenging biases in public tax and credit policies as well as in demonstrating alternatives to energy, soil, and water management appropriate to moderate-size family farms.

Marvin L. Hayenga is a professor of economics at Iowa State University and has taught and/or conducted research at the University of Wisconsin-Madison, Michigan State University, the University of California-Berkeley, and the University of Illinois. His work involves the changing structure, price behavior, and risk management in the food sector. In 1980, he received the American Agricultural Economics Association Award for Professional Excellence, Distinguished Policy Contribution.

Eric O. Hoiberg is a professor of sociology at Iowa State University. His research and teaching interests include the adoption/diffusion of innovative agricultural technologies, environmental sociology, agricultural sociology, and the sociology of the community. His most recent studies center on the application of the adoption/diffusion framework to porcine somatotropin, coupled with an investigation of the likely social and economic impacts of the widespread diffusion of this technology.

Rachelle D. Hollander coordinates Ethics and Values Studies at the National Science Foundation. EVS supports research and education projects examining ethical or value issues of significance to U.S. science or engineering. Dr. Hollander has authored articles on applied ethics in numerous fields as well as articles on science policy and citizen participation. She recently was elected a Fellow of the American Association for the Advancement of Science for her work in fostering this kind of research and its integration in professional practice.

David T. Kingsbury is the assistant director of the National Science Foundation (NSF) for Biological, Behavior, and Social Sciences and an adjunct professor of microbiology at the George Washington University in Washington, D.C. Previously, Dr. Kingsbury served on the faculty of the University of California-Berkeley. He currently chairs the Biotechnology Science Coordinating Committee of the Federal Coordinating Council for Science, Engineering and Technology, which has been responsible for developing a coordinated federal framework for regulating the biotechnology industry.

James B. Kliebenstein is an associate professor of economics at Iowa State University, and has taught and conducted research at the University of Missouri-Columbia and the University of Wisconsin-Platteville. His research interests are in farm-level adjustments and economic impacts of emerging biotechnologies as well as in the economics of animal-health preventive medicine.

A. David Kline is an associate professor and chair of the Department of Philosophy at Iowa State University. He conducts research in philosophy of science, philosophy of technology, and modern philosophy and has lectured extensively on ethical issues related to technological advancement. Kline was recently appointed to the U.S. Department of Agriculture's Agriculture Biotechnology Research Advisory Committee.

Bruce M. Koppel is a research associate at the East-West Center Resource Systems Institute. He is interested in the social consequences of technological change with special reference to Asia. He has conducted assessment studies on biological nitrogen fixation and has written on ethical problems associated with knowledge utilization.

Paul Lasley is an associate professor and extension sociologist at Iowa State University as well as director of the Iowa Farm and Rural Life Poll. His research involves changes in the structure of agriculture, agricultural policy, rural development, and adoption and diffusion of innovations. He has also been involved extensively in survey research among state-wide samples of farm families. Lasley's awards include the Outstanding Extension Educator Award and the Iowa State University Excellence in Applied Research and Extension Award.

Peter J. Nowak is an associate professor of agriculture and extension soil and water conservation specialist at the University of Wisconsin-Madison. He is researching why farmers adopt or reject various agricultural technologies. In addition to his academic authorships, he contributes frequent articles to the popular press on both the obstacles farmers must overcome to adopt new technologies and the consequences of their decisions.

Brian J. Reichel is a doctoral candidate in sociology at Iowa State University. He has coauthored two books related to biotechnology and has researched and coauthored numerous monographs dealing with public-private sector alliances.

Bernard E. Rollin is a professor of philosophy, a professor of physiology and biophysics, and the director of Bioethical Planning at Colorado State University. He has written and spoken extensively on animal rights and genetic engineering as well as on medical approaches to the dying aged. Dr. Rollin developed the first course offered in the United States on ethical issues in intensive agriculture for animal science students and was a pioneer in developing philosophy-related courses for premed students.

Hope J. Shand is the research director of the Rural Advancement Fund International, a subsidiary of the Rural Advancement Fund/National Sharecroppers Fund, which is a fifty-year-old nonprofit organization based in North Carolina. Prior to joining RAFI, she was director of RAF's Agriculture Accountability Project. She wrote *Uncertain Harvest: A Report on N.C. Agriculture* and is currently editor of *RAFI COMMUNIQUE,* a newsletter concerned with genetic resources and the socioeconomic impact of agricultural biotechnology.

Mack C. Shelley II is an associate professor of political science and statistics at Iowa State University. His research interests are in international studies, economics, political science, and applied statistics, with a focus on public policy decisions and their consequences. He has been active in research on institutional impacts and societal concerns arising from biotechnology, including university-industry relationships and economic development.

Seung Y. Shin is a graduate assistant in the Department of Economics at Iowa State University. His master's thesis dealt with the impact of Bovine Somatotropin on dairy farm profitability. He is currently working on a U.S. livestock model at ISU's Center for Agricultural and Rural Development.

Matthew Shulman is a researcher and analyst for the New York State Office of Rural Affairs. He has served as editor of *R.F.D.: Vermont Farm News,* and has helped design a state-wide extension system for the Federal Agricultural Coordinating Unit of Nigeria's Federal Department of Rural Development. He became interested in the agricultural, economic, and social impacts of bGH on American family farms through his experience in observing the impact of exogenous technologies on agriculture and society in West Africa.

Dennis M. Warren is a professor of anthropology at Iowa State University and chair of ISU's Technology and Social Change Program, an interdisciplinary academic and research program that focuses on relationships among science, technology, and social change. His research assesses the impact of technology in the developing world.

Gladys B. White is an analyst at the U.S. Congressional Office of Technology Assessment (OTA). She joined OTA to work on an assessment of life-sustaining technologies for the elderly and was study director for the special report, *New Developments in Biotechnology: Ownership of Human Tissues and Cells.* Dr. White is a bioethicist and a nurse.

William F. Woodman is a professor of sociology at Iowa State University. He has served as principal or co-principal investigator for projects in the areas of transportation, public policy, biotechnology, and program evaluation in human services. Dr. Woodman also has been a consultant for regional planning agencies, human and social service organizations, and the state of Iowa (during the state's reorganization).

Susan Wright teaches history of science at the Residential College at University of Michigan, where she also directs the Science and Society Program. She has written widely on the history of genetic engineering, particularly on the policy process, the assessment of hazards, and the development of military interest in the field. Currently, she is completing "Molecular Politics," a comparative study of the development of national policies for genetic engineering in Britain and the United States.

Faye S. Yates is the assistant director of Iowa State University's Institute for Physical Research and Technology, where she is in charge of public relations and marketing. Her interest in agricultural bioethics began when she helped coordinate ISU's first agricultural biotechnology workshop in 1984. She served as the first editor of the *Ag Bioethics Forum,* a quarterly newsletter dedicated to agricultural bioethics issues, and is an active member of ISU's Agricultural Bioethics Committee.

Daniel J. Zaffarano, until his recent retirement, was vice president for research and dean of the graduate college at Iowa State University (ISU). Prior to assuming this dual role, he was Physics Division chief of the Ames Laboratory of the Atomic Energy Commission. Since 1969, Dr. Zaffarano has been a consultant-examiner for the North Central Association of Colleges and Universities. His interest in bioethics began with the inception of Iowa State University's Agricultural Biotechnology Program, and he was instrumental in initiating and establishing the university's Agricultural Bioethics Program.

Index

DATE DUE